Professional and Practical Considerations for Landscape Design

Professional and Practical Considerations for Landscape Design

Steven L. Cantor

With illustrations by Vicky Chan and Richard Alomar

Oxford University Press is a department of the University of Oxford. It furthers
the University's objective of excellence in research, scholarship, and education
by publishing worldwide. Oxford is a registered trade mark of Oxford University
Press in the UK and certain other countries.

Published in the United States of America by Oxford University Press
198 Madison Avenue, New York, NY 10016, United States of America.

© Steven L. Cantor 2020

All rights reserved. No part of this publication may be reproduced, stored in
a retrieval system, or transmitted, in any form or by any means, without the
prior permission in writing of Oxford University Press, or as expressly permitted
by law, by license, or under terms agreed with the appropriate reproduction
rights organization. Inquiries concerning reproduction outside the scope of the
above should be sent to the Rights Department, Oxford University Press, at the
address above.

You must not circulate this work in any other form
and you must impose this same condition on any acquirer.

Library of Congress Cataloging-in-Publication Data
Names: Cantor, Steven L., author.
Title: Professional and practical considerations for landscape design / Steven L. Cantor.
Description: New York, NY : Oxford University Press, [2020] |
Includes bibliographical references and index.
Identifiers: LCCN 2019017399 | ISBN 9780190623340 (paperback) |
ISBN 9780190623333 (hardback) | ISBN 9780190623357 (updf) |
ISBN 9780190909666 (epub)
Subjects: LCSH: Landscape design.
Classification: LCC SB472.45 .C35 2020 | DDC 712—dc23
LC record available at https://lccn.loc.gov/2019017399

For Thomas von Foerster

I always console myself with thinking that I have done my best.
—Charles Darwin, Autograph letter: January 21, 1891[1]

NOTE

1. Charles Darwin, Autograph letter: January 21, 1891 written at Down House, Beckenham, to Sir George Jackson Mivart; from *The Magic of Handwriting: The Pedro Corrêa do Lago Collection*, exhibition at J. P. Morgan Library Museum June 1–September 16, 2018.

CONTENTS

List of Problems   ix
Preface   xi
Acknowledgments   xiii

1: Introduction   1
2: Project Management and Office Administration   16
    Part 1: Project Management   16
    Part 2: Billing   43
    Part 3: Office Standards and Metric System   49
3: Contracts   62
4: Marketing and Human Resources   76
    Part 1: Marketing   76
    Part 2: Human Resources   88
5: Drawings   100
    Part 1: Surveys   100
    Part 2: Drawings: Conceptual, Schematic, Design Development and
        Master Plan   108
    Part 3: Construction Drawings/Contract Drawings   124
    Part 4: Planting Design   142
6: Building Code, Specifications, and Cost Estimates   170
    Part 1: Building Codes and Related Planning Issues   170
    Part 2: Specifications   187
    Part 3: Cost Estimates   201
7: Areas of Practice   213
    Part 1: Site Planning   213
    Part 2: Single-Family Residential Design   224
    Part 3: Public Works   251
    Part 4: Roof Gardens and Green Roofs   273
    Part 5: Environmental Assessment   305
    Part 6: International Work   332

8: Construction Materials, Details and Site Inspections   348
    Part 1: Pavements, Steps, and Walls   350
    Part 2: Lighting, Irrigation, and Water Features   376
    Part 3: Metals, Wood, Plastics, and Site Furniture   387
    Part 4: Site Inspections   412
9: Ethics and Sustainability   422
    Part 1: Ethics   422
    Part 2: Sustainability   433
10: Resources   460
Appendix: Projects and Contact Information   473

Index   483

LIST OF PROBLEMS

1. Rumor Mill (following Chapter 2, Part 1) — 40
2. Project Management Writing (following Chapter 2, Part 1) — 41
3. The Most Beautiful Place in the World (following Chapter 3) — 74
4. Topics in Landscape Architecture (following Chapter 4, Part 1) — 84
5. Resumes (following Chapter 4, Part 1) — 85
6. Job Offer (following Chapter 4, Part 1) — 86
7. Proposals for Various Project Types (following Chapter 4, Part 1) — 86
8. Details (following Chapter 5, Part 2) — 97
9. Plant List and Cost Estimate (following Chapter 5, Part 3) — 166
10. Zoning Variance (following Chapter 6, Part 1) — 185
11. Historic Building and Americans with Disabilities Act of 1990 (following Chapter 6, Part 1) — 186
12. Specifications Review (following Chapter 6, Part 2) — 200
13. Outline and Detailed Specifications (following Chapter 6, Part 2) — 201
14. Detailed Cost Estimate (following Chapter 6, Part 3) — 212
15. The Shifting One-Hundred-Year Flood Plain (following Chapter 7, Part 1) — 223
16. Residential Site and Fee (following Chapter 7, Part 2) — 248
17. Residential Upgrade (following Chapter 7, Part 2) — 249
18. Role-Play for Residential Design Fee and Services (following Chapter 7, Part 2) — 250
19. Public Housing Apartment Landscape and Infrastructure Proposal (following Chapter 7, Part 3) — 270
20. Co-op Apartment Landscape and Infrastructure Proposal (following Chapter 7, Part 3) — 270
21. Green Roof as Habitat for Endangered Species (following Chapter 7, Part 4) — 304
22. Quick Environmental Assessment (following Chapter 7, Part 5) — 331

23. International Work: Resort Planning, Botanical Gardens,
    Transportation Systems (following Chapter 7, Part 6)   343
24. Set of Construction Details (following Chapter 8, Part 1)   372
25. Water Feature (following Chapter 8, Part 2)   387
26. Site Furnishings and Lighting (following Chapter 8, Part 3)   411
27. Wood and Metal Gate (following Chapter 8, Part 3)   411
28. Site Inspection (following Chapter 8, Part 4)   417
29. Project Management Writing (following Chapter 8, Part 4;
    same as Problem 2)   420
30. Sustainable Design (following Chapter 9, Part 2)   458

Please note the following about the problems and exercises:

1. Many of these problems require the participant to create sites of varying size with varying difficulties, such as challenging topography, trees in poor condition, and so on. This is an act of the imagination based on the experience of the creator. Once the base map has been created, it's important to follow these existing conditions carefully and be bound by them, and also to share your information with other participants, whose sites likely will be quite different from your own. Learn from one another. Sometimes the creation of a site with a number of challenges results in a clearer design response.
2. In all of the problems, any *terms in italics* may be changed to suit the particular instructor or team leader or your own personal, studio setting. Typical changes might be people or clients or contractors, plant materials, or the locations of sites. It's fine to make the sites and conditions overflow with local color. The exceptions are problems in which the focus is on something necessarily outside one's typical experience or something which changes the degree of difficulty, such as in the chapter on international design.
3. With many of the multiple-choice and short-answer questions, there are not necessarily right or wrong answers. At least some questions are suggested to generate discussion. The answers to some questions may change depending on the group of people responding to them, particularly depending on their age and where they are located.
4. The problems cover a range of difficulty, and the difficulty may be increased or diminished depending on how the sites are created. At the same time, it's not necessary to work through all thirty problems. Some are much more challenging than others.
5. In drawing and presenting graphics to other colleagues, students, or just for yourself, prepare a series of sheets with title block and clear labels. Ensure a continuous sense of organization over the series of sheets you use to present and solve the problem(s).

PREFACE

Long careers run in my family. One of my sisters is nearing retirement as a social worker after forty-five years. The other probably wrote fiction in the womb, continues to do so, and has already had two novels published. I earned my degree in landscape architecture in the bicentennial year and started having short essays published not too long afterward. What I've learned in a long career as a landscape architect is that I'm often my own worst critic, there's never enough time to get everything right, and the simplest method is often the best. I've also enjoyed a parallel career as a pianist, occasionally as a soloist, most frequently as an accompanist in chamber music and opera, usually as a collaborator with solo singers or instrumentalists. Music is rewarding in that if there are mistakes in performances, there are frequent opportunities to try again, whereas if you direct the transplanting of a specimen tree or stake out the location of a new one, nature is not always that forgiving if you don't find exactly the right spot (and the client may not be either). Therefore, the reason for this book. *How do you document the process and give yourself and your firm the best chance to get things right?*

Aside from a great deal of writing, some experience as a professor, and some design projects on my own, most of my career in various firms has been in project management and professional practice. Professional practice involves a good deal of design, and design requires a good deal of professional practice. Although individuals may choose to specialize in one area, you can't divorce one subject from another. I have often been caught up with the nitty-gritty aspects of practice—verifying that the design is being executed in the field as shown on the drawings, that it is being done within budget, that the staff in the office and the contractor in the field communicate with one another, and that there are no surprises or fireworks, so that when the project ends, everyone is on an even keel, is pleased with the results, and wants to hire you for another project.

I intend the audience for this book to be advanced students in landscape architecture curricula who are taking a professional practice course or design studio in which they need some exposure to such issues. The topics are presented in the sequence that I believe will be most useful but are also independent of one another, so readers are welcome to dive in wherever appropriate. The variety of styles and formats is intended to keep you engaged. The book is also meant for emerging professionals, people who have been in an office for a few years and may be considering starting their own firm or want a study guide for the registration exam. Throughout the book are thirty problems and exercises; some are quite challenging, others are multiple-choice questions, a number are intended to stimulate discussion, and some involve design explorations. There are no answer guides, because often there are no clear-cut answers. For problems in which you are asked to draw, each person must create a site; if you are part of a group in a studio at a firm or within an educational setting, all the sites will be different, so you should be able to review one another's work without worrying about copying ideas.

ACKNOWLEDGMENTS

I thank Vicky Chan and Richard Alomar for their selfless dedication to and generous collaboration on this project.

I thank Sarah Humphreville, Abigail Johnson, Bronwyn Geyer, and all the others at Oxford University Press who have supported this project from its inception through publication.

I thank Asish Krishna & Rathna Bhavya, project managers, responsible for the production process of the book at Newgen in Chennai, India, along with Sue Warga, the copyeditor and Mr. Prabhakar, the art director, and their support staff.

I thank landscape architect Elizabeth J. Kennedy, for stimulating me to write.

I thank Gail Addiss, Deborah and Tom Bauer, Diane and Sally Cantor, Joseph M. Dunn, Thomas Echols, Cedric Tolley, Lucas Heston Sandford, Stephen Rapp, Gary Evans and Bill Gabello, Geoffrey Rogers and Wallace Sanders for their encouragement.

I thank Mohammad Waris Safi and Miron Dion-Arivas for their computer expertise and wise counsel on the intricacies of software glitches.

I thank Caroly Wilcox and Frank Misiurski for their generosity and "Anything Goes."

I thank all of the contributors to this book—landscape architects and designers, contractors, artists, project managers, architects, designers, directors, project managers, clients, owners, and so on, in all their diversity—who have trusted me to write about and sometimes to photograph their work. Some of you I've never met in person, and the wonders of technology have connected us. May you all continue in rewarding lives and careers.

I thank Dean Anderson for his thorough review and patience in vetting the section on metals.

I thank Mark Davies for his expertise on plant materials and landscape contracting.

I thank Scott Weinberg for maintaining academic links.

I thank Heidrun Eckert at Zinco for transatlantic patience.

I thank Linda and Aramis Velazquez for continuing to link me with Greenroofs.com.

I thank Karen Brown, Gary Ramsey, Dr. Astaire Selassie, and others for supporting my health.

I thank the New York City Public Library's Dyckman Street Branch, Columbia University libraries and the Indian Road Cafe.

I thank all my employers in landscape architecture for providing an environment which has stimulated me to study, to create, to learn and to grow, and to teach others for over four decades, including the following: Ralph Liss Associates; Edward L. Daugherty; University of Georgia School of Environmental Design; University of Colorado College of Environmental Design; Bruce Kelly/ David Varnell; Clarke + Rapuano, Inc.; New York Botanical Garden; Madison Cox Design Inc.; Edmund D. Hollander, Landscape Architect Design P.C.; Anhalt University of Applied Sciences; Thomas Balsley Associates; Stantec Consulting; Greenroofs.com, and others.

Professional and Practical Considerations for Landscape Design

CHAPTER 1

Introduction

The professional practice of landscape architecture consists of the methods, approaches, and documents that landscape architects use and prepare in order to apply for and find work; to hire staff with appropriate salary, training, and benefits; to sustain a project while it progresses through the design and construction stages; to carry out such projects at a profit; and to archive projects when they are finished. This book describes what is required to turn drawings of designs into works of landscape architecture: interpreting contracts, directing the development of conceptual and preliminary drawings, preparing contract documents (contract, plans, and specifications), creating cost estimates, and administering projects. If such skills are not emphasized, then a firm risks a pattern of producing designs that may be beautiful and effective but nevertheless force layoffs of staff because the firm could not do the work within the allowed budget. Diversified staff with varying levels of experience also need to have a common understanding of how projects are designed and executed. When creating practice guidelines, keep in mind not only for those with technical skills but also administrators and designers, all of whom want to achieve the best result possible in the field. Also, in a period of technological innovations and advancements, landscape architects must be fluent in software applications from AutoCAD to Facebook and use them appropriately. Finally, just as landscape architects must know what different materials cost so that they can achieve a successful and aesthetically satisfying project within budget, so they must also understand how people are paid for their skills.

Figure 1.0 An intimate garden with a recirculating water feature in Little Rock, Arkansas.
Design by Gary Evans and William Gabello
Photo by Steven L. Cantor

With each major topic in this book there will be discussions of professional and practical considerations that apply to human resources (that is, the diverse people who make up the office), the various processes that are followed, the documents that are produced, and the projects that result. Professional considerations include specific guidelines and methods of achieving successful projects within the practice of landscape architecture and design. Practical considerations include shortcuts to save time in the production of drawings or other documents or in the implementation of a project, ways to achieve clarity with graphics in the specific challenges of particular types of drawings, and quick methods to estimate how many hours different important tasks take to complete. Practical considerations can be thought of as ways to respond to the question of *how* something is achieved, whereas professional considerations are more a response to *what* needs to be achieved within a set budget. For example, professional considerations for a landscape plan for a residence include all of the requirements that enable the designer to create and implement a landscape that will satisfy the owner's program requirements, while practical considerations include ways to streamline the process while still achieving high-quality results. Within this framework, there should be a checklist of what to include and also an acknowledgment that some factors may be unnecessary or too complex to consider for a particular application; often these considerations overlap. Also, each landscape designer or other professional must decide and choose what works best for specific types of projects and employees within a particular office setting.

This book demonstrates one overarching approach to professional practice, but it is hardly the only one. Particularly since this volume is being offered as an ebook, I am hopeful that over time it can be expanded and revised to meet the evolving needs of landscape architecture education and new areas of practice. At the end of each section and chapter, problem statements and other exercises are included for use by anyone reading the book, including students and faculty, groups of professionals, and staff in an office setting. These problems are most helpful when users—students in a class, say, or staff within an office—can compare their results to others' results in the same setting and learn from one another. In some cases there may be clear right and wrong answers; in others what's most important is understanding how to proceed with a particular problem or how to have a wide-ranging discussion with colleagues or classmates in response to a question or problem and use that as a catalyst. Some of the exercises involve setting up a role-play scenario. Have fun with these, but also use your imagination to understand how a particular personality or character would act. At the same time, suspend your disbelief and see what happens and what you can learn. Other problems describe specific sites with particular properties and characteristics—such as slopes, vegetation, and floodplains—that each participant must draw on a scale map of the property, which you must create from your imagination in order to proceed with the problem. The more specific and varied your creation is, the more satisfying your results may be. By creating your own sites, different conditions and concerns will arise for each of you, and it's easy and useful to share results with other students or colleagues. Similarly, you can challenge yourself by creating more unusual site conditions within the parameters of some of the problems, perhaps based on what you are studying in a specific studio setting or what you may be considering for a future project. It is not necessary to complete all thirty problems in this book; a professor may focus a course on the topics that best fit the curriculum. A professional may find fewer topics or situations that match what the office is experiencing.

Each section and chapter also include a series of **dos and don'ts** for that specific topic and a few problems or questions that enable the reader to further experiment and explore. Occasionally topics will appear in more than one chapter, so that someone reading a particular section does not have to flip back and forth a lot between chapters. There is a variety to the formats of the various sections: some topics lend themselves more readily to checklists and others more to discussion. It is hoped that this diverse approach will create an interesting book.

The remainder of this chapter gives definitions of key terms used in the book. Chapter 2 is divided into two parts, the first of which is a discussion of some principles of project management with numerous examples.

Each chapter begins with a color image showing either exceptional designs, often with elegant, sustainable design components, or design challenges. The goal is to encourage the reader to delve into the subjects of this book. The appendix provides more information on some of these projects.

The second part of Chapter 2 is a discussion of three topics: billing, the development of office standards, and the metric system. It is essential to bill in a timely and efficient manner and to share design standards throughout an office. Even though the metric system is not used in the United States, it's important to understand it, since its use is required almost everywhere else in the world.

Chapter 3 discusses the essential requirements of contracts as well as various contract details to include and to avoid for particular types of projects or clients. Sometimes consideration of contractual issues sets the tone for a project, as a good or bad contract can make or break a project. The material is presented in an outline format for ease of organization, and examples are provided for each case to highlight the points the author wishes to emphasize. A design problem is presented based on a reading from the ecologist Edward Abbey's classic work *Desert Solitaire*, in which there is a provocative abstract connection to contractual relationships between the landscape architect and client—a good exercise in thinking outside the box.

Chapter 4 covers marketing and human resources. Marketing addresses the process and methods of a firm seeking and applying for new work, while human resources is concerned with all aspects of hiring employees, keeping track of their benefits such as health insurance, setting policies for social media use, and other aspects of employment that apply to all employees. Also included are some of the factors that graduating students and emerging professionals must consider when they are seeking a job in landscape design and embarking on a career in the field. There are a considerable number of questions that help the reader imagine different scenarios involving email systems and the use of social media, since they are ubiquitous in contemporary office settings, from small firms to large corporate interdisciplinary design firms.

Chapter 5 deals with surveys, design drawings, and construction drawings—in short, the entire series of drawings landscape architects undertake, from the earliest sketches to the final bid documents. Each section has a summary of checklists of dos and don'ts, plus examples. The aim is to emphasize what should be included and excluded from each type of drawing. It is important not to repeat major information from one type of construction drawing in another type; at the same time, there should be clear connection and flow from one type of construction drawing to the next in the sequence. A system of two-letter binomials is presented for use in planting plans that help readily identify plant materials as trees, shrubs, or perennials; this is a time-saving device for large projects and can help both designers and contractors.

Chapter 6 is divided into three sections: applications of the building code, specifications, and cost estimates. There is an emphasis on applicable sections of the Americans with Disabilities Act of 1990, with its impacts on accessible design and the importance of checking and verifying local building codes despite the use of the International Building Code, which is applicable in many jurisdictions. Specifications are defined, and examples of each type of technical specification—descriptive, performance, proprietary, and reference—are presented, along with the circumstances and conditions in which each type is preferred. Guidelines for writing styles are discussed, as well as how to organize specifications and coordinate them with a set of construction drawings and a cost estimate. The final section of this chapter focuses on cost estimates, including how to gradually make them more precise as the drawings they are based on become more exact as well, and how to coordinate them with specific details in the construction drawings.

Chapter 7 describes and gives examples of six major types of landscape architecture practice. Site planning and residential design focus on perhaps the two most traditional areas of practice from which the profession has evolved over many decades. Public works, roof gardens, green roofs, environmental assessment, and international work sometimes have traditional origins but suggest where practice may continue to evolve in the future. International practice is discussed from two perspectives: that of people living in the foreign country and that of individuals traveling to it for work. Examples of projects are given throughout, as well as sample problems to help provide the reader, student, and/or practitioner a clear sense of some of the issues involved in each type of practice. Examples of projects shown include two award-winning green roofs in New York City and a residential estate in Easthampton, New York, plus an array of projects from other settings. The goal is to tantalize the reader with diverse examples and snapshots of practice so that he or she may investigate further. The designers range from students to acclaimed practitioners, so that the reader can appreciate both the variety and the challenges of landscape architecture. This section concludes with some images of Freshkills Park in Staten Island, a remarkably complex achievement involving public works, environmental assessment, and site planning.

Chapter 8 describes all the nuts and bolts of practical detailing of the tools of the trade, so that the landscape designer has a sense of what's involved in working with pavements, steps, walls, metals, wood, plastics, and site furniture, and incorporating lighting, irrigation, and water features into design. Each category of detail has its own concomitant dos-and-don'ts list. Since designs can be durable only if they are built to last, there is a thorough section on doing a detailed site inspection of a project under construction. The chapter concludes with several problem statements on this topic. Illustrations range from drawings showing the integration of several types of details to achieve a solid wall or pavement or other construction to some photographs of as-built installations.

Chapter 9 concerns issues in landscape architecture and design practice about what is right and wrong, and segues directly into sustainability, which gains in importance at a time when resources are stretched to the limit and being affected by climate change. Several definitions of sustainable design are offered, one more for a lay audience and another based on measurable scientific results. A project is presented from South Korea in which sustainability was a parameter of the design process. A range of questions is offered as means of generating discussions about ethical issues, and a problem is offered with sustainability as a theme. Throughout the book, several other projects are highlighted in which sustainability is at least one goal of the design. The LEED process is defined; to be most effective, it must be followed in a detailed and methodical way from beginning to conclusion.

Chapter 10 provides a bibliography of books and a list of websites. Many of the books are references on topics covered in the previous chapters. There are also a few works in which scientists have tried their hand at fiction. The websites listed are government organizations, not-for-profits, and trade organizations. I have included a few manufacturers because they are so widely used in a particular field for a particular product that it feels to me better to include them than not. If you investigate most of the websites further, you can find links to free materials such as publications, specifications, and opportunities for classes (some free, some at reduced cost for members). Some trade organizations provide AutoCAD details and other useful information if you are willing or wanting to specify their particular product. Many of the organizations that have "American" or some other specific nationality in their official title are, in fact, international and have offices, membership, and products overseas.

DEFINITIONS

The following definitions, in alphabetical order, apply for terms used in this book. The approach is to give the reader the essence of each term. In some other sections of the book, such as the description of specifications, more detailed information is provided, as appropriate for that particular discussion. Only terms discussed in this book are included; for definitions of other terms, check some of the references listed in Chapter 10.

1. *Addendum*. An addition made to a set of contract documents before a project is under contract but after it has gone out to bid. Late changes to documents prior to the issuing of bid documents are simple updates to drawings, specifications, or cost estimates and may be called out by the in-house landscape architect's staff by use of a revision date. However, once the project is out to bid, the landscape architect must be certain that all bidders receive addenda in a uniform manner and acknowledge receipt of them.
2. *Bid form*. A "fill-in-the-blank" document given to any contractor planning to bid on a particular project, set up in a format that requires the contractor to provide a breakdown of the bid in such a way that

a fair comparison can be made between different bidders. Typically, a bid form includes the name and address of the project, a listing of the documents on which the bid is based (including addenda), a breakdown of phases (if any) of the work, the anticipated work schedule, evidence of insurability, and the complete name, address, and signature of the bidder. The bidder lists a breakdown of the bid in a standardized format convenient to the purposes of the landscape architect and owner. All bidders use the same form to ensure a uniform basis for comparison. In order to encourage a fair process, there is usually a due date and time by which all bids must be received.

3. *Bond.* A financial guarantee of the execution of the contract and the payment of bills. The four most common types are the bid bond, the performance bond, the payment bond, and the labor and materials bond, which are all described in Chapter 6.

4. *Change order.* A revision to a set of contract documents after a project is under contract. The landscape architect must document what is included in each change order, such as drawings, specifications, or other items, and include a transmittal form to verify that the contractor has received the change order and is adjusting the work accordingly. Often change orders result in a change in the price of the work under contract, so the landscape architect must give the contractor ample time to check on the impact the change order will have on the project, and must submit to the client or at least discuss with the client an estimate of the change order's cost.

5. *Client.* The person or entity for whom the landscape architect is doing the work. Sometimes the client is a married couple, such as the owners of a residence for whom the landscape architect is hired to do designs. Of course, the client can also be an individual. Sometimes the client is the representative of the owner and serves as the project manager or point of contact for the landscape architect. For residential projects and smaller projects, the client and the owner are often the same. The landscape architect is often referred to as the owner's representative, as her professional role is to act on behalf of the owner/client to ensure the satisfactory completion of the work,

6. *Consultant.* A specialist, such as an engineer or swimming pool designer, who provides professional services, such as design or engineering drawings and specifications, for some specific aspect of the project. A landscape architect may be a consultant to another designer or firm, or may be the principal to whom others are consultants.

7. *Contract.* A legal agreement in which two or more parties describe an agreed-upon scope of work to be done for a certain price and according to a certain schedule.

8. *Contract documents.* The contract, plans, and specifications for a particular project.
9. *Contractor.* The firm or individual doing the work on the project. Just as the owner or design firm can have a project manager or representative acting on their behalf, so can the contractor. Furthermore, there is often a site superintendent on the job site at all times directing the work for the contractor and in charge of all work and communications at the site. Subcontractors are contractors working under the direction of contractors on specific phases of work; for example, a subcontractor may have a contract for landscape construction under the direction of the general contractor for a project.
10. *Disability.* As per the Americans with Disabilities Act of 1970, a disability is an impairment that must be accommodated to enable an employee to enter and leave the work environment, function, and interact comfortably with colleagues.
11. *Ethics.* The set of principles governing right and wrong conduct that apply to the professional responsibilities and actions of the landscape architect and other individuals or groups with whom s/he interacts.
12. *Imagination.* "The act or power of forming a mental image of something not present to the senses or never before wholly perceived in reality; creative ability; ability to confront and deal with a problem: resourcefulness; a creation of the mind."[1]
13. *Insurance.* Coverage by contract in which one party, such as an insurance company, provides protection (to the owner, client, landscape architect, or contractor) against the potential actions of another and/or guarantees against loss from a specific action, inaction, or peril. It is typical for landscape architects to have *liability insurance* against claims that might be brought against her as a result of design errors or actions by others, such as the contractor. The contractor must usually provide liability insurance protecting the owner and landscape architect against damages from bodily injuries, occupational sickness or disease, or death, as well as damage caused as a result of the work being performed. The owner must usually have in place or purchase *property insurance* as well as his/her own liability insurance. Sometimes the cost of property insurance is shared between the owner and the contractor. This insurance provides protection against damage to property as a result of fire, theft, or vandalism and any other property-related damages the parties agree on. Finally, the contractor must usually verify that s/he has in place *workmen's compensation insurance*, providing protection to the entire staff as a result of illness, disability, or sickness or disease that occurs during the project. *Certificates of*

insurance are usually provided to the owner by the insurance company to verify that all required insurance is in place and that all key insured individuals or companies are listed as being beneficiaries or insured under each appropriate policy.

14. *Joint venture.* An agreement between two entities to work in tandem to provide complete design services or specific expertise for a project under terms they negotiate.
15. *Landscape architect* and *landscape designer.* A professional, trained and experienced in the design and construction of landscapes of all kinds, with varying degrees of specialization. The term *landscape architect* usually implies that the individual has graduated from an accredited university program in the field and has passed a licensing exam. Often a landscape architect registered in one state may readily obtain reciprocal registration in an adjacent state as long as education and licensing information can be shared. The term *landscape designer* is more generic, implying that the individual is skilled in some aspects of landscape architecture but does not necessarily have the full range of skills and expertise as a landscape architect, such as in grading, drainage design, and detailing.
16. *LEED.* Leadership in Energy and Environmental Design (LEED) is an environmental rating system originally developed by the U.S. Green Building Council to help identify and rate sustainable design. The Canadian Green Building Council has a similar system. In either system points are awarded for beneficial or positively rated environmental accomplishments in six study areas: sustainable site development, water efficiency, energy efficiency, materials and resources selection, indoor environmental quality, and innovation in design. There are four LEED ratings:

Certified	26 to 32 points
Silver	33 to 38 points
Gold	39 to 51 points
Platinum	52 or more points

It is now standard practice for federal and state agencies, and some cities and local governments, to require a minimum level of LEED certification in their new buildings and related landscape projects. LEED also provides certification for consultants in guiding a project through a LEED process.[2]

There are consultants specialized and certified as LEED trained who are expert at guiding a project through the LEED process, submitting the test results to the appropriate American or Canadian authority, and helping to gain as many points as is possible for each step of the design and building process.

17. *Lien.* A legal hold or charge against either personal or real property to satisfy a debt or other obligation. Usually the placement of a lien prevents continuation of work on a project until the lien is satisfied according to the legal requirements.

18. *Liquidated damages.* A clause inserted into contracts when it is urgent that the work be finished by a certain deadline. The contract amount may be somewhat higher than usual, in return for which the contractor agrees to pay a penalty per day if the work is not completed by the deadline. There are usually exclusions for extreme weather or other elements beyond the contractor's control.

19. *Minority-owned business enterprise (MBE).* An enterprise, such as a landscape architecture or design firm or contractor, in which at least 50% of the ownership is held by a person belonging to a minority group. Some federal, state, and city contracts require that a certain minimum percentage of the work under contract for a project be given to firms that are MBEs. Such firms run the gamut of the industry, from those specializing in printing and graphics to major design firms.

20. *Owner.* The person or company who has title to the property for which you are hired to provide service as a landscape architect or designer. Sometimes the owner is also the client. At other times there is an owner's representative who functions as the client and acts on behalf of the owner but must periodically meet separately with or otherwise communicate with the owner to be certain of approval at all key phases of the project.

21. *Shop drawing.* Detailed drawing, usually to scale, or diagrams and related information prepared to show the proposed execution of a specific aspect of the design, such as a unique pavement pattern, that cannot be fully prepared until the design, as proposed, is being built in the field. The dimensions, materials, and other detailed information shown in the shop drawing are the contractor's means of verifying that measurements in the field and the specific materials proposed for use will work within the design parameters shown in the contract drawings and specifications. The shop drawings must be carefully reviewed by the landscape architect and approved prior to fabrication and construction. Mistakes or inconsistencies noted by the landscape architect during review must be corrected prior to final approval and fabrication. To facilitate careful review, it is typical to stamp and sign the drawing in ink, and distribute copies for appropriate revisions, if necessary, to all parties (landscape architect, fabricator, contractor, etc.). (See Figures 1.1 and 1.2).

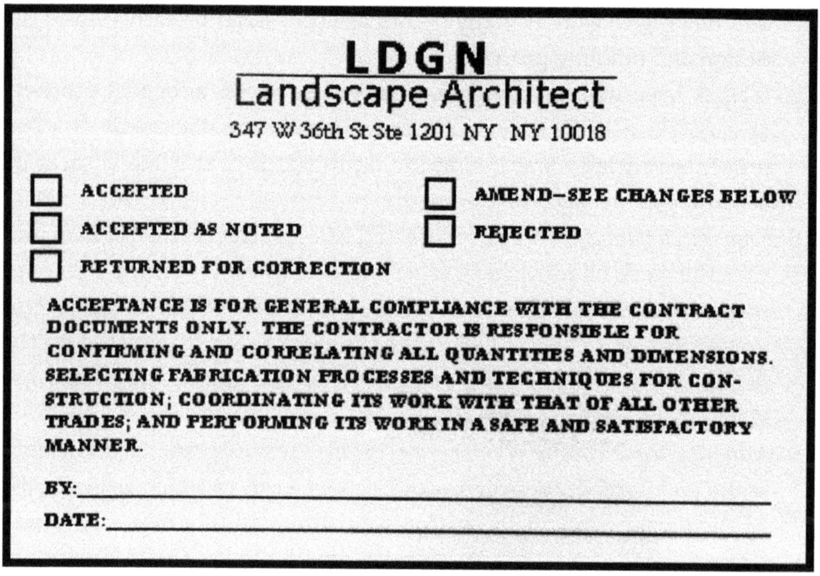

```
                    LDGN
              Landscape Architect
              347 W 36th St Ste 1201 NY NY 10018

    ☐ ACCEPTED                  ☐ AMEND—SEE CHANGES BELOW
    ☐ ACCEPTED AS NOTED         ☐ REJECTED
    ☐ RETURNED FOR CORRECTION

    ACCEPTANCE IS FOR GENERAL COMPLIANCE WITH THE CONTRACT
    DOCUMENTS ONLY. THE CONTRACTOR IS RESPONSIBLE FOR
    CONFIRMING AND CORRELATING ALL QUANTITIES AND DIMENSIONS.
    SELECTING FABRICATION PROCESSES AND TECHNIQUES FOR CON-
    STRUCTION; COORDINATING ITS WORK WITH THAT OF ALL OTHER
    TRADES; AND PERFORMING ITS WORK IN A SAFE AND SATISFACTORY
    MANNER.

    BY:_____
    DATE:_____
```

Figure 1.1 Typical shop drawing stamp from landscape architecture office.
Courtesy of Jeff Dragan, LDGN Landscape Architects, DPC

22. *Specification.* A set of technical directions describing a specific item of work or various requirements necessary for executing or constructing or installing a specific item of work shown in a set of contract documents. See Chapter 6 for a detailed discussion.

23. *Sustainable design.* Design "that uses and recycles all materials and systems so there is no net loss to the environment as a result of implementing the design, and that, once complete, functions without a net drag on the resources and systems on which it depends."[3] See Chapter 9 for additional definitions and further discussion.

24. *Unit prices.* The bid cost for an item in a bid list, such as the cost per square foot for a type of pavement or the cost for a specific size of a specified plant material. Usually the bid requirements state that these costs must hold for the life of the project, so that if change orders or subsequent phases of work occur within a reasonable time frame, the unit cost for these items is predetermined. (Should the bid form not include this requirement, it's sound policy for the specification to include it.)

25. *Variance.* Legal permission from a zoning board to implement construction or other activity at odds with building code standards, such as setback requirements or location of outbuildings, driveways, or other structures.

26. *Women-owned business enterprise (WBE)*. An enterprise, such as a landscape architecture or design firm or contractor, in which at least 50% of the ownership is held by women. Some federal, state, and city contracts require that a certain minimum percentage of the work under contract for a project be given to firms that are WBEs. Such firms run the gamut of the industry, from those specializing in printing and graphics to major design firms.

NOTES

1. *Merriam-Webster's Collegiate Dictionary*, 11th ed. (Springfield, MA: Merriam-Webster, 2009), 620.
2. U.S. Green Building Council, www.USGBC.org; Canadian Green Building Council, www.CGBC.org See also Steven L. Cantor, *Green Roofs in Sustainable Landscape Design* (New York: W. W. Norton, 2008), 35–37.
3. Cantor, *Green Roofs*, 34. See also Stephen R. Kellert, *Building for Life: Designing and Understanding the Human-Nature Connection* (Washington, DC: Island Press, 2005).

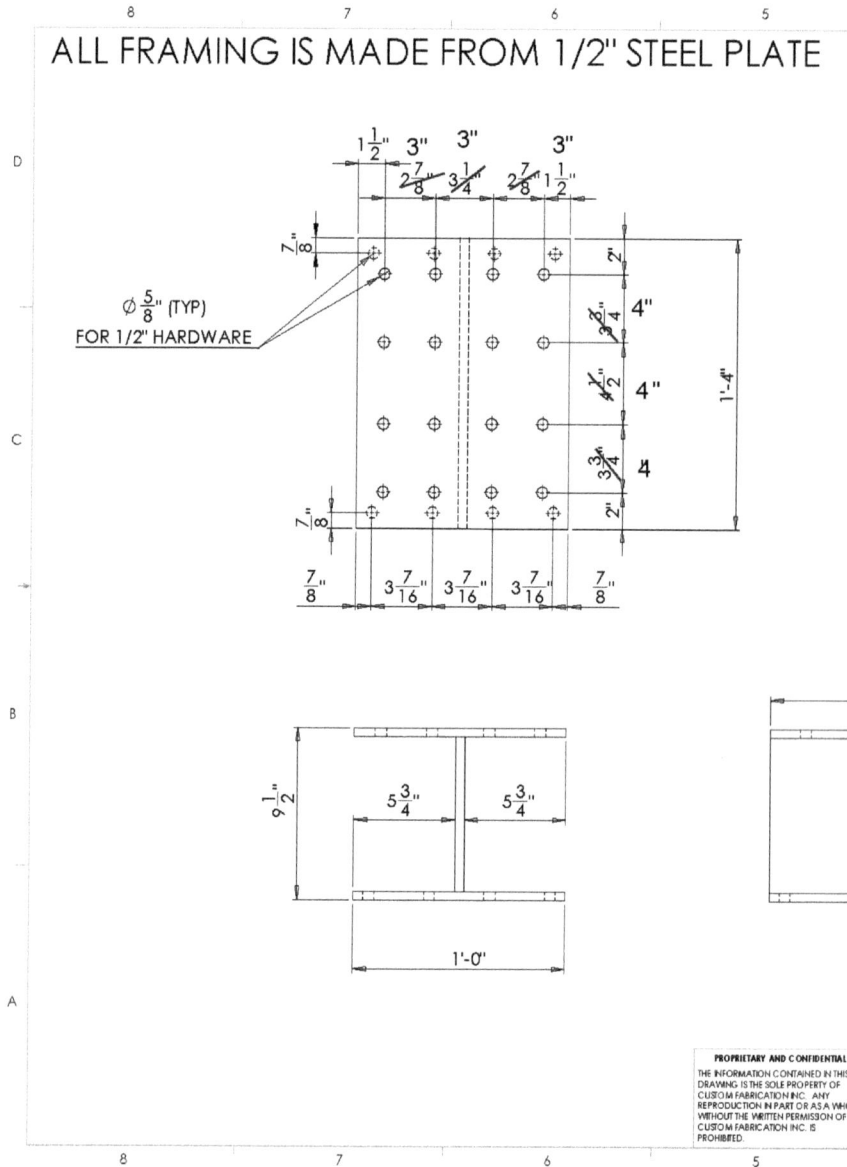

Figure 1.2 Shop drawing example.
Courtesy of Jeff Dragan, LDGN Landscape Architects, DPC

CHAPTER 2
Project Management and Office Administration

PART 1: PROJECT MANAGEMENT

I look at Hurricane Katrina, and I think if four days before landfall you gave a movie studio autonomy and a 100th of the billions the government spent on that disaster, and told them, "Lock this place down and get everyone taken care of," we wouldn't be using that disaster as an example of what *not* to do. A big movie involves clothing, feeding, and moving thousands of people around the world on a tight schedule. Problems are solved creatively and efficiently within a budget, or your ass is out of work.[1]
—Film director Steven Soderbergh

A goal of this book is to give the reader a sense of some of the personal, political, and social processes that contribute to the success of a project. A good design, if it is well constructed and in tune with the dynamics of its site, is self-evident. However, there is something mysterious, even magical about the process whereby a number of diverse and often egotistical design professionals join forces to produce a creative, cogent result that satisfies both the articulated and unstated desires of a client, either an individual or a group of people. This process is the art of project management: certain factors, attitudes, and guidelines expedite a good design process and contribute to a quality result. Some of the following guidelines may work better for you than others; please revise them as you see fit.[2]

Even though wonderful designs are created and constructed in violation of these guidelines, such results are usually a credit to the dedication, professionalism, and talent of the designers in overcoming obstacles, some of

Figure 2.0 The towering curved steel forms of *"Wake"*, massive yet graceful sculptures by Richard Serra, enhance the landscape at Olympic Sculpture Park in Seattle, Washington. Landscape architecture juxtaposes contrasting scales with dramatic effect. Part of the Seattle Art Museum, Washington. See https://www.seattleartmuseum.org/Documents/OSP%20Map%20and%20Guide.pdf.

Photo by Steven L. Cantor

which are created by the people who employ them. Project management in a vacuum is exhausting and debilitating for everyone, from principals to staff. In the landscape architecture and design professions, in which there is usually a narrow profit margin, some project managers and even principals may unconsciously undermine their own best intentions. Since uncertain economic conditions in landscape architecture intensify the competition for a limited number of potential projects, a natural tendency is for employees to be on the cusp between being full-time and part-time, or between being part-time and laid off.

Communication

To communicate clearly in all ways, whether written, oral, computer-aided, or graphic, is an essential skill. It is rare to have equal talents and skills in all such areas; therefore, it is crucial to know one's limitations.

Someone must communicate with the client consistently, clearly, and directly. All design and contractual aspects of the projects must have organized documentation. Depending on the size of the firm and the role it is playing in a particular project, the communicator may be as crucial as the designer. One must communicate with consultants, colleagues, and those being supervised. One must deal with a designated member of the bureaucracy, as high up as possible, without stepping on anyone's toes. The results of all of these efforts must be carefully organized in correspondence files, most of which should be saved in the office computer network.

An important aspect of this documentation often overlooked is the need to photograph projects prior to, during, and after construction. A photographic record should document the existing conditions of the survey and start to point out problems emphasized in a graphic such as a site analysis drawing. Photographs of ongoing construction activities document that the details are being built as specified. Photographs of as-built conditions form a record to show the completion of the work, and also can represent the baseline for considering future phases of work. A complete and accurate photographic record is crucial in publicizing the work and recalling what was done in order to inform contemporary practice. It is far preferable to be organized about photographing in a timely manner during the progression of the work than to do it in a frantic rush to meet a publisher's deadline or to gather evidence in a lawsuit.

In larger firms and complex projects, the role of communicating is often assigned to a person deemed the project manager. She or he may also be a designer, but her or his principal role is to keep the project on schedule and within budget. This person is responsible for all communications with the client, discussing and negotiating contractual issues, arranging meetings, distributing documents, and trying to free the time of the design team members to do what they do best. If there is friction between the different participants, the project manager needs to instill confidence in insecure personalities, assuage bruised egos, and tantalize other members with the project's goals. The project manager may be a capable designer and may contribute in major ways to the design, yet it is not likely that credit will be given for that function. Instead, other skills are what are most needed: business sense, common sense, and an ability to listen and communicate. (For the balance of this discussion I will use the term *project manager*, although the role may be taken on by the principal-in-charge, a landscape architect, or someone entirely different.)

Listening

Although there are times when a project manager must make recommendations to a client, give clear directions, and tell people, usually the staff, what to do, there are also times when it is important to listen. Active listening in a nonjudgmental manner is an aspect of listening often overlooked. This refers to evincing clear expressions from others, rather than asking leading questions. For example, there is a tremendous difference between asking "What do you think?" and "Don't you agree?" The difference in responses is likely to be even more compelling. For example, in the initial conceptual phases of a project when a team is becoming familiar with the site and design issues, a design process often used is brainstorming. Random ideas are generated without any evaluation. A project manager participating in this process should be helping participants to generate as many ideas as possible, not short-circuiting it with narrowing strictures before it even begins. (See Figure 2.1).

Another setting where active listening can be crucial is at public hearings, where a project manager may be one of those representing her firm and presenting its designs to the politicians and the public. Although part of the presentation may well involve demonstrating and illustrating the project in as rosy a manner as possible, to show that the design meets all the assigned criteria, often this process is followed by intense public comment and discussion. The more that the project manager or landscape architect can ask and respond to nonjudgmental questions, the more likely it is to develop a complete sense of the public reaction, and start to develop a sense of how to proceed.

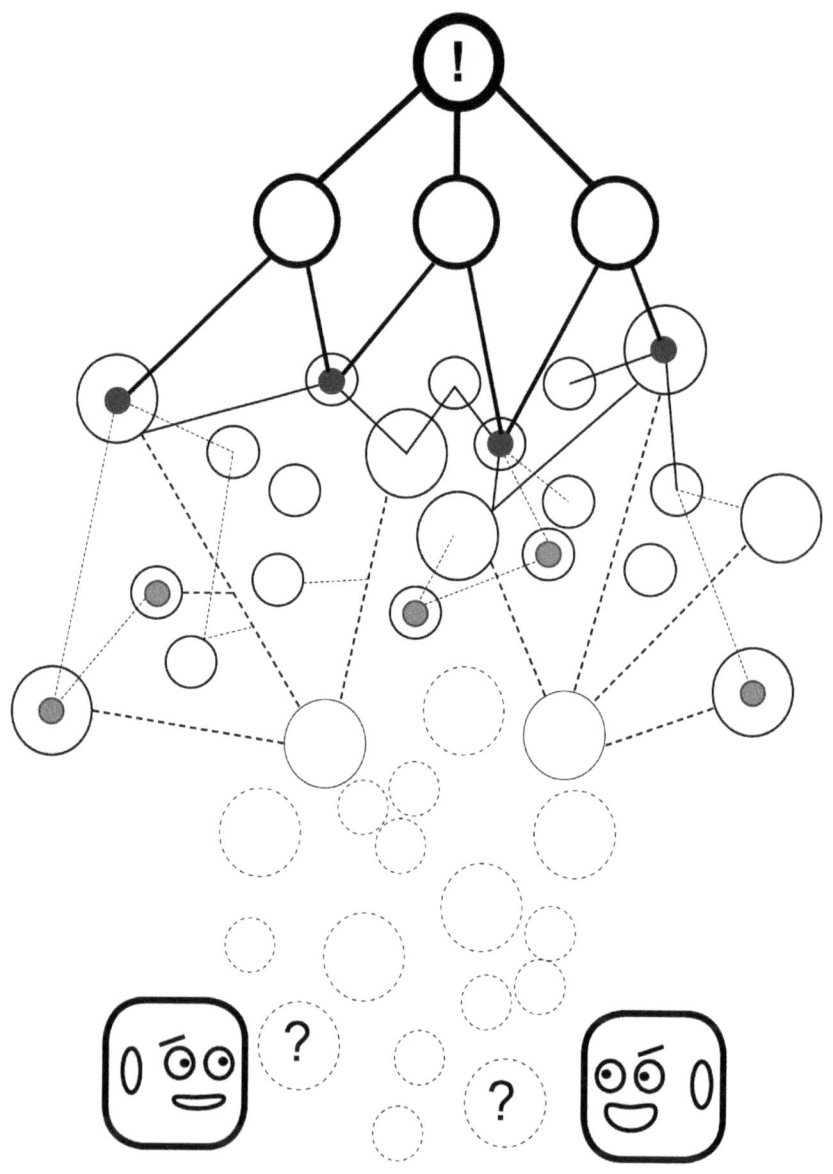

Figure 2.1 Brainstorming.
Richard Alomar

Organization

The primary way to keep a project on schedule is to organize the time of everyone participating. For a large project involving many consultants, this may not be possible, but the project manager can help consultants organize their time by being certain that they have all the information, documents, and schedules in hand and are completely aware of deadlines and requirements.

The most important aspect of organization is planning the production of the work in increments that do not involve repeating or redoing steps once they are completed. Obviously, doing this maximizes the efficiency of the operations and also boosts morale. No employee, whether at the highest or lowest levels, finds satisfying the prospect of redoing a plan because critical information available at the time it was done was somehow not distributed promptly and properly. In the end there can be no excuse for this omission. Due to circumstances beyond anyone's control and the nature of the design process, there are occasionally times when the design evolves and the drawings must be revised. Therefore, every effort should be made to limit rework to those situations in which it is absolutely necessary.

Organization according to critical path is important whether documents are being produced by manual drafting or by computer-aided drafting. The project manager learns to prioritize those *critical* tasks which, in turn, have the greatest impact on a sequence of other tasks which are all required to be carried out in a timely manner in order for the entire project to be completed on time. Which drawing comes first, and what work must necessarily precede it and follow it? How much time should it take to do each particular drawing, and which other drawings in a set are dependent on the completion of the earlier ones? (See Figure 2.2).

Particularly on larger projects, the project manager may be responsible for seeing the larger picture, while civil or structural engineers or other employees, for example, may focus on particular details or problems to such a degree that they have less sense of the overall picture. It is therefore essential that the project manager know what everyone on the design team is doing at any given time, in order to be certain that what is being produced reflects the most current design decisions. One of the most difficult aspects of project management is the challenge of being certain that all staff are working from the most current plan sheets and using the correct, standardized language and symbols. Therefore, the organization of this information for all members of the design team is a critical part of an efficient design production process. Although for complex projects it is not always possible for every member of the design team to be aware of how his

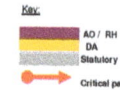

Figure 2.2 Critical Path Chart for K-farm, urban farming project, Hong Kong showing a completion date of May, 2019 (see also Chapter 9).
David S. K. Au and Associates Ltd. and Vicky Chan

or her efforts are contributing to the overall results, if the project manager can communicate some sense of the concept, strategy, and framework in which these efforts fit, it is always helpful.

Trust

Just as it is crucial to communicate with clients and colleagues, it is equally important to trust one another. Everyone on the design team, from the most senior to the most junior person, must have equal access to basic information about the project. There is no room for controlling personalities or insecure people who secretly hoard information with the fear that they would become obsolete if that information were available to those who needed it. By contrast, since every member of the design team has a different role, each functions synergistically when the lines of communication are smoothly connected.

Lawrence Halprin, in his book *RSVP Cycles*, introduced the idea that design is a cyclical process, and that the four major functions of the design process as he defined it, research, scoring, valuation, and performance, may occur anywhere within the cycle.[3] Design is not a linear process. Similarly, imagine being a musician in a chamber music ensemble. At the first rehearsal of a new work, all the musicians have the music for their instruments, but usually only a conductor or pianist has a complete score showing all the parts. At any point in the rehearsal process, any musician may stop in order to orient herself or himself to the cues in the parts for the other musicians, or to discuss how to phrase passages or interpret score markings or dynamics. This is a joyous but often tedious and challenging process, made possible by everyone trusting one another. Even in song recitals it is common for singers to perform from memory and rely on knowing their entrances from thorough review of the score and rehearsals with their pianist and collaborator. It is unfortunate that sometimes in design offices people are asked to perform without ever having seen the score, so to speak, or are expected to contribute to the whole even though they have no context for where their part fits. Pinning the tail on the donkey while blindfolded is not a game employees should engage in. (See Figure 2.3).

Clients can also be caught in these situations. It is not uncommon for a client to hire a consultant for his or her expert advice and then find reasons to ignore every recommendation that is made. This seems not so much a result of the client's expertise in areas unknown to the consultant as a matter of the client not trusting the consultant to do what she or he was hired to do in the first place. Often this problem can be solved by patient communications with the client in which clear examples are pointed out and explained.

Figure 2.3 Piano/vocal music score.
Source: www.imslp.org: from Franz Schubert's "Der Lindenbaum,"
"The Linden Tree," from his song cycle, "*Der Winterreise*," ("Winter Journey") Breitkopf and Hartel, publisher, www.imslp.org

Delegation

It is only human to have gaps in one's knowledge about any subject, particularly complex areas of design. When faced with such situations in the course of a project, a design team member must have enough self-confidence and trust in others to acknowledge that she or he does not have all the answers or even know the right questions. Therefore, it becomes critical in moving forward with the work to delegate tasks to members of the design team, the staff, or other consultants. The tasks one delegates to others are not only those one is ignorant or uncertain about but also those that are very simple but time-consuming and would better be accomplished by a person in a lower position in the firm.

Some offices are organized vertically, in which each person is expected to carry out all duties relating to a job, while others prefer a horizontal organization, in which each person has one principal area of responsibility. In complex projects, it is more efficient to assign people to what they do best, if that is at all possible. Occasionally this means that someone who is making a valuable contribution to the project may not be as stimulated as if she or he were assigned a totally different task. On the other hand, landscape architecture projects are diverse. Any two projects are distinct enough from each other that grading plans, specifications, construction details, and graphic techniques vary widely. A person with a specific area of responsibility can learn a great deal and maintain continuity by carrying out similar tasks on several projects. One way to avoid developing staff who are too specialized is to be certain that each has at least one major area of responsibility plus a secondary area. Ideally, each person should be working in both areas on the same or different projects. (See Figure 2.4).

Responsibility

The project manager must take responsibility for the success and failure of all communications and results. Regardless of the reasons for problems, the only way to solve them is for one person to have the responsibility of answering to the client and facing difficult situations. Complex projects can involve incredibly torturous and convoluted problems, with ramifications that take weeks or months to sort out. The client and the staff respect someone who is willing to take charge of a fraught situation. The project manager must be the sounding board for the whole office and the contact person for the client, regardless of the circumstances.

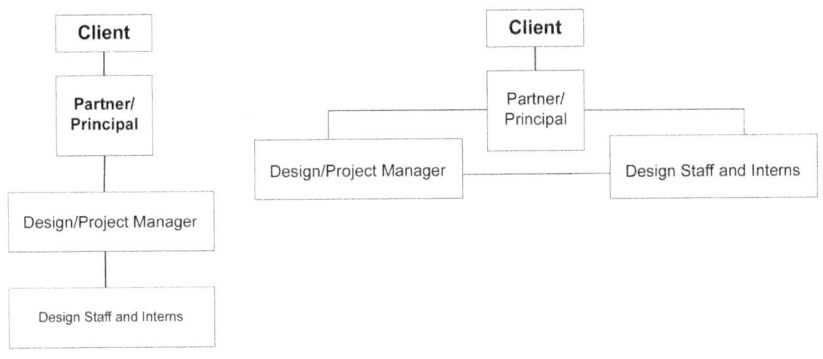

Vertical vs. Horizontal Organization

Figure 2.4 Vertical vs. horizontal organization.
Richard Alomar

Another important aspect of this principle is to hold others to a standard of accountability. If the project manager is responsible for direct, clear, and timely communications with the client or the client's representatives, then reciprocation is essential for a successful result. It can feel quite satisfying indeed to tell a client, "Yes, we submitted the design developments for review as you requested. We did not incorporate the changes we received from you last week, because as was agreed at the last review meeting two months ago (see the minutes of that meeting), we would need responses from you within a month's time in order to incorporate them into the next submittal." This response is not being combative, but merely encouraging the client to be timely and responsive.

Practicality and Common Sense

Practicality is a design issue as well as a project management issue. Just as the designer should not promise a client the reincarnation of Versailles when the budget only permits a color garden, the project manager cannot promise the client the delivery of a complete set of working drawings in two weeks when at least five to six weeks are necessary. The results of such efforts are disastrous: drawings full of inconsistencies that incite the frustration of the client and inevitably require time-consuming revisions. How much better to determine from the beginning a realistic design for a realistic budget with a realistic delivery date!

Flexibility and Adaptability

Landscape architecture deals with fluid subjects, anything but the immutable hard-line drawings that some draftspeople and project managers insist on producing. One of my first jobs was with a landscape architect, Edward L. Daugherty, of Atlanta, Georgia, who insisted that planting plans (and many other drawings) be done in pencil, rather than ink, to underscore and mimic the softness of the palette the drawings represent. He also sought to emphasize that just as designs evolve, so can the drawings to represent them. It is not a crime to change one plant material to another in deference to the client's wishes, nor to erase a line that is incorrect or to correct a misspelled word without replotting an entire drawing. Not is it inappropriate for the computer plotter to draw on a grading plan or layout plan a chain-link fence as a simple line symbol, rather than as some perfectly exact literal symbol with the precise number of posts required to scale. (See Figure 2.5).

A basic requirement of layout drawings is to include some give and play at beginnings or ends of long strings of dimensions. In this way, if there are occasionally site conditions or other factors that do not match precisely with proposed conditions, there are still ways to accommodate them in the field without having to delay the work and make costly revisions to the drawings.

Similarly, just as the drawings and design process must be flexible to reflect the nature of design, so must the process be adaptable when the design is applied to the site. Expect to make adjustments on the site. Develop construction

Chain link fence symbol (scale independent)

Chain link fence scaled drawing

Scale independent and scale dependent objects

Figure 2.5 Chain-link fence as symbol vs. exactly to scale.
Richard Alomar

drawings that have inherent starting points that can be adjusted to reflect site conditions or unexpected new conditions that may arise. Often the drawing that appears most precise may fit the site the worst. The layout of a centerline for a road or driveway, or a cluster of five major trees, may be adjusted in the field to best suit what the landscape architect encounters on the site. The drawings that he or she works from should allow this adaptability.

Honesty and Respect

More than a matter of following the golden rule, honesty is also a matter of self-interest. On complex projects that can last for months and even years, the project manager cannot be in a position of rehashing more than one account of important events. If careful notes are taken at meetings and incorporated into the contractual records, these minutes and notes must be used to construct chronologies and develop responses to questions from the client or the staff. If one is in the habit of being truthful, even if it may occasionally be embarrassing, then one does not have to remember more than one version of any important event. Project management built on fabrications, no matter how skillful, will eventually indict itself.

Designers draw upon complex information about how people behave in the environment and what types of facilities will make them comfortable and evoke the results desired by the client. It is hypocritical to think of ourselves as erudite designers if we do not treat one another with respect, acceptance, and dignity. There is no room for bigotry against any group—women, ethnic groups, immigrants, lesbians, gays, transgender people, people with disabilities, the elderly—in our profession. It insults everyone and undermines what we do. The person who has great expertise in grading plans that meet the standards of the Americans with Disabilities Act but then tells jokes about cripples does not belong in a landscape architecture office, nor does the designer who works on a garden for an AIDS hospice but uses homophobic slurs. Acceptance, respect, and courtesy are essential traits of well-adjusted people and will be appreciated by colleagues and the client.

Tenacity

Tenacity in the face of lethargic bureaucracy is sometimes required. Similarly, one must be clear on the basic principles governing the design and management of a project and be prepared to fight hard if they are being compromised. The more one develops good rapport with staff and the client on routine matters, the more natural it becomes to challenge the client or staff on critical matters. One must learn how to push without pushing too hard. One cannot be afraid to push or to say no.

Tenacity requires persistent communications with clients and consultants. Some would criticize and call this tendency pigheadedness, single-mindedness, or stubbornness. However, the results of such efforts count. There are simply situations in which it is much better to accept such criticism if, as a result, an important decision is reached with a client—for example, a client whose schedule is so busy that the consultant must call frequently in order to communicate the basics. My aunt Shirley once confided in me, "To try and fail is to learn; not to try is to suffer the inestimable loss of what might have been." Condensed into a project management principle, it is important for project managers to push the envelope and prod the design team into the best possible results. An experienced project manager will not win every argument but will be appreciated for her thoughtful persistence in doing her best to advance the design's best interest in an articulate manner.

If the Timing's Wrong, Nothing Else Matters

If the project manager is both a good communicator and a good organizer, the timing falls into place. However, the design team can be so excited about the work being done, so distracted by pressures of production, and sometimes so irresponsible that they forget to bill the client. Similarly, no matter how wonderful the product, if it is submitted a month late without careful explanation to the client, the job may be over.

Billings, submittals of milestone documents, and major communications must be prepared and submitted in a timely manner. Clients, particularly government agencies, often have specific requirements for billings. They may occur on the first of the month, with appropriate backup of employee time sheets, for example. A description of the work performed should accompany any bill as a frame of reference for both the client and the landscape architect. If there is a question about a bill months after its submittal, it is helpful to see a summary of the services for which the client was being billed.

The time to ask for payment for extra services is before they start, or at worst just as they are beginning, but not after they have been completed and submitted. It seems obvious, but too many designers overlook the simple notion that a client, no matter how generous and fair-minded, will be less likely to pay for extra services that she or he needs if the complete drawings are submitted before there is any discussion of those services in contractual language.

The submittals of milestone documents can sometimes become delayed as a result of numerous last-minute changes. Except for final contract documents, it is important to remember that a particular drawing is only as final as the revision date noted on it. For example, it is often better to submit design development drawings on time rather than withhold the submittal while waiting to resolve two or three issues. If these are major issues, it is likely that extra services will be involved. In addition, important issues require careful study. To include sketchy information about them in an otherwise comprehensive submittal may create problems during the review, as if the client is being insulted rather than accommodated. Therefore, it is better to submit the documents and discuss the extra services in a separate communication. Too often, if the project manager withholds a submittal indefinitely on a complex project while the last issues are being resolved, additional problems will be discovered by the client or the designers that will further complicate the design, delay its completion, and end any hope for a review in a timely manner.

Review a Project's Payroll Regularly

"Good" design that results in a huge monetary loss to the firm is *not*, in fact, good at all, but inept. The human costs in staff layoffs, deadline pressures, and overtime without compensation are very real and very significant to the people subjected to them. It must be the responsibility of the project manager to know, at any time, where the project is in relation to the budget. In addition, she or he must be able to anticipate what this relationship will be in the future. When financial problems arise, they must be addressed immediately, whether this means a conference with the staff to discuss how to work more efficiently, a discussion with the client about the fee structure, or both. There can be difficult situations in which the fee agreed upon is found to be low but there are no options for financial salvation. These are the types of projects in which it can be most crucial to be able to predict what is left to do and how it can be done as efficiently as possible while still serving the client's needs.

I am sadly familiar with a project in which the negotiated fee was too low, and even though the scope of work increased by millions of dollars, the client refused to acknowledge that the fee should be increased. The project manager, horrified at the situation, was too embarrassed about the negotiations and so dedicated to doing a perfect job that he neglected to inform the principals. Years later, the client is still asking for the completion of the work for the same fee, the drawings for this wonderful design are still not finished, and hundreds of thousands of dollars in payroll are gone and have never even been billed.

On the other hand, situations arise in which a design team, after careful consideration, may decide that a project is so prestigious that it justifies being carried out at a loss. For example, this could be a pro bono project that will greatly benefit the community or city in which the firm is located and might generate additional projects for which fees could be charged. What if a firm were asked to develop prototypical low-cost housing and community gardens for three sites under the expectation that if the designs are successful, considerably more work could be funded? Or perhaps, on a rare occasion, the firm might be approached by a very wealthy patron for a signature landmark building that would generate so much publicity that, even if implemented at a modest loss, it could reasonably be expected to lead to considerable additional work. There are few absolutes except not to go blindly forward without regularly reviewing payroll.

Don't Promise More than Your Team Can Deliver

Based on the contract, the landscape architecture or design firm is providing specific design documents for a negotiated fee. It is important that both the client and the designer accept these terms. During the design process the lead designer and the project manager may be tempted to provide a more complex design than can be developed within the fee structure or to specify materials too lavish for the client's budget. Unless there are negotiations between the client and designer in which there is an agreement to provide more than originally intended, which will cost more than the agreed-upon construction budget, keep such ideas and notions in abeyance, and save them for future discussions at an appropriate time.

Similarly, the schedule of the project is important. The client may want the landscape finished by the time of an event or party being planned. If this is an agreed-upon element of the contract, that's fine. But if it's not, it can be quite risky to give in to pressures from the client for a completion date ahead of schedule or incorporate extras without careful consultation with all the team leaders for the design.

Separate Contractual Matters from Design Issues

Project management is about both design and contractual matters. Although the distinctions between the two areas are often clear, at times they can be fuzzy. The project manager must draw the line—that is, know what is required under the terms of the contract, and be certain that what is being produced is within those parameters. At the same time, if the design process uncovers new directions that should be pursued, it is the responsibility of the project manager to communicate with the client at the proper time for additional compensation under the terms of the contract. It is important to rely on previous experience, particularly when working with a regular client. If a client has responded favorably to a particular approach, then such a strategy merits repeating when a similar situation arises. On the other hand, don't repeat mistakes. Finally, if contractual matters become difficult, it is important not to make idle threats. For example, never write or tell a client that you will have to stop work if payment is not made or an agreement on extra services is not reached unless you are absolutely certain that you will be able to direct your staff to do so. Bluffing can be disastrous.

Some firms have the policy that the project manager is responsible for both the design and management of a project. If practical, this is the preferred method, because she or he can immediately be aware of changes in the scope of work and fee structure, for example, that will have a bearing on the design. In larger and complex projects, this may not be possible; therefore, there must be clear communication between the principal designer and the project manager. The designer does not deserve to find, for example, that there is no more fee to finish the work, nor should the project manager endure a situation in which a great many beautiful design drawings have been prepared that have nothing to do with the listed items in the scope of work in the contract.

Acknowledge What You Don't Know

No one is omniscient. In most complex projects, and even in simpler ones, there will be issues in which your opinion or knowledge is not the most informed. Your abilities as a listener and manager behoove you to seek out solutions and other input from members of the design team who know more than you, have a different point of view and appreciate that you value their opinion. The discussion process by which such information emerges should not be suppressed. The project manager must acknowledge areas where expertise in allied or overlapping fields is appreciated.

The time to seek out diverse points of view, to gather knowledge about subjects relevant to the project, is in the beginning, when there is the most flexibility to consider diverse points of view. If some questions emerge for which you don't know the answer, don't hesitate to seek responses from those who might contribute a solution. If there is something about a design that is wrong—that is, it just doesn't fit—any member of the design team should be able to point this out, and the project manager should train all staff to come forward with solutions. Perhaps not every suggestion and improvement to the design can be implemented immediately, but the discussion process by which such information emerges should not be suppressed; in fact, it should be encouraged.

Share the Credit

When things go right, include everyone in sharing the credit. No complex landscape architecture project is conceived entirely by one person, nor is it ever realized by one person. It is so easy for a project manager to thank everyone who has participated in an award-winning project. Be certain that the names of all design team members are included in publications, press releases, articles, and other media contacts. The goodwill and appreciation that such inclusive policies generate more than compensates for the time spent learning how to spell and pronounce everyone's name.

Publicize the Work

Occasionally a project's timeliness and appropriateness light up the world. It is great but rare to celebrate such occasions. There are other successful projects in which the landscape architecture firm is bound by agreements with the client, who may be famous or wealthy, to avoid publicity and keep all details of the design private or secret. However, between these two extremes are myriad projects for which publicity is a godsend. Project managers must take every opportunity to bring attention to works whose execution and results demonstrate the expertise and sensitivity of the firm in solving problems and meeting the needs of the clients. Social media (such as Facebook), websites, and standard print media and broadcast networks are invaluable resources for such efforts.

Don't Reinvent the Wheel

Landscape architects often want to stamp a project as uniquely their own design by incorporating signature forms or other unique items that are instantly recognized as being associated with a particular designer. We have the Breuer chair, still recognizable decades after the architect's death. What about Cantor Concrete? Would this serve any purpose? Probably not. It is laudable to propose a design that solves identifiable problems and serves future needs. Yet care should be exercised before substituting a new design for a well-established traditional standard. That is not to say it should never be done, but be careful. (See Figure 2.6).

For example, consider planting details. There is considerable variation in methods of planting trees, shrubs, and perennials, and the specific methods are necessary details for any planting plan. If the planting is for a local parks department or other bureaucracy, there may well be established standards for installation. It's best to incorporate and comply with those standards, although sometimes adjustments may be recommended for specific plants in specific settings. For example, in a very windy site, it may be necessary to specify a thicker, heavier mulch layer to prevent evaporation of moisture and the mulch from being blown away.

There are standardized details for pool or tennis court fencing and similar items. Many pool manufacturers and tennis court manufacturers recommend fences of set dimensions with openings to prevent a baby's head or a tennis ball from becoming stuck in the mesh. Although there may be local variations in the height permitted or required, and some variation in the materials permitted or required, be particularly careful in adjusting the actual opening size of the mesh.

Riser/tread ratios for steps have been studied by landscape architects for decades. There are charts and formulas that will indicate the desirable tread or riser dimension when only one is known. Unless you're designing an unusual, nonstandard setting, such as just for children, or if you're trying to impede comfortable pedestrian movements, be cautious about ignoring these standards. The Louisiana Museum of Modern Art, 21 miles (35 km) from Copenhagen, features a 3 story children's wing with all dimensions scaled just for children including interior stairs, so that adults feel uncomfortable. Immensely popular activities for children are scheduled there.

Specifications for many materials are standardized as well. For example, over many years of testing, certain proportions of materials have been shown to result in the best quality of concrete for particular uses, so

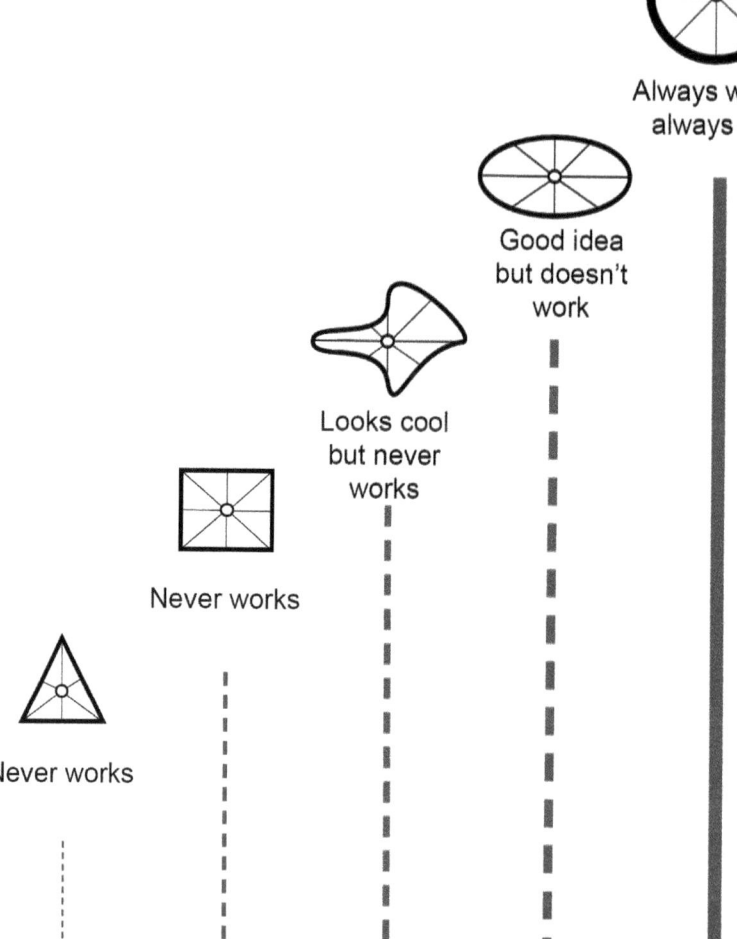

Figure 2.6 Don't reinvent the wheel.
Richard Alomar

every time that concrete is proposed for a construction detail, this same mix is usually applicable. It's rare that it makes sense to change or fine-tune standard definitions and recipes incorporated into long-established specifications.

A more complex example might be a unique play equipment installation. In the United States this is quite challenging, because there are so many potential liabilities arising from the use of nonstandardized components, like rubberized safety mat elements to absorb the impact of falls, that designers must be particularly cautious and rigorous in complying with all sorts of safety standards before embarking on or developing a unique design. Perhaps the sculpting of unique landforms and the massing and choices of plant materials are more rewarding with less liability. (See Figures 2.7a and 2.7b).

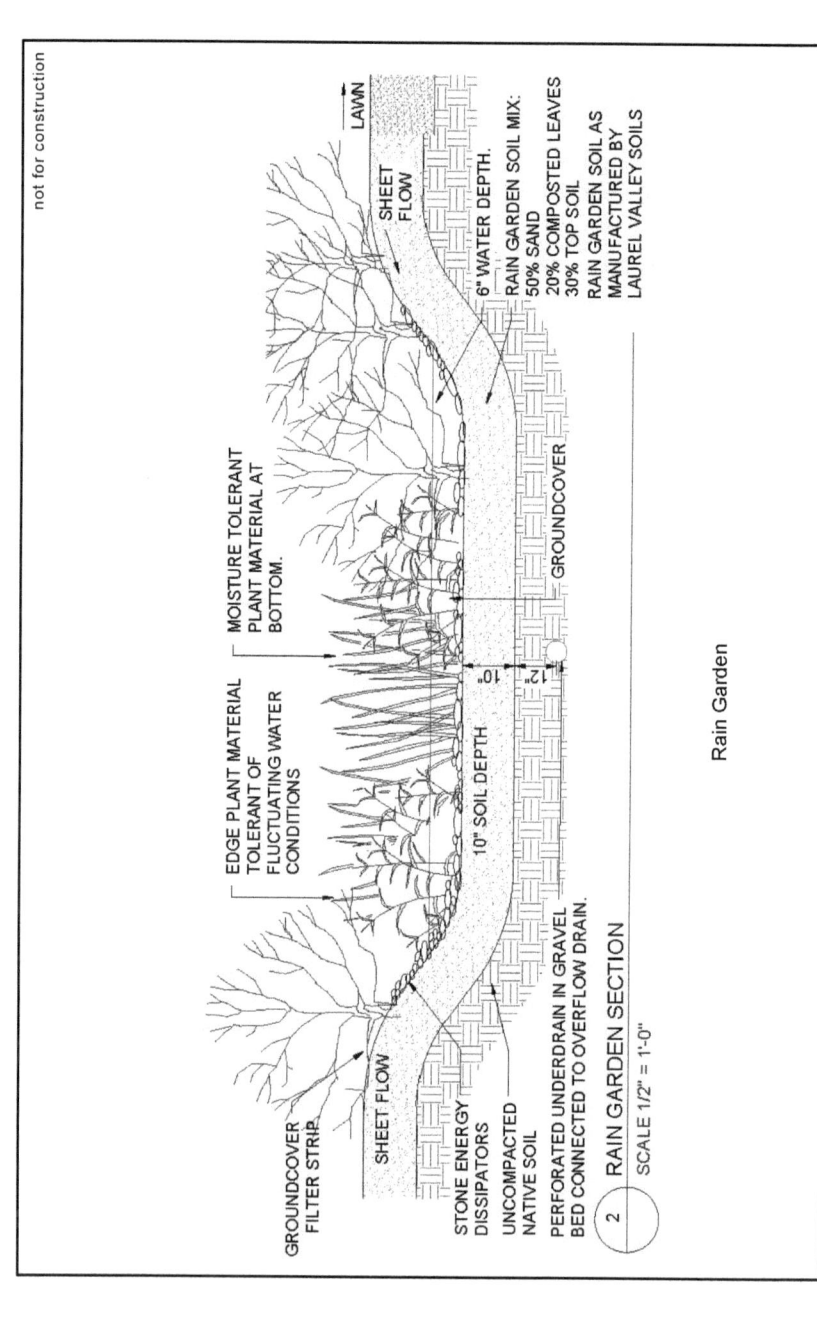

Rain Garden

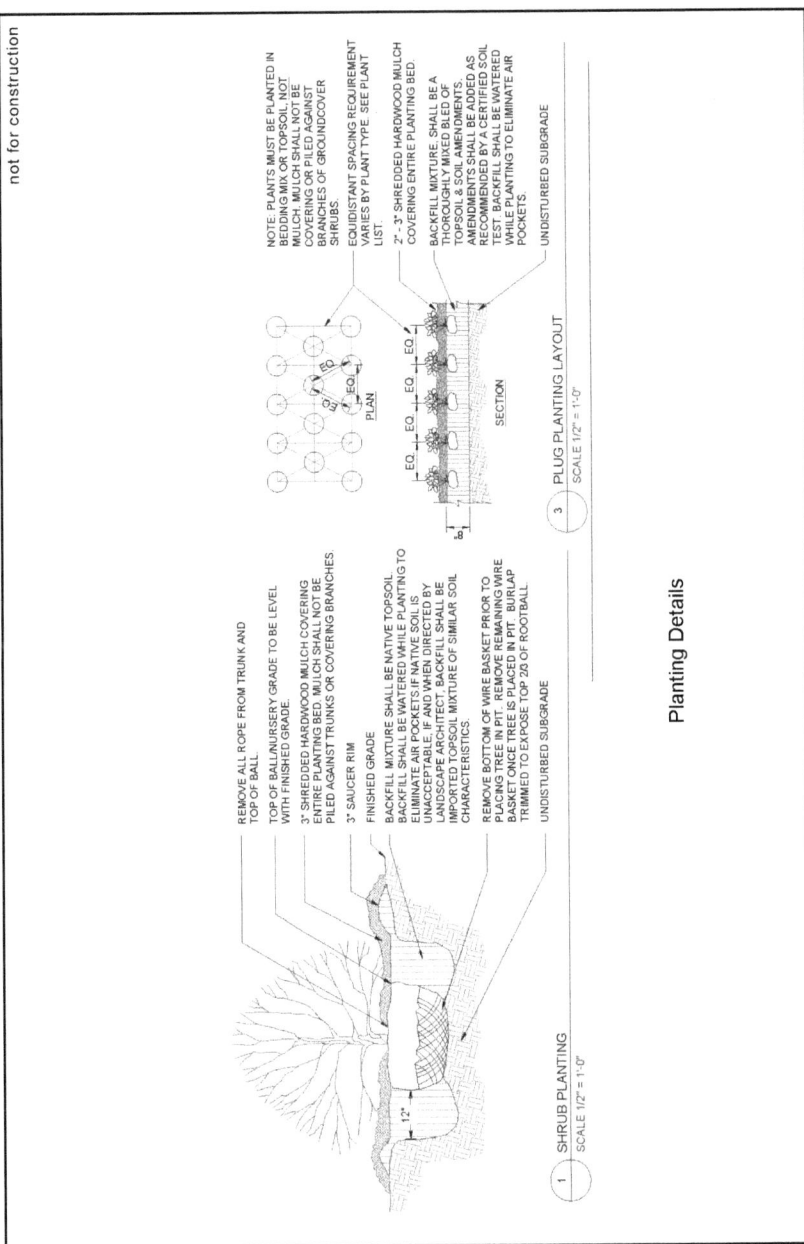

Figures 2.7a and 2.7b Example of something worth inventing as a unique design element, in this case a rain garden at the Floriculture Greenhouse Outdoor Living Lab at Rutgers University, New Brunswick, New Jersey (see also Figures 5.11 to 5.16).
Neil J. Werket and Richard Alomar

Instant Design

With the advent of the digital age, virtual reality, and videogames, don't overlook the possibility that some clients will expect almost instant results and instant feedback. There are tools that allow clients to pick and choose instantly from an array of design choices, or at least to react to design choices very quickly when speed of decision is a priority, in the same way that email and smartphones made faxes almost obsolete. Although ruminating about design decisions—that is, to draw something up, write up a concept, and put it aside for a weekend—may still be worthwhile, designers are often pressured to make major decisions quickly. There are software packages like BIM, for example, that convert two-dimensional drawings into a three-dimensional model or animation sequence, enabling contractors to estimate costs faster and anticipate how to place equipment and stage the sequence of operations necessary for complex projects in a more concise manner, all of which saves time and money for the client.

Look Before You Leap

In spite of all the modern ways to streamline the design process, including all of the tools that enable the quick generation of reams of analysis, myriad concepts, and diverse designs, take enough time to be convinced that you have the correct final design and detail before you make a major submission to a client. We live in a pressure-cooker world, but moments of relaxation are still important. Routinely demand a final review by a colleague of an important set of design documents, but also know the difference between a final review and anal retentiveness. Let them go on time.

Problem 1

RUMOR MILL

Depending on the size of the class, you may want to divide it into two or three groups. If there are twelve or fourteen students or colleagues, divide them into two groups of six or seven, who should sit in two separate rows side by side. If there are fewer than twelve, then students may sit in one continuous row, end to end. If there are more than eighteen, then divide them into three groups of six students each (or more) in a line.

The teacher should hand out to the student at one end of each row one of the following statements. That student should read it carefully and speak it clearly into the ear of the neighboring student, who repeats the statement until the last student at the end of the row has heard the statement. Then the last student should clearly state what s/he feels to be the statement and compare that with the written statement the first student still holds in his hands.

1. *Acer rubrum* "October Glory" is a very adaptable cultivar of red maple.
2. Plant fourteen *Forsythia suspensa* along the length of the retaining wall at the toe of the slope.
3. Use mortar mix Type C with raked joints for a brick retaining wall 3 feet 6 inches (1.1 m) high.
4. Do you prefer *Hemerocallis* 'Miss Sunshine' or 'Daisy Mae' and at what spacing for the spring garden on the south side of the house?
5. The driveway is flanked by an allee of 18-foot-tall (5.5 m) sugar maples planted 20 feet on center.

Problem 2

PROJECT MANAGEMENT WRITING
(SEE ALSO CHAPTER 8, PART 4)

(In the following problem, any terms in italics may be changed to suit the particular instructor or studio leader, or your own personal, work or studio setting.)

1. Write a letter to *a client* in which you recommend a solution to a particular problem relating to design drawings you have prepared.
2. The letter may not exceed two pages, double-spaced, word-processed.
3. You may include with the letter one sketch, at 8½ × 11 inches (same size as letter).
4. Address the reason for your site visit and describe the problem.
5. Indicate *the drawings and specifications* that are relevant to this problem.
6. Explain how what you've encountered on the site is different from what is required by the drawings and specifications.
7. Provide your recommended solution.
8. Some examples could be:
   a. *Plant materials* are not in compliance with drawings and/or specs.
   b. Material or construction is not in compliance with drawings and/or specs.

    c. There is some other issue with lack of compliance.
9. Additional work. If there is a significant change in scope of work that would result in a change order, describe the reasons this is appropriate (and not something for which you should be held responsible), explain why it is necessary, and indicate what the anticipated cost increase is going to be. Call it Change Order #1 and any drawings relating to it SK-1a, SK-1b, etc. This change order text should not exceed one to two pages. You may include one additional sketch with it. If you submit this, then the maximum size of the document will be three pages for questions 1–8 and three pages for the change order.

Please do the basics immediately.

Set up a sheet size with office name, address, phone, and email.
Include the *name and address of the client*.
Write in a business letter format.
Be certain that all pages are in the same format.
Sketches must be the same size as the letter (8½ × 11 inches).
Number the pages (the better format is "page 1 of 6," "page 2 of 6," etc., so that the person reviewing it knows the total number of pages included).

PART 2: BILLING

One of the most important but often overlooked aspects of project management is billing. No matter what the quality of the design work created, detailed, and built, or other tasks performed, and no matter what the accolades received from appreciative clients, if the firm or individual in charge does not bill in a timely manner, all sorts of problems emerge. Most design firms run with tight profit margins, so in order to pay salaries when employees expect them, the inflow of cash received from invoices is critical. If an invoice is sent out late, it becomes harder and harder to demand timely processing and payment. It's not that clients or government agencies are punitive; it's simply that they must follow specific protocols upon receipt of an invoice, usually involving a step-by-step procedure that takes a minimum of days or weeks to execute, so patience is demanded. The time to realize that an invoice should be sent is immediately upon completion of the task being billed, and not when it's time to issue bimonthly payroll checks and the funds to pay everyone are not available. Depending on the nature of the projects under contract and the typical time frame a particular client takes to process an invoice and send payment, the most important task for the accountant is to keep checking politely on the status of every outstanding invoice to be certain that any delays are not the firm's responsibility. The best way to check is to call directly. Emails may establish a record of communication, but direct conversations are the timeliest way to ensure that concerns are expressed and understood. Occasionally such a call may help expedite payment, as the representative of the client may have minor or major questions that the accountant could easily answer without having to resubmit the invoice. Also, don't overlook that simple, repeated human interactions with another individual build rapport in a way that impersonal communications do not.

Importance of Billing

The following discussion underscores how important billing procedures are to the long-term success of a design office and its smooth functioning. When a new contract is signed and a project begins, the project manager must meet with the office accountant (sometimes this is the same person as the project manager, and s/he has dual functions) and set up the tasks to be performed and the invoicing that should occur during or at the conclusion of each task. Each project and client may be different, so it's urgent to set up a task number and budget, as per contract, for each item of work to be performed and to be billed. The project manager assigns people to work on each task and is responsible not only for managing the design of the project but also for keeping track of employees' time so that the work to be performed does not exceed the allocated budget. If s/he senses that the amount of time spent on a specific task is taking up a greater percentage of the time (and fee) allocated to that task, then it's essential to review the work done, study how it can be completed more efficiently, and continue to keep track of it. The importance of this responsibility of the project manager cannot be overstated. Often it can take a week or two to track and print out the records showing each employee's allocation of time. Principals may review such documents perhaps twice a month. If the project manager is not fully engaged in reviewing the work in progress, both from a design perspective and from a financial perspective, then the firm could well be in a big financial hole by the time s/he brings it to the attention of higher-ups in the firm. By that point, it could be a crisis with little hope of corrective actions. Occasionally, overruns (that is, spending more time/salary on a certain task or phase of a project than budgeted) occur due to a learning curve—the client is new and has certain ways of working, and it takes time for the team assigned to a project to adjust to each other and the project requirements. But by the second or third task within the contract, everyone is working efficiently. At other times, overruns may occur because the task to be performed is more complex than originally conceived or the client starts to ask for items of work that the project manager judges to be outside the scope of work of the specific task. When this happens, it's important to document the concern to the client and attempt to negotiate a "change in scope" or "extra work" addition to the contract. If such patterns emerge several times during the course of a contract and the client is responsive, so that both the client and owner feel that an equitable solution is reached, then this is a promising sign that future projects with this client or agency should be encouraged and sought. However, if every such discussion of extra work involves long delays or protracted negotiations, if they result in fee increases that hardly cover the extra time involved, or if approval for payment for extra services is denied altogether, then this is a sign to be very cautious in undertaking additional projects with this client.

Timely Billing

To bill in a timely manner requires a lot of coordination between the project manager and the accountant. Together they should review what the client requires for submittal of each invoice. The requirements vary tremendously depending on the nature of the work, the type of client, and the specific requirements of the contract. For example, assume that a contract has five phases:

1. Schematic	15% of fee	$ 7,500
2. Design Development	25%	$ 12,500
3. Construction Documents	35%	$ 17,500
4. Bidding	10%	$ 5,000
5. Site Construction	15%	$ 7,500
TOTAL CONTRACT FEE		$ 50,000

Then, for each phase of work, the project manager should determine how many person-hours of time assigned to the staff s/he has in mind will add up to these fees. Since changes may occur, it's prudent to include a safety factor, say, 10%, so that when 90% of the fee is used up, a red flag registers; then the work can be reviewed and a method can be found for completing the work as efficiently as possible. When the client has approved a phase of work and authorized payment, then the project manager must communicate with Accounting to be absolutely clear what must be submitted. The contract may be simple enough that all that needs to be stated and verified on the invoice is that, say, Task 1 or 2 is complete and the fee is $X. However, it's often the case that clients require backup, so the Accounting Department may need to print out time sheets and mark the hours each employee has spent on the task being billed. Sometimes at the beginning of a job the client must approve the job title, function, and hourly rate for each employee assigned to the job. Often government agencies have a particular invoice form, frequently in Excel or other spreadsheet format, that must be used to calculate the total due. Again, this form should be reviewed carefully by the landscape architect and project manager when the contract is signed, so that it's absolutely clear to them how and when it's to be used, and what a complete submittal consists of. The project manager must be certain that all of these preliminary tasks are accomplished and approved by the client *prior* to starting significant work on the first task. Once this has been done, then subsequent tasks should run smoothly.

Verifying Receipt of Invoices

It's still the case that most invoices are sent by surface mail or courier, rather than email. It's important for the accountant to verify receipt—to know that a client has received the completed invoice and that it was sent to the correct department and is not sitting around in the wrong office or department. After the accountant is certain that the invoice has been received, it's important to follow up within a few days in order to receive verification that all is in order and the client doesn't notice anything missing, such as a stray time sheet or one page that may not have printed clearly. Once the accountant verifies that the invoice is received and is complete, then s/he should call every week or two to check on its status. Gentle pressure is helpful, as some clients are besieged by invoices at certain times, so politely reminding them that you're waiting for payment keeps you at the top of the list for processing. Yet even if you feel that payment is overdue, it's important to be polite and firm in all communications and not risk antagonizing anyone.

Delays in Being Paid

Sometimes it's a challenge to get paid. One factor that may complicate the process is that the scope of work for which you are billing is a minor part of the contract because you're a consultant or subconsultant to the principal firm in charge. This firm may take the attitude that they can't pay you until they are paid. Minority business enterprises, women-owned businesses, or any consultant whose scope of work is somewhat limited (5% to 20% of a budget task), are often in this situation, and it can be daunting. Perhaps the most important aspect of handling such financial issues is to establish rapport with the client and principal firm from the beginning and emphasize that timely payment is urgent. If you find that an effort is made to pay you promptly, then this is a good sign of a solid working relationship. If, however, most efforts to be paid are thwarted, and you're in a situation where payment is weeks or months overdue, this is the time to consider finding other firms that need your expertise as a consultant. Although it's often a standard practice in the industry for the prime (the principal firm in charge of a project, for which all other firms are consultants or subconsultants) to tell firms working under their direction that they will get paid when the prime gets paid, this can create difficult problems for small firms. The small percentage of the contract for which they are responsible may still be their major income for several months.

Ancillary Costs

Pay strict attention to all contract requirements that involve costs. For example, sometimes the contract will spell out how many copies of construction plans, specifications and cost estimates must be submitted, and what scale these drawings must be. It may be necessary to take a lot of photographs and print some images for site analyses or other purposes. Even with advanced digital photography, the time spent photographing and the cost of printing images for review can be considerable. If many different individuals or agencies must review sets of drawings and documents, the logistics and costs of printing and delivering can add up quickly. If the client requires that half-size drawings must be delivered as well as full-size sets, it's important for the project manager to establish graphic standards that will be clearly legible at both scales.

Reviews, Approvals, and Permissions

Reviews and approvals are sometimes spelled out clearly in contracts, sometimes not. In complex projects involving governmental entities, it's often necessary to have reviews and approvals by various government agencies at different phases of the work. Some approvals processes require public hearings, presentations to the public or to agency personnel, or both. Prior to starting work on a project, the project manager should verify with the client exactly what reviews and approvals will be necessary. For complex projects, this may need to be a task unto itself. For example, a historic cultural landscape project may need review and approval from a local landmarks commission or public design commission. Their reviews may require submitting not only documents but samples of proposed materials as well. The landscape architect may need to present to a design commission, appear at a public hearing, and present a prepared PowerPoint presentation, all of which take considerable time and coordination, so the contract for services must cover these items of work. Often it is assumed that the landscape architect is responsible for all permissions and approvals, because such permissions and approvals tend to be standard and most firms are well educated and informed about them. Negotiations must occur in order to agree on a fee for all such aspects of the job. Similarly, it's urgent to know when each permission or approval should be applied for. In some cases, the same agency might require approvals at both a preliminary stage and the final stage.

One solution for firms that are primarily subconsultants for public works (depending on the size and expertise of the staff) is to find and seek out one niche for which the firm can be the prime or leader rather than follower. Such a position may seem at first to be untenable and require a lot of time and deliberation to bring about, but to be in a situation at least occasionally in which you are directly in charge of all major design, management, and invoicing decisions may help to stabilize cash flow and sustain the firm during difficult times. This may also be helpful in an increasingly competitive environment in which employees, even at a senior level, don't always know the entire context of every project on which they are working. To feel that one has control over and clarity on at least a few projects makes others much more palatable.

A sad reality in the modern world is that with international competition, and with the delegation or farming out of some aspects of the work to drafting pools or other specialists, landscape architecture practice may be less profitable. Flexibility in working hours and the application of benefits can help in navigating the vagaries of working life. For example, employees might be asked to work extra hours during a crunch period to finish a project, and then take comp time for several days after the project is finished, while being paid the same amount week to week.

PART 3: OFFICE STANDARDS AND METRIC SYSTEM
Word and Excel

Even small design offices need standards for drawings and documents so that anyone needing to find a particular file will know what procedures to follow in order to retrieve it and open it and will have a clear sense of how to proceed if revisions are required. Often someone must work on a drawing or finalize a letter that was initiated by another colleague. The following discussion touches on overall organization of Word documents, Excel spreadsheets, and AutoCAD drawings.

For each major project, it's useful to have a folder in the computer system that is labeled with the client's name, usually last name first, followed by first name or initial. Documents within a project may be divided between those "sent" (or "out") and those "received" (or "in"). Within a folder marked "out" or "sent" the staff member may wish to create subfolders for all the types of documents that can be expected to be included during the duration of a project. Finally, if it's anticipated that the project will take several years, some firms choose to further subdivide the project by creating separate folders for each year that work is being done. Assume that a new client's name is Martha Spruce. Therefore, the basic organization could be as follows for a labeled project. The examples indicated are for Word files, with the ending .doc. Excel documents would be the same except with the suffix .xls.

BASIC FILE ORGANIZATION

Spruce-Martha	
IN or RECEIVED	OUT or SENT
2016	2016
Correspondence	Correspondence
Drawings	Drawings
Estimates	Estimates
Photographs	Photographs
2017	**2017**
Correspondence	Correspondence
Drawings	Drawings
Estimates	Estimates
Photographs	Photographs
2018	**2018**
Correspondence	Correspondence
Drawings	Drawings
Estimates	Estimates
Photographs	Photographs

Since dates in Word automatically organize alphanumerically, it's practical to start the file name of any document in Word with the date as a string of eight numbers, beginning with the year, then two digits each for month and day, respectively. Then develop some additional nomenclature to describe the purpose of the document and perhaps a few additional letters to suggest the recipient.

YEAR-MONTH-DAY-RECIPIENT-SUBJECT
20190319-Parks-peren.doc (a letter sent on March 19, 2019, to the Parks Department about perennials)
20190423-DEP-drnge.doc (a letter sent on April 23, 2019, to the Department of Environmental Protection about drainage)
20190806-Jones-cost-est.doc (a letter sent on August 6, 2019, to Jones about cost estimate)

This basic string length doesn't explode into impossibly long file names and it's easy for anyone scrolling through the folders on this project to identify the date and purpose of the document without having to open and review it. Some firms and project managers develop elaborate codes and abbreviations for each project name, each project participant, and other detailed data. However, this is only useful to the extent that it's applied consistently and becomes the template for everyone to share in an office. If some people find it too tedious or time-consuming to follow such a highly detailed model, then the risk is that some older documents will disappear from the system and require advanced search options to locate. Therefore, if several months after discussion of drainage with the Department of Environmental Protection it is necessary for a landscape architect to review this topic again for the same project, the file is easily found under:

20190423-DEP-drnge.doc.

Some organizers prefer to use the underscore line (_) instead of a hyphen (-) to separate elements within the naming system. I prefer the hyphen because it stays readily visible even if the entire file name might be underlined and because it's too easy to forget to use the underscore and end up with a bunch of files in which some have hyphens and some underscore lines, a confusing result. Therefore, 20190319-Parks-peren.doc and not 20190319_Parks_peren.doc. At the same time, the underscore line has its proponents because it's less corruptible when copied by some software. If the firm for which you work has an IT professional, she should review and develop guidelines for all such standards. But it's good to have an office discussion that includes her and experienced employees, because everyone may have ideas about glitches to avoid and shortcuts that help in naming and saving files and organizing folders.

AutoCAD

AutoCAD conventions are equally important. Staff highly experienced in AutoCAD should develop conventions for all standard drawings, so that anyone working on a drawing, whether initiating it, updating it, or archiving it, knows exactly how to proceed: what layers should be uniquely assigned to a color, and be plotted at a particular line weight so that a layout plan printed now will have the same legibility and graphic character as an earlier version of the same drawing printed two years ago. Developing and enforcing AutoCAD standards within the office is time-consuming but pays off in terms of efficiency of the work and consistency of results.

If the firm or individual is a principal or prime on a project, then each consultant who is under contract or joins the design firm must be informed about these standards as well. Within AutoCAD (or other software programs) the work accomplished by the surveyor, lighting or irrigation consultant, or any other consultant must be clearly distinguishable from the base data or other design information shared or provided by the prime. If standards and criteria for achieving this are not emphasized from the beginning of a project, it can become very difficult, even impossible, to identify any work done by a consultant or subconsultant on a job. All AutoCAD drawings initially received by consultants must first be saved as undisturbed digital originals, and then *referenced* to appropriate design and construction drawings being developed. If any questions or conflicts occur about the consultant's data, the landscape architect can refer to the original with confidence and contact the consultant with specific questions if necessary.

It is important to occasionally have tutorials or continuing education sessions, either outside the firm or within it, to update staff on new standards or important changes to existing office standards. At such meetings, comments can be solicited for suggestions on how to improve office standards. If all employees working regularly in AutoCAD or other programs feel that they are contributing to the ongoing development of the firm's standards, then it sustains their interest and keeps them involved and creative.

Several principles are critical with AutoCAD standards. A record copy of any incoming AutoCAD drawing received from a consultant (surveyor, architect, engineer, landscape architect) must be saved as a unique reference document before it is edited in any way, so that if any questions arise about the information provided on it, a staff member can easily check the original and verify information. In assembling complex drawings based on such reference documents, there must be a common X-Reference, usually established by the standards of the office with an insertion point so that each drawing reads correctly in relation to another in model space and prints correctly in paper space. In AutoCAD colors are assigned to elements based on the density (thickness) of the line that will eventually be plotted for that element; for example, wall coping will be assigned to a color representing a thicker plot line than the color for a wall base. Over many iterations of projects and drawings and plots, a firm will develop and perfect a set of layer/color assignments that work best for its particular designs.

Various offices have different requirements about how to manage the preparation, labeling, and digital filing of a set of drawings in AutoCAD.

Even what to do in model space compared to what to do in paper space may vary from place to place, so the critical factor is to learn the standards for whatever office that you join. Rules and directions will cover such requirements as how to date drawings, how to label and update details, where to create match lines, the steps to follow before starting to work on a new drawing (such as freezing unwanted layers in a base sheet received from a surveyor or architect), the creation and use of blocks in planting plans or other special features, plotting standards, and the archiving of a project when it's finished. Especially valued are employees who studiously learn the requirements and follow them, yet also make occasional suggestions at appropriate times to the CAD manager about how some methods might be improved.

Minutes of Meetings

As just one example of important documents generated by an office, that might be produced by many different people over a long period of time, consider minutes of meetings. They are perhaps one of the most important summaries an employee produces because they provide a continuous record of the progress of a project over its entire calendar. Also, for anyone becoming part of the design team well after the project has started or needing to review its history, to sit down with the sequence of design drawings and minutes of meetings in chronological order gives a clear sense of the progression of the decisions on all aspects of the project and how the design evolved. If disagreements occur or disputes arise, the minutes may form a basis for resolving them.

A considerable level of skill is required to write and edit them, and important guidelines follow. For example, one purpose of minutes is to hold people accountable for what they are responsible to do within the roles they are assigned in a project. Often a set of minutes ends with a list of items of work pending or critical tasks remaining. If the person within the group who is responsible for each of these tasks is identified (and it may be more than one person), then there is a way to follow up with each team member. For a busy principal in an office who needs to review the status of a number of projects in a short period of time, he or she should be able to skip to the end of each set of minutes and know what's going on. Therefore, a good set of minutes concludes not only with a final summary (such as "tasks completed," "key tasks remaining," and action items such as "schedule a meeting for coordination with Agency PDQ within ten days") but also, and equally important, *with the date and main agenda for the next meeting.* Finally, all minutes should be submitted for review to all attendees by the person who prepared the document (usually, but not always, this is someone who attended the meeting). Anyone who attended the meeting and notices an error, needs another copy, or has questions about the contents must be able to contact someone immediately to submit or request additional information. Usually this contact person is the project manager for the job, or at least someone in the position of being able to facilitate such a request.

Stylistically, minutes should be written in a clear, concise style. Many of the rules of specification writing apply, with the exception of not writing in the imperative voice. Many firms have their own templates and requirements, so be certain to review your firm's or client's requirements. Long prose should be avoided; crisp, shorter sentences are preferred, as they tend to evoke the least amount of disagreement or confusion, particularly when complex design issues are being itemized or discussed. Bulleted (or, better, numbered) summaries and lists are fine. For meetings that last a long time, it may or may not be suitable to describe events chronologically; sometimes it's more important to order them by importance, yet still include everything. Rely on your professional judgment: if ten critical topics were covered, include that as a boldface sentence at the beginning of the minutes, and then devote adequate coverage to each.

DOS AND DON'TS OF MINUTES OF MEETINGS

1. Have the person taking minutes identify him- or herself, so that people tend to speak toward that person and give him or her any documents that need to be copied and distributed to others during the meeting or after it concludes. If others know that someone else is in charge, it frees them up to focus on other matters. At the same time, if they take notes too, they can give them to the person taking minutes, for incorporation and corroboration. At meetings at which complex issues are discussed, it's helpful to have more than one set of notes.
2. If you are responsible for writing the minutes, take careful notes during the meeting, whether on a smartphone, on a laptop, or by hand.
3. If possible, write the minutes on the day of the meeting. Even with accurate note-taking, one forgets very quickly what happens, so accuracy or a sense of urgency can be lost very quickly.
4. Follow required formats. Many public agencies or firms have standardized templates that must be used, and they automatically reject

minutes that do not comply with these formats, no matter how accurately they are written.
5. Be certain to identify everyone who attends the meetings as well as those who usually attend but might have been absent and should receive a copy of the minutes. There might also be a list of people who do not attend but should still receive copies of minutes of meetings about projects relating to subjects or departments of which they are in charge. There should be a clear way, such as with single and double asterisks, to distinguish between those who attended and those who did not.
6. Give the full name, title, and contact information (usually phone number and email address) of each attendee. Be certain to give the correct contact information, and not private email or phone numbers that the attendees *do not* want shared with others. List the names of presiding officers, principals, or other key people first. For large groups, sometimes it helps to boldface the information for key people.
7. If someone within your firm or organization attended the meeting with you, even if he/she was not responsible for the minutes, ask this person to review your first draft of the minutes. Often another person will remember important details or points of emphasis, or the presiding mood or tone on key issues that you might miss just because you are so focused on all of the details. Two heads are better than one.
8. Follow through. Often minutes of meetings cannot officially be distributed before key representatives of the sponsoring organizations have approved them, so corrections must be made within a short period of time.
9. Conclude the minutes with the date, time, and location of the next meeting; the work to be discussed; and any incomplete work that must be brought to a conclusion in order to be discussed at this meeting so that the project can stay on schedule.
10. Don't embarrass anyone. At the meeting itself, for example, it's fine to have a discussion about a particular item of work that is late, needs work, or must be further studied. However, in the minutes, indicate the item of work clearly, and give the name of the firm or firms responsible. Often it may be a joint responsibility of a firm and an agency. *At least at first, don't single out an individual.* The only reason to start listing or calling out the individual would be if this problem persists across several meetings and no one at the firm or agency responds. The lack of a response might require the project manager to call the individual(s) on the team within the particular firm and/or agency and ask for an explanation for why the work is delayed. It might even be necessary to have a separate meeting on a particularly challenging issue. Design cannot occur in a vacuum. Yet

if everyone feels terribly pressured or singled out, it may not occur at all.

11. Use the previous minutes of meetings as a "table of contents" for the next meeting, to be certain that there are no loose ends and that the project is moving along steadily.
12. Don't share minutes of meetings or accompanying documents with people not associated with the project, as the listings of participants with contact information and the detailed discussions of design, construction, and other issues might occasionally contain information that should remain private.
13. In your role as project manager, use the minutes as a step in the budgeting process. After each meeting, use the minutes to calculate what proportion of the budget has been spent and what remains, and make a rough estimate of where you are within the overall budget. If you are situated about where you need to be, congratulate yourself. If you have had three meetings and anticipate that you need four or possibly five more, yet have spent half the budget already, then have a discussion with the principal to see where savings can be made and whether any of the work remaining could be construed as extra work. You will likely *have* to manage with four meetings, not five. It might still be possible to manage within the budget.
14. Beware the contract that allows for an open-ended number of meetings, which you can assume that you will have to attend, possibly with a colleague, and produce a set of minutes. There is a limit to how much discussion and interaction with clients and consultants is necessary or desirable in order to produce a sound design. Yet if half of your design fee is spent on meetings and minutes, distributing the minutes, and making corrections to them, how will this lead to results on the ground?

Metric System

Almost all countries outside the United States use the metric system. As of this writing, there are only two, Liberia and Myanmar, that do not. Although some state and federal agencies in the United States require plans in the metric system or drawings and documents containing metric equivalents for imperial measurements, it is difficult to predict when, if ever, the entire United States will make such a major jump. However, if any landscape architect or other designer wishes to engage fully with work in a foreign setting or even to collaborate from a U.S. office via email, computer, and video links with a foreign-based company doing design projects for overseas settings, then a basic knowledge of the metric system is essential. Most software will automatically convert or adjust to the metric system, so what sets one employee apart from another is having a working knowledge of the metric system and knowing the basic use of the measurements in a particular country and culture: what's cricket in one is baseball in another.

The following measurements are useful for understanding the metric system; approximate equivalents are provided, along with a more precise equivalent. Even if one becomes fluent in the metric system, to have memorized or be able to easily recall some approximate metric/imperial equivalents is helpful.

METRIC AND IMPERIAL MEASUREMENTS

Distance	
Approximate	**More Precise**
1 inch (in.) is about 2.5 centimeters (cm)	1 in. = 2.54 cm
1 cm is about 0.40 in.	1 cm = 0.395 in.
1 foot (ft.) is about 31 cm	1 ft. = 30.96 cm
1 yard (yd.) is about 1 meter (m)	1 yd. = 0.9144 m
1 m is about 39 in.	1 m = 39.372 in.
1 m is about 3 ft.	1 m = 3.281 ft.
1 mile (mi.) is about 1.6 kilometers (km)	1 mi. = 1.61 km
1 km is about 0.6 mi.	1 km = 0.6214 mi.
Area	
1 hectare (ha) is about 2.5 acres (ac.)	1 ha = 2.47 ac.
1 ac. is about 0.4 ha	1 ac. = 0.405 ha
Liquid volume	
1 liter (L) is about 2.0 gallons (gal.)	1 L = 0.265 gal.
1 gal. is about 4 L	1 gal. = 3.79 L
Solid volume	
1 cubic yard (CY) is about 0.8 cubic meter (CM)	1 CY = 0.765 CM
1 CM is about 1.3 CY	1 CM = 1.31 CY
Weight	
1 pound (lb.) is about 0.5 kilograms (kg)	1 lb. = 0.454 kg
1 kg is about 2.0 lb.	1 kg = 2.2 lb.
Scales for drawings	
1:1	Full-size
1:2	Half-size
1:5	3" = 1' 0"
1:10	1½" = 1' 0"
1:20	¾" = 1' 0"
1:25	½" = 1' 0"
1:50	¼" = 1' 0"
1:100	⅛" = 1' 0"
1:200	¹⁄₁₆" = 1' 0"
1:250	(1" = 20' 0")
1:500	(1" = 40' 0 ")
1:1000	(1" = 80' 0")

All these equivalents notwithstanding, if you are converting drawing details in the imperial system to details in the metric system, there are two options: to provide precise equivalents, which is helpful with a limited number of details in which absolute precision is required, or rounded equivalents, which is suitable for most examples. So for a planting detail in which the tree pit is to be excavated to a depth of 2 feet and the mulch will be a 3-inch layer of shredded hardwood bark, it's fine to translate the depth of the tree pit to 60 or 62 cm (or 600 to 620 mm) and translate the mulch to a thickness of 8 cm (or 80 mm) or even round farther up to 10 cm (or 100 mm). Use rounding to create metric equivalents that are practical. The same types of approximate equivalence might be applied to depths of various layers for a pavement cross section or excavation. On the other hand, if details must show dimensions of metal hardware or other elements of precise dimensions, then it's important to give an exact metric equivalent. (See Figures 2.8a and 2.8b).

NOTES

1. http://www.vulture.com/2013/01/steven-soderbergh-in-conversation.html (Interview originally published Jan. 27, 2013). Also quoted by Maureen Ryan, Chair, Film Program, Columbia University, May 16, 2018, Miller Theater, Recognition of Graduates, Film MFA and Film and Media Studies MA.
2. Some of this section originally appeared in Steven L. Cantor, *Innovative Design Solutions in Landscape Architecture* (New York: John Wiley & Sons, 1997), 271–284, and is incorporated (revised and updated) with the publisher's permission,.
3. Lawrence Halprin, *The RSVP Cycles: Creative Processes in the Human Environment* (New York: George Braziller, 1969), 2–3.

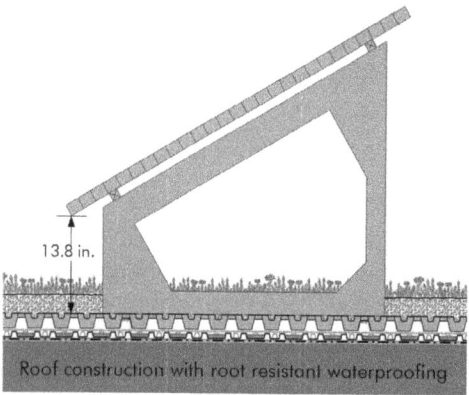

Figures 2.8a and 2.8b Zinco is one green roof/photovoltaic manufacturer that provides products and services on both sides of the Atlantic, so it indicates both metric and imperial dimensions. Note that the dimension comparisons are approximate, rather than precise.
Courtesy of Zinco, copyright/trademark name, Amsterdam, Netherlands)

System Build-up
"Solar Green"

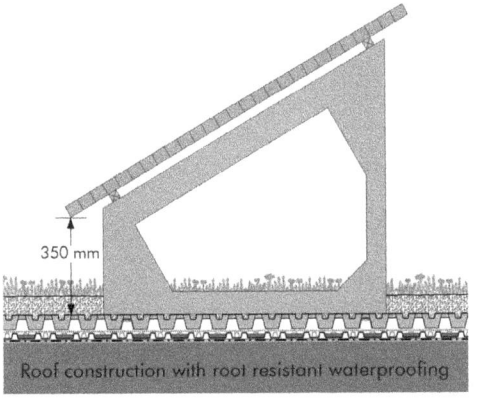

Solar Panel

Solar Base Frame SGR
(inclination between 5° and 45°)

Plant community "Sedum Carpet"
Growing Media "Zincoblend E", depth depending on structurally required load (project specific)
ZinCo Solar Base SB 200 with infill
Drainage Mat Fixodrain® XD 20

Roof construction with root resistant waterproofing

* The required superimposed load and the resulting weight of the system build-up need to be determined depending on the location and building geometry according to the structural calculation.

Technical Data of build-up:

Height:	min. 127 mm
Weight, dry:	min. 126 kg/m²
Weight, saturated:	min. 161 kg/m²
Water retentin capacity:	ca. 35 l/m²

CHAPTER 3

Contracts

For this purpose let us maintain as far as possible an objective point of view, examining the more technical and tangible characteristics of melody, and leaving for more specialized and philosophical consideration those elusive but nevertheless important values such as emotion, tension and relaxation. This is not to belittle the force of aesthetic qualities.[1]

—Walter Piston, *Counterpoint*

Contracts set the tone for design practice. Although there are many types of people who may enter into a contract with a landscape designer or landscape architect, the terms of the contract must be fair to permit and encourage the designers to develop and execute a design for a specific scope of work for a reasonable fee, resulting in the satisfaction of both the designer and the client. There may be a contract between a landscape architect and a private client, between a landscape architect and a public client or agency, or between the landscape architect and a contractor. Sometimes the landscape architect is on the outside looking in, and as a courtesy introduces the client to a contractor who executes the landscape architect's design, subject to his/her review. The contractor may receive payment upon the landscape architect's certification that the work in the field is in accordance with the designer's contract documents.

When my own career began, the American Society of Landscape Architects (ASLA) had standard fill-in-the-blank contract forms that could be used by landscape architects and landscape designers. However, as landscape architecture evolved as a profession and took on more and

Figure 3.0 Net Zero competition entry for site planning and sustainable design in Montpelier, Vermont, https://netzerovt.org. The plan is to allow people to live, work, and play in downtown Montpelier without the need to use motor vehicles. See also Figure 4.0.

more complex roles, it became standard practice to use contract forms from the American Institute of Architects (AIA), which have sections in which specific contractual information may be added to suit a particular project and a particular group of people. The principal or project manager, sometimes in consultation with a lawyer, fills in the blanks with the names of the specific clients and designers, the scope of work, the time frame for execution, and the fees to be paid for the execution of the work. Some government agencies and some private corporations have their own contracts in which they have incorporated language favorable to their own interests, so an experienced project manager and/or lawyer must review carefully the contract and the proposed language, and negotiate whatever changes to the language are felt to be necessary in order to proceed with the work.

Often, in hindsight, the most important reason a particular landscape architecture project was not as successful as it could have been may be that the contract was not thoroughly reviewed and understood before the actual design work began. It is optimal for a landscape architect to take an active role in every contractual aspect of the work, from the initial marketing to the final agreement with a client to perform design services. The more complex the project, the more critical it is to thoroughly review its contract during the process of negotiations, often with the assistance of a lawyer or someone expert in contract language. Where possible, the landscape architect should always aim to limit open-ended requirements and imprecise language, so that there is a finite number of deliverable documents based on the expenditure of a finite number of hours of labor. All of the following may take on urgent importance, depending on how the project unfolds. In the examples that follow, the *lettered text in italics* is intended as examples of helpful revisions to protect the landscape architect:

1. How is the "scope of work" defined?
    a. It is essential to limit the language used to define the scope of work as much as possible, so that the landscape architect does not end up in a situation where anything the client wants, regardless of when it is requested or discovered, is defined as standard services to be performed.
    b. *The scope of work for the Belladonna Landscape Development shall be within the limits of the demolition, layout, grading, drainage, construction, and planting shown on the construction drawings titled "Belladonna Landscape Development" and dated June 17, 2019, and the accompanying specifications.*

2. What happens if there is disagreement between the client and landscape architect as to the exact nature of the work to be implemented?
   a. It may be appropriate to add another phase of work, typically called "scope development," in which time is taken to clarify the exact nature of the program and site, so that there is a clear consensus prior to the implementation of any design services.
   b. *Prior to the conclusion of design development (or conceptual design) and after site analysis is complete, the landscape architect shall engage in further program development to clarify the best proposed use for the site.*
3. What standard support services are to be included within the basic scope of work?
   a. This can vary considerably from contract to contract. For example, it is not uncommon for a client to require that all printing and delivery costs of progress documents and final contract documents to myriad individuals or agencies be included within the total fee structure of the contract, without any additional compensation for the landscape architect.
   b. If many iterations of the project are required before the documents are finalized or many different individuals or agencies must review and approve the documents before they are finalized, then significant costs may be incurred. To the extent possible, the landscape architect should limit the number of sets of contract documents that may be considered required deliverables during the course of the contract, and have an agreement that additional sets are a reimbursable expense.
   c. *The landscape architect acknowledges that for each of the four phases of the contract, one full-size set and four half-size sets of drawings and related documents (specifications and cost estimates) shall be sent by courier to the client agency, and three other representative team members within the city (names and addresses to be determined). It is agreed that additional half-size sets will be printed and delivered for a cost of $25 each, and full-size sets for a cost of $40 each.*
4. Does the contract include a list of tasks or activities that are specifically excluded from the scope of work?
   a. With the inclusion of such a list in the contract, if the client requests that one of these tasks be implemented, it provides the landscape architect with a clear justification for categorizing such tasks as extra services. The landscape architect must not overlook any prudent and practical means to focus the scope of work described within the contract to the exclusion of any task or item of work that might conceivably be included but is not.

  b. The landscape development work under contract is for Parcel A on the set of drawings and specifications dated November 15, 2017, and does not include the adjacent Parcel B.
5. How are "extra services" defined?
  a. Who makes the determination that certain tasks may be beyond the original scope of work and thereby entitle the landscape architect to additional compensation?
  b. Clear language in the contract governing change orders will often give protection to the owner, contractor, and landscape architect, and provide initial guidance on how to proceed if/when there are additions, deletions, or other revisions to the scope of work that must be discussed and the specifications adjusted accordingly.
6. Are limitations placed on the number of meetings or locations of meetings for each phase of work for the total extent of the project?
  a. What happens if a phase of work continues indefinitely because the resolution of complex tasks takes many months? The landscape architect must find a way to limit the number of meetings and site visits that are covered by the initial design fees to an agreed-upon maximum, and have a way of billing for additional meetings or site visits should they occur.
  b. This can be as simple as spelling out a set number (or range of numbers) of meetings for each phase of the project. At least, there should be some anticipation that certain phases of the work, depending on how the scope is defined, may require many more meetings than others. If the landscape architecture office is some distance from the site or from the client's office, these meetings can become a major expense.
7. How is the landscape architect to be paid?
  a. There is a range of possibilities. Sometimes fixed fees based on a percentage of a cost of construction are assigned to each phase of work. The landscape architect must anticipate what happens if a fixed fee for a particular phase of work is exhausted through no fault of his/her own but just as a result of any number of factors. There should be some fair means described within the contract to allow for additional compensation should this become necessary.
  b. Each contract must spell out clear milestones, at the end of each of which the landscape architect is paid a progress or milestone payment commensurate with the work completed.
8. Is there agreement prior to starting work on the maximum and minimum hourly rates for each employee to be assigned to the project?
  a. Sometimes contracts may be more open-ended and allow for all or most of the landscape architect's time to be billable, based on submission of

invoicing identifying the task and job title of all work performed. The landscape architect may be expected to submit to the client a list of all employees, their job descriptions, and resumes or justifications for their inclusion in a particular class (or salary rate) for the tasks to be performed.
   b. The preceding example is typical for some public agencies; from the beginning of the project the landscape architect must be aware of the billing criteria and requirements for each employee assigned to a new project.
9. Does the contract include means of increasing hourly rates after an agreed-upon base period, which may typically be two to three years?
   a. It is often standard practice to escalate hourly rates for the next calendar year based on criteria set by the U.S. Department of Labor.
   b. Read the contract and be aware of what your firm or individuals are entitled to. Sometimes an agency representative may inform you of such additional entitlements, but it's better to be aware yourself of what benefits you may deserve.
10. Who determines that the work is finished?
    a. Often final payment is not given for design services until completion is certified. Again, landscape architects should avoid contract language that gives the client all of the authority to make this determination.
    b. This is usually determined by an inspection jointly carried out by the landscape architect and the client or owner's representative in several phases, according to terms spelled out in the specifications. Just be certain that the landscape architect's interests are represented.
11. Does the contract indicate a calendar in months or years by which time the work is expected to be completed and a corresponding calendar of compensation?
    a. The landscape architect should be careful that the dates by which each phase of work is expected to be completed are reasonable.
    b. Furthermore, there should be clear language explaining what is to happen if a particular phase of work lasts considerably longer than anticipated.
    c. If design development is scheduled to last twelve months but drags on indefinitely, there is no way for the landscape architect to stay within the budget while working steadily to complete the task. Therefore, there should be means spelled out within the contract for the consultant to receive additional compensation.

12. What permits and approvals are required and who is responsible for securing them?
    a. Although these tasks may seem simple at first, the landscape architect must be certain that such language is not so open-ended as to require myriad submissions or to present conflicting requirements—for example, where one agency might require something for approval that another agency rejects automatically.
    b. It is not uncommon in large cities with many different bureaucracies for complex projects to require a wide range of permits and approvals from many different agencies, from local to state and even federal jurisdictions.
    c. The number, range, and scope of permits and approvals vary dramatically. If significant preparations, review, presentations, and meetings are required for some permits and approvals—in short, as if they are a phase of work in of themselves—then clearly some compensation is required. Other permits or approvals may perhaps be considered as routine administrative procedures. The landscape architect, in reviewing the contract, may compile a list and estimate the hours of time that will be required to prepare, present, submit, and revise all applications for permits and approvals and use this estimate to negotiate a fee with the client.
13. What happens if the project may involve multiple sites?
    a. In such contracts, the landscape architect may be expected to carry out typical phases of work, such as schematic design, design development, final design, and construction administration, for a range of different sites—for example, the design of five neighborhood parks at different locations within a community. A fee for most public projects is based on a percentage of the cost of construction, and is typically calculated based on a sliding scale where the higher the cost of construction, the lower the percentage that is awarded as a fee. It could be important in a project like this to have the fee for each park based on the cost of construction of that park, rather than a fee based on the combined cost of the five parks. In that way, the landscape architect can maximize the fee potential for the total project. Also, this could protect him/her from long delays in payment if the design schedule for a few of the parks lags behind because of unforeseen issues, such as the need for an updated survey or community approvals. Then, if each park is treated independently, the consultant does not wait indefinitely to bill for the completion of a particular task, such as final design, when it might be completed for most of the parks but not all.

b. It is rare that what works for one site is an exact fit for another site, even if standard details are being installed (say, in parks or schools), because site conditions, such as grading, drainage, existing trees, and pavements, will likely be different from site to site. Therefore, it's much preferred, if at all possible, to develop a separate fee for each site, even if some leeway is allowed for applying similar standard details to different sites.
14. Who will own digital images and other records when the project is finished?
   a. Often it is required and quite appropriate for the client to have a full set of photographic digital images of site conditions and finished construction, as well as sets of construction documents. However, upon the completion of the work, the landscape architect should still own or have full use of these same materials as resources for future work and as archival documents.
   b. This issue is still evolving. Many digital applications quickly become obsolete as software manufacturers issue updates that make earlier versions harder or impossible to use. At a minimum firms and clients should save legible record copies of prints (full-size and half-size) and a current version of the software at the time the project is closed out.
15. Does the contract spell out the names and job titles of the key personnel for the client and the landscape architect who are to be the key contacts throughout the project?
   a. These individuals should be experienced, should possess the ability to communicate clearly, and should have the authority to make decisions with the full support of the client and the landscape architect. There should be a backup person in each category in the event that urgent business must be agreed upon or enacted when the principal contact is unavailable.
   b. This information harks back to the human resources aspect of the firm. All personnel must know where and how to record information on key personnel or from project managers and other important people on major projects, and in what format.
16. Does the contract spell out the initial existing conditions and the documents the landscape architect (and contractor) is expected to use?
   a. One crucial document is an accurate topographic survey. Prior to implementation of even schematic design, there must be a way of verifying the accuracy of the survey information and a means for securing additional accurate information if the existing data is inaccurate or incomplete.

b. For some projects, even if the topographic survey is reasonably accurate, if it is not current and signed and sealed by a registered surveyor, there may be problems in securing approval for infrastructure construction and other improvements.

c. See the section "Surveys" in Chapter 5 for detailed information on the dos and don'ts of surveys.

17. Does the contract describe accurately any special conditions unique to the project or site?

　a. Sometimes there are strict guidelines, based on historic preservation or other criteria, that must be followed. The contract must spell out the applicable guidelines or resource documents that apply.

　　i. Certain species of trees may be forbidden due to their susceptibility or vulnerability to a local insect or disease.

　　ii. There may be restrictions on the use of construction materials—for example, the exclusion of tropical hardwoods due to the potential damage from harvesting them.

18. Is LEED (Leadership in Energy and Environmental Design) certification included?

　a. LEED certification has gained more and more influence, particularly in public works projects for city, state, and federal agencies. If gold, silver, or platinum LEED certification is required, then the contract should stipulate the key LEED contacts for both the client and the landscape architect.

　b. Coordination on LEED issues must be scrupulous and start at the very beginning of the design process, or there is a risk of falling behind in verification and certification procedures and not being able to catch up. (See Chapter 1 for definition of LEED and Chapter 9 for discussion of LEED certification process.)

19. How will disputes be resolved?

　a. Many procedures are standard and acceptable, such as arbitration or some other typical review by an independent third party.

　b. Be wary of any contract in which all disputes are resolved by and at the discretion of the client or the owner, with no room for compromise or balance. Contracts involving public agencies or bureaucracies in which disputes are resolved "at the discretion of the executive director" should also be carefully vetted.

20. Does the contract allow time for the simple administration of a project?

　a. Even if a project is in a slow period, time is still required by the project manager or occasionally other personnel. Unless work stops altogether for periods—while a phase of work is being reviewed, for example—the client must expect a certain minimal ongoing cost so that the landscape architect can make key staff available should questions arise, to update

or correct drawings as review comments are received, and to facilitate accurate record keeping.
b. Whenever possible, a project manager should try to build into the schedule enough time for a few of the staff to review their work, check drawings, coordinate them, and compare them to cost estimates and specifications. This could be the time when you have a staff member not familiar with the project take a look at a set of drawings and documents and evaluate them for consistency and accuracy.

21. Is there coordination with other disciplines, such as civil or structural engineering?
    a. If such disciplines are required as part of the work to be implemented, the landscape architect must be certain that the coordination of these services is spelled out in the contract, including how it will be paid for. This coordination is critical whether these other disciplines are represented within the landscape architect's own firm, are independent consultants themselves, or are subconsultants to the landscape architect. The individual roles should be carefully spelled out and agreed upon to avoid time-consuming delays to determine which firm has the principal role.
    b. Any certification or registration of subconsultants should be completed prior to beginning any of the work of the contract.
    c. Some firms with separate landscape architecture and engineering departments have elaborate time-sharing arrangements, so it's important to understand how to charge for a subconsultant's time. Of course, if you're hiring a consultant from another firm, it's all the more important to have a clear agreement on the scope of work, which should include an approximation of the number of hours required for various tasks in each phase of work.

22. How are special services or contingencies defined—for example, tasks that may be required but cannot be predicted or known until design has already begun?
    a. Such services could include borings, additional topographic survey work, property line verification, asbestos abatement, and so on. An allowance should be set up within the contract budget provided by the client to cover the cost of such services so that funds are immediately available when such work is necessary and there are not long delays while waiting for these services to be approved and budgeted.
    b. It is not uncommon for totally unexpected needs to arise or situations to occur that could stop work indefinitely if there are not already allowance budgets set up to accommodate such circumstances. The landscape architect must be certain that the contract spells out the range of services that could conceivably be covered by allowance budgets.

23. Are clear procedures described for project documentation?
    a. There should be clear guidelines agreed upon for all project documentation: minutes of meetings, records of important telephone conversations, AutoCAD or other drawing standards, and so on.
    b. Neither the client nor consultant can afford to be far along in early phases of work and find that they are not in compliance with standard requirements, or that minutes of earlier meetings in which key decisions were made and for which there should be clear records are not, in fact, complete and approved.
    c. Of course, it's easiest if you are the prime and all other participants are consultants to you, so that your office standards are the template that must be followed. However, if you are working for a public bureaucracy, it's just as important from the beginning to review all of that agency's standards and be certain that the key personnel of each subconsultant are aware of all of the requirements before the work begins in earnest.
24. How will the project be bid?
    a. Will there be a requirement for at least three competitive, approved bidders, or will there be negotiations with a preapproved contractor selected by the client or jointly selected by the client and the landscape architect?
    b. Competitive bidding requires the most complete contract documents, because until a preferred bid is selected, it is difficult to discuss details of the work, so everything must be self-evident within the contract documents.
    c. If a preselected bidder is used, often there can be ample discussion even during earlier phases of the design process. The contract for design services should stipulate which approach is favored.
    d. Whether public or private bidding, it's incumbent upon the landscape architect that the process be fair. You can never afford even a shadow of a rumor that you've showed preference to a bidder in a private setting. In a public setting such rumors can lead to criminal charges.
25. Does the contract include a separate fee for construction services or services during construction?
    a. At a minimum the contract should set aside a predetermined portion of the total fee, perhaps as much as 25%, to cover these costs. Often the hours of construction services required of the consultant depend on the complexity of the design, the degree to which custom design elements are included, the distance between the site and the design office, and how much coordination with other contractors or agencies is required.
    b. If any of these factors are not clearly established when design services begin, then there should be language in the contract to allow for

renegotiation of the fee for construction services. This fee might also change if the construction budget is significantly more or less than originally anticipated.
   c. The hardest fee to determine is for construction services, so shy away from a predetermined fee if it is not substantial, particularly if it set before the design has been done. There are too many factors that cannot be predicted that can cause this to be a very time-consuming and labor-intensive process. It's also critical to the success of the design that the landscape architect devote adequate time to ensuring that what is shown in the construction documents is faithfully and accurately executed, so it puts a great deal of pressure on the design office if the fee for such services is not commensurate with what is appropriate for this purpose.
26. What is an appropriate use for a liquidated damages clause?
   a. A good example would be a school or university with a construction project that must be built within a tight schedule—say, between the time classes end in May and when students return to campus in early September. A contract could include a liquidated damages clause fining the contractor for every day past September 5 that the work is not complete.
   b. The advantage for the contractor is that in temperate climates the summer normally offers fair weather. The disadvantage is that because it's a limited period of time, the contractor must be highly organized and focused regarding all operations on the site.
   c. Do not consider a liquidated damages clause for a contract starting in mid-fall, just as long periods of inclement weather begin. Periods below freezing, high winds, or blizzards could give the contractor all sorts of reasons to be able to avoid penalties if work is delayed.

DOS AND DON'TS FOR CONTRACTS

1. Use www.aia.org as a primary source for contract forms and asla.org for excellent information within its professional practice network on specific topics of interest and expertise.
2. Although it is standard to save and archive old contracts and documents, take the time to periodically save old sections of specifications that are written in clear language, as they may be quite helpful in later projects when a frame of reference is desired.
3. Keep track of the history of fees charged for projects of similar scopes of work and be certain that you can justify increases over time. You may need a better explanation than the accountant who simply adds a set amount each year to the charge for doing your tax return.
4. Don't start working on a contract if you are uncomfortable with any of its terms, or if your decision to move ahead is based on some sort of compromise; be clear about why and how you are proceeding.
5. Open-ended clauses allowing someone to bill you without an upset limit are almost always dangerous.
6. Verify the experience of any contractor who will have a major role in any work on a project you are designing.
7. If a client will not pay you for construction administration, and you take great pride in the quality of your design services, be very cautious about whether you want to do the design work without any control over its execution.
8. Do not trust a handshake agreement. Too much is at stake.
9. Don't review a contract when you are tired.
10. Include a good attorney as part of your project management team.

Problem 1

MOST BEAUTIFUL PLACE IN THE WORLD

Pair off in teams of two. If there is an odd number of students in the class, have one team of three.

Read the opening chapter, "The First Morning," of Edward Abbey's book *Desert Solitaire: A Season in the Wilderness* (New York: Ballantine Books, 1968). There are many editions in paperback. Discuss among the team members how the author manages to communicate so vividly his impressions of the beauty of the desert environment in which he resided or hiked.

Each person in the team should develop a list of what s/he feels describes his/her teammate's Most Beautiful Place in the World. Invent strong, concise phrases modeled after and/or inspired by the language, style, and format in Abbey's book. The place may be real or imaginary. Your final list may be printed out in Word or handwritten, but it should be legible (and large enough) for presentation to the class. Give as clear a depiction as possible of what you envision.

Meet once to discuss your list with the other team member(s). The partner should question whatever is not clear, in order to clarify and draw out as accurate and complete a list as possible. Revise the list as necessary and submit it to your partner. The team partner should approve the final list as being accurate.

Based on your partner's revised list, draw the Most Beautiful Place in the World. You may draw a sketch in plan view with or without a few cross sections or elevations and/or a perspective. Label key elements. The drawings may or may not be to scale. Show your preliminary results to your partner, and encourage him/her to critique your work. Revise your drawings as appropriate. How pleased is your partner with the revised efforts? Is s/he satisfied that you have depicted her/his concept of the Most Beautiful Place in the World? The list, sketch, and any other related drawings and sketches should read as a group—for example, they should be on the same size sheets with similar labels, title blocks, and so on.

Critique each other's work. Who has done the best job of developing a list? Who has done the best job of drawing what's on the partner's list? How does this exercise compare to the practice of landscape architecture?

NOTE

1. Walter Piston. *Counterpoint*. New York: W.W. Norton, 1947, p.13.

CHAPTER 4

Marketing and Human Resources

PART 1: MARKETING

Many a small thing has been made large by the right kind of advertising.[1]
—Mark Twain, *A Connecticut Yankee in King Arthur's Court*

It can be argued that marketing is the most important function of an efficient landscape architecture office. A marketing department or director must allocate its resources to search for work opportunities, decide which of the requests for proposals that are received merit a response, coordinate and prepare submissions in response to these requests, enter competitions, and maintain accurate and up-to-date records of all key personnel (organized according to areas of expertise, states of registration, and qualifications) and projects (organized according to public or private and type of work). This must all be accomplished with limited financial resources that do not make a serious dent in budgets of projects already under way. Larger firms may have the resources to have a specialist whose full-time job is to handle all of these tasks, but smaller firms must often share these tasks among various employees, from principals and graphic specialists to accountants and secretarial staff. Needless to say, if the marketing functions of an office do not succeed, then the firm folds, regardless of the skills and expertise of its personnel.

When my own career in landscape architecture began, multidisciplinary firms abounded. They often included many professionals in different fields—landscape architecture, architecture, engineering (civil, structural,

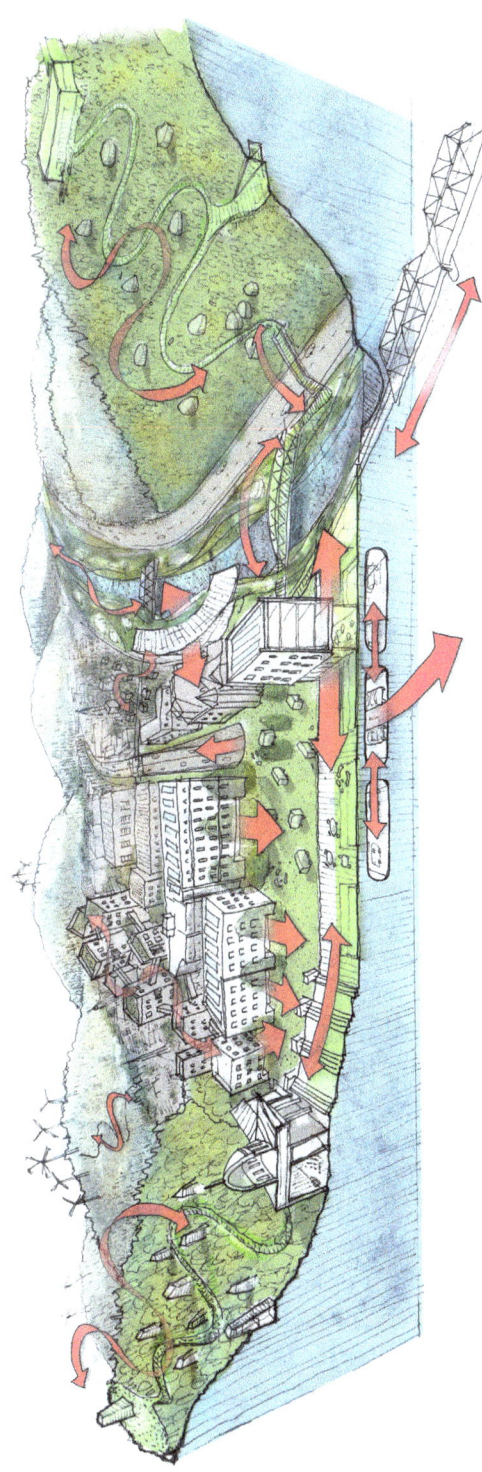

Figure 4.0 Companion entry for Net Zero competition entry for Sustainable Vermont community https://netzerovt.org. A series of programmatic and urban strategies is used to increase the city's performance on collaboration, livability, walkability, energy, water, and air. Vicky Chan, Richard Alomar, and others

mechanical, and electrical), transportation planning, irrigation design, lighting, graphics, support services, and so on—and often all of these professionals would have experience with a wide range of project types and clients in both the private and public sectors. Such firms are much rarer now because specialization is often required. A firm that focuses on residential landscape design may have little experience with public parks; a firm emphasizing environmental assessment will have little experience in detailed, as-built projects. Marketing requirements are another reason that such segregation and separation occur. A public agency or other government entity will have totally different submission requirements from residential firms as to experience and qualifications (sometimes even preselection requirements) for those seeking to be awarded a contract, so firms must decide which areas of practice to focus on.

The rapid advancement and sophistication of social media is, of course, another major trend. Email systems, apps, and websites all may offer nearly instant information, which results in the demand or expectation for almost instant responses. The sheer magnitude of what's available perhaps adds a criterion to marketing that did not exist some decades ago: how does your package, application, or resume emerge at the top of the list out of all that noise? Perhaps one way is for a response to a proposal to be greater than the sum of its parts—for it to suggest a story line, whether hidden or overt. This matches what has occurred in television advertising. A certain brand of car is associated with such a quiet and stable ride that a diamond cutter in a hurry can slice a raw stone into perfection while traveling in the car; certain cable TV packages guarantee exciting relief from a boring suburban lifestyle. The challenge for a landscape architect is to be perceived as current, dynamic, and skilled in what the prospective client requires. A landscape architecture firm may need to advertise and promote itself based on any or all of the following: an ecological approach to design, sustainable design practices, a site-driven design process, the unique expertise of particular employees at specific tasks, a design team with a record of solving problems through innovative design approaches and methods, or any clear documented achievement that helps the firm's aura soar above those of its competitors.

It is now common in design firms with offices in various locations, whether regionally or internationally, to have meetings via videoconferencing: groups of staff members or individuals share information and discuss projects on which they are collaborating. Such methods are promoted as being efficient and necessary in the fast-paced modern world. Certainly, with clear communication and sharing of the agenda of the meeting beforehand and concise minutes of the meeting afterward (in which what was decided and agreed upon is described), these methods can be appropriate.

In such settings, it's important to have adequate computer power to share and review complex drawings. In local offices where everyone knows one another or at least many colleagues, the participants' roles in any project are reinforced daily by communications from people who are part of a design team. By contrast, in videoconferencing, where people may not know one another except as joint workers on a remote project, it's important to have clearly defined roles.

The following guidelines are intended to give you a list of basic requirements that are essential for applying for new work and helping keep everyone in a firm fully employed, no matter its size.

Cover Letter/Narrative

A strong cover letter is critical. It functions as a narrative to tell the story of what the firm and its collaborators are offering, identifies key members of the proposed design team, and summarizes the approach to the project. Anticipate that the reviewer will only skim this document, so the occasional but consistent use of **boldface,** <u>underlined</u>, or *italic* fonts and various styles can help the reviewer see points of emphasis quickly without reading every detail. Numbered or bulleted summaries may also be helpful in organizing and revealing key materials for the prospective client.

Resumes

Resumes are ubiquitous. The best ones are concise and custom-tailored to the job for which an application is being made. A flexible Word or other format is needed, so that different versions of each resume can be created in which at least five relevant projects are cited. It is helpful to have one or two sentences highlighting the qualifications and experience of each employee proposed as part of a design team, so that the reviewer understands this person's role and contribution.

Project Summaries

Each landscape architect or firm should develop a list of short descriptions (a few paragraphs long) of key projects, organized by type, size, budget, or other required criteria, that demonstrate the team's experience in relevant projects. It's helpful for each document to include a few images, perhaps a plan graphic of the design accompanied by an image of the completed as-built project, or some other combination (such as before-and-after images) useful for the application.

Often at least a few photographic images of as-built projects are critical in convincing prospective clients, particularly those less fluent in reading drawings and graphics, of the firm's capabilities. Therefore, in developing contracts for design services it is crucial to be certain that your design firm has permission to photograph and use images of your finished designs for publicity purposes, although you can negotiate omitting street addresses and other elements to protect the client's privacy. There was a period during which my career focused on high-end residential design, and negotiation for design and construction services was severely constrained by clients who demanded stringent limitations on what could be photographed. Since it was almost impossible to show drawings or photographs of work, reputations had to spread by word of mouth, but that was not a good way to gain recognition or gain more work.

Hierarchy of Employees and Project Management Timeline

Although the cover letter is useful to identify key personnel, a chart to show the organizational framework of key personnel—who answers to whom—is quite useful. Similarly, a graph or bar chart indicating the proposed timeline for carrying out the proposed work can be critical in helping those assembling the data and coming up with a final proposal to estimate what the fees should be and to be certain that key personnel are not being over- or underutilized.

Breakdown of Fees

At some point a critical factor is the fee for the project being proposed and the breakdown of this fee. Sometimes the reviewer, whether an agency or individual, requires a lump sum fee, instead of a range with an upper and lower limit. In such cases the marketing manager or other key people assembling and participating in the application process must do a best-guess estimate of what they feel each phase of the project should cost, and reply on that basis. Since so much may remain unknown to them until or unless they actually win the award of the project, it is important to try to retain some flexibility. It may be possible to exclude certain items from the proposal. For example, construction inspections are one of the most difficult items to estimate, because there can be so much variability depending on the complexity of the work, its proximity to the location of the office that would send employees to inspect the work, the quality and experience of those executing the work, and its time frame. Therefore, with some proposals, it may at least be possible to give a precise fee proposal for the first phase of a project, but then offer a range of fees for subsequent phases, or simply say that these phases of work will be determined by mutual consent as the earlier phases are nearing completion, so that it's then quite clear precisely what tasks remain to be executed.

For some types of work, particularly public works for city, state, federal, or other governmental agencies, the fees are dictated by long-established standards. Payment is often based on a percentage of the estimated cost of construction, and is further broken down into percentages for each phase of work, such as 20% conceptual, 30% preliminary (or design development), 35% construction documents and bidding, and 15% construction implementation. Everyone applying to be awarded that scope of work will be paid the same; therefore, the materials submitted by the marketing department are the first step in distinguishing one firm or project team from another. Sometimes the prospective client interviews representatives of the short-listed firms and then makes a final decision; at other times there is no such intermediate step.

One advantage of private sector work is that there is more flexibility in the fee structure. A company located in a city or neighborhood with higher overhead costs may need to charge higher fees, yet demonstrates in its marketing proposal an efficient operation, relevant experience, and a clear talent for the specific scope of work, and therefore has a chance of being selected over a less experienced firm with lower overhead costs. For such projects, it may be possible to negotiate fees for the initial phases of work and then have a clear understanding that fees for subsequent phases will be determined after there is greater clarity, based on conceptual and design development studies, of the final phases of the job.

Design Sketches

It's often typical that a proposal for design services may require the team to submit some sort of preliminary graphic of a proposed design. If this is required, such a graphic should be stimulating and suggestive; yet it is not possible to do more than indicate conceptual ideas, outlined in written and graphic form.

Be as Open-ended as Possible

Being open-ended does not mean being evasive; rather, it simply acknowledges that it's very difficult to predict the amount of time required for every task in a complex project, in which later tasks depend on the successful completion of earlier tasks that are not as yet fully defined.

Required Schedules and Specialized Documents

Often some standardized forms developed by bureaucracies must be filled out in order to apply for certain types of work. The forms may require a breakdown of the firm's experience according to type of project and total cost of construction. Such forms can be quite tedious to complete, but once done, they can easily be edited for future applications to the same or a similar agency. So an applicant, whether an individual, a firm, or a team of collaborators, must have accurate records of all existing and past projects from which to draw upon in completing such applications. Often such applications will require that the list of relevant projects and budgets must be from work completed within the last decade or so. Even when work has been archived by the firm, with most records in storage, a basic summary of each completed project, together with its costs, some relevant images, and summaries, should be created and saved for this purpose. The time to do this is while the project is under way and soon after its completion, not years after its implementation.

Red Flags

Some requests for proposals may include requirements or directions that in and of themselves suggest a lack of understanding of a fair contractual design process or suggest that there would be such difficulty in moving forward with the work that it is best not to participate in the application process. Insist on being able to review a standard contract that would be incorporated into the project should you be fortunate enough to be awarded the job. See how disputes are resolved, and how requests for extra services are handled in a project when something new arises not covered by the defined scope of work. It's also useful with government and other agencies to ask to review the format for invoicing and understand when invoices may be submitted and how long it will take to receive payment once the client has the invoice.

It is rare in government work and more common in private sector work to be given, upon signing of the contract, an advance against which to charge. As an example, think of a book author who signs a contract. Some publishers pay only upon receipt and approval of a complete manuscript and illustrations, with all approvals signed and delivered, so the author must absorb all costs, perhaps even building up a level of debt, until notification by the publisher that everything is complete. Other publishers may pay an advance against royalties, so the author is able to pay for reasonable expenses incurred, such as travel and supplies, while the work is well under way but not yet complete. A final advance is paid upon receipt and approval of all manuscript, illustrations, and images. Royalties are then paid depending on the type of publication and agreed-upon schedules listed in the contract. A landscape architect, firm, or design team faces similar issues. If upon the award of a contract no fees are paid, then the office accountant and/or principal must be certain that there are reserve funds available to pay all the employees while they are working on the first phase of the project, for which billing is not allowed until a certain minimum percentage of completion or other milestone is achieved.

Diversify the Work of the Office

One way around this conundrum is to diversify the work in an office. Consider having some private clients, such as homeowners or developers, in some area of landscape architecture practice, such as residential design, roof gardens, or environmental assessment, in which the payment of an advance upon signing of the contract is an accepted standard. Then the availability of such funds may be used as an antidote to public contracts in which there is still a long waiting period between finishing the first phase of work, being able to bill for it, and receiving payment.

As landscape architecture has diversified and matured as a profession, one encouraging trend is that landscape architects are valued within both the private and public sectors for their diverse skills and abilities to manage complex issues. It's increasingly common for a landscape architect, instead of an engineer or architect, to be the principal-in-charge of a design team. Landscape architects may work as site planners, problem solvers, and coordinators to direct teams of designers in executing challenging programs on large or small sites, whether urban design, residential, or some combination of scales. Or a landscape architect may take on a traditional role but have the trust of the client to such a degree that the architect is directed by the landscape architect (rather than the other way around) and the architecture is first presented to the client only after its relationships to the site have been tested and approved by the landscape architect. In such cases the landscape architecture firm gains traction in the design profession and must skillfully manage the responsibilities of project manager and designer. At the same time s/he must hire as consultants engineers, architects, and other professionals who are comfortable in a role that may be somewhat different from their traditional expectations.

Pro Bono Work

Most offices should consider one or a few pro bono projects a year in which services are provided for free to benefit a cause or organization, in return for which the firm may expect to receive a considerable amount of good publicity, which in turn might lead to a paying job in the future. Of course, there is the possibility that no profit-making work will be generated, so one must be cautious in what work is done. Yet such projects can be beneficial in terms of promoting office morale, giving people in the office who normally do not collaborate an opportunity to work together and sometimes just good hands-on work in the field.

Problem 1

TOPICS IN LANDSCAPE ARCHITECTURE

Research a study area of your choice. The following list suggests broad topics, from which you should find a specific example. After you have chosen your area, prepare a detailed critique of a completed landscape architecture project. Why and how was it successful? What were the limitations? Can you guess at any issues that may have occurred with the design fees and management of the project?

1. Healing gardens
2. Historic preservation
3. Cultural landscapes
4. Invasive species
5. Grant applications
6. Butterfly gardens
7. Fragrance gardens

8. Gardens for the visually impaired
9. Landscape architecture education
10. Memorials
11. Cemeteries
12. Parkways
13. Roads
14. Resorts
15. Master plans
16. Elderly populations
17. Children's playgrounds

Problem 2

RESUMES

Sometimes the best way to understand one subject is through the understanding of another. Think of the two time-honored versions of the popular childhood ditty "Row Your Boat." One is thought to be for common folk and the other for more highbrow folk, yet each has a certain charm.

Row, row, row your boat
Gently down the stream
Merrily, merrily, merrily, merrily
Life is but a dream

Propel, propel, propel your craft
Rapidly down the solution
Ecstatic, ecstatic, ecstatic, ecstatic
Existence is but an illusion

Write two versions of your resume, one that would appeal to someone who would be charmed by the first version of the song, the second to someone who is entranced by the companion version. In each of these versions of your resume, try to use a vocabulary and style consistent to that version. (It need not be set to music.)

Listen to a recording for solo piano of Twelve Variations on "Ah, vous dirai-je, Maman?" (Oh, shall I tell you, Mother?) by Wolfgang Amadeus Mozart (1756–1791), known in English as "Twinkle Twinkle Little Star" or "Baa Baa Black Sheep." On www.youtube.com, you can easily find 3 or 4 fine renditions. Does this change your interpretation or approach in any way?

Problem 3
JOB OFFER

(In the following problem, any *terms in italics* may be changed to suit the particular instructor or team leader, or your own personal, work or studio setting. Typical changes might be specific people, such as clients or contractors, plant materials, or the locations of sites, people, institutions or companies.)

Write a letter to a prospective employee offering him/her a job and explaining why your firm would provide the best environment for her/him to be in. Limit: two pages double-spaced. The letter should include *the name and address of the firm* of which you are president or human resources officer and *an assumed address of the prospective employee* to whom you are writing, both anywhere within *North America*, or *Europe*. Use a business letter format.

Problem 4
PROPOSALS FOR VARIOUS PROJECT TYPES

(In the following problem, any *terms in italics* may be changed to suit the particular instructor, or the personal, work or studio setting. Typical changes might be specific people, such as clients or contractors, plant materials, or the locations of sites, people, institutions or companies.)

1. Divide the class or studio group into teams of three students. Each team should meet and designate one person to be the *client*, another to be the *landscape architect*, and the third to be the *contractor*. (If class size is not divisible by three, create one or two teams of four students and have two participants as the client instead of one.) All sites to be within 25 mile radius (40 km) of *your location*.
2. Assign one project type to each team:
    a. Single-family residential estate on 4 acres (1.6 ha), with a budget of $150,000, not including design fees, licenses, permits, etc.
    b. Public park recreational facilities on 5 acres (2 hectares), with a budget of $350,000
    c. A garden on 2.5 acres (1 ha) for hospital patients and their families, with a budget of $500,000

d. Twenty-five existing multifamily housing units on 10-acre site (4 ha) in need of erosion control, beautification, and plantings, with a budget of $250,000
e. Upgrade (access ramps and railings) of five entrances for public housing high-rise apartment buildings on existing square-block urban site to comply with Americans with Disabilities Act, with a budget of $175,000
f. Environmental impact assessment for a proposed park on an abandoned 7.5-acre (3 ha) industrial site, with a budget to be determined
g. Renovation of a 3-acre (1.2 ha) historic garden that has fallen into disrepair, with a budget of $300,000
h. Site planning for 10-acre (4 ha) elderly housing development on rolling topography, to include gardens, seating, amenities, parking, etc. in compliance with Americans with Disabilities Act, with budget to be determined
i. Project of size and scope different from the above
3. For each project, the team should discuss and decide the following:
   a. What other professionals, consultants, or individuals should be added to the team?
   b. What can be achieved with the budget indicated? What restrictions might need to be considered in order not to exceed the budget? What should the minimum budget for construction be, if not indicated? What should design fees be?
   c. How long will it take to complete the design and install it on the site?
   d. The client/owner should emphasize five to ten items of the highest priority (that is, program or bucket list) to be implemented. The landscape architect should determine what drawings will be required, and suggest the number and titles of an overall package of drawings and specifications. The contractor should research costs and develop an estimate of how much time will be required to implement the wish list provided by the owner. Reach an agreement about what will be done over how long a period of time at what cost.
4. Present your results to the class, or whatever group you are a part of at work.
5. The class or studio as a whole should evaluate the results of each team. Which team has the best design strategy for reaching the most important goals in a reasonable amount of time and at an acceptable cost?

PART 2: HUMAN RESOURCES

Human resources (HR) is the department within a firm that focuses on the records of employees and provides them with information on benefits, sick leave, health insurance, pensions, and references. In smaller firms and organizations, the HR department may include marketing and accounting; in larger settings, the latter two are often separate departments. In either case, updated resumes must be maintained for each employee so that proposals for new jobs include information that shows off the accomplishments of the firm and the qualifications and experience of its relevant employees in the best possible light. HR must also stay familiar with all regulations governing the conditions of the work space for all employees, including the fair treatment and protection of people with disabilities.

An important consideration becomes how much time a firm should devote to HR activities: updating records, researching changes in health insurance or other benefits, scanning timesheets and developing accounting statements for each project within the firm, responding to requests for references or validating the employment of each person, and so on. Individual employees are typically responsible for updating their own resumes according to company-provided criteria, but someone in HR or marketing must review the resumes for accuracy, consistency, and compliance with the prescribed format, so that they may readily be incorporated into proposals.

New Graduates' Short- and Long-Term Goals

Upon graduating from a program in landscape design or landscape architecture, an eager person typically starts to look for employment. In our modern world there are vast possibilities, from small firms to large corporate entities, from educational institutions and conservancies to for-profit organizations, from firms that are directly involved in design to contractors who execute designs and industries that manufacture various products such as benches or light standards. However, one basic consideration underpins all of these possibilities: do you want to work for yourself or would you rather be employed by another?

Many new graduates have as a goal to eventually have their own firm, yet they seek to gain experience within a variety of settings prior to embarking on their own adventures. Some people are just so independent-minded that they chafe at being someone else's employee. Still others much prefer giving up the longer hours, increased stresses, and other problems that ownership entails in return for a regular salary or set hourly rate, time off during the evenings and on weekends, and so on. As one starts to look for work in the field, it's important to gauge one's needs and interests. A young family with children must allow ample time for childcare and joint activities. Single people who may want to eventually meet a partner or spouse must allow for time during the workweek or at least on weekends to enjoy social activities. Sometimes individuals, no matter how committed they are to a career in landscape architecture, may have other interests, such as music or theater, to which they devote time outside work hours. Some people, whether due to temperament, personal, family, or other reasons, are simply not good candidates for full-time management or ownership. Often firms over a certain size have a package of benefits that new employees find particularly helpful, including health insurance, disability protection, and 401(k) plans. It can be much more difficult for smaller firms or a family or individual to find comparable coverage or benefits.

Compare Benefits

Any prospective employee fortunate enough to have several job offers and the time to analyze and compare them must pay attention not only to the location of the office, the range of work, and the types of projects that the prospective firm performs but also to the package of benefits. With the wide variation in health insurance benefits offered by different employers in different states, some of which have incorporated plans authorized by the Affordable Care Act, it is important for the prospective employee to be able to determine what, if anything, s/he will pay for health insurance (if it is available through the employer) and what the benefits will be both for the individual and for his or her family. A job offer $10,000 to $15,000 higher from a firm with limited benefits may actually be less competitive than a lower offer from a firm that has a more complete benefits package. If a firm pays less in direct salary but offers a 401(k) plan in which the employee's contribution is matched up to a certain maximum percentage, includes disability protection, provides more vacation time, and offers tuition reimbursement for courses deemed relevant to the job, then the lower-salary option may be the better choice, even for someone quite ambitious.

Years ago there were a limited number of university curricula offering degrees in landscape architecture. A great majority of graduates were white men. As the field has gained in breadth, applications, and importance, educational opportunities abound, and the diversity of students and graduates has greatly increased. This seems a natural extension of the demographics of major cities, where many landscape architecture school graduates work and play, but also is typical of the complex urban design, assessment, infrastructure, and landscape projects that attract people to this area in the first place. African Americans, women, gays and lesbians, transgender people, foreign-born students, and immigrants, among others, all have a role to play in the profession. Of course, the vast richness of these groups' experience is a great resource to draw upon in the design of the modern world.

Some firms will give each employee a manual that lays out all policies of the firm regarding personnel matters—for example, if overtime is paid and how it is calculated; the rates at which sick time and vacation time accrue; restrictions on the use of company equipment for personal use; how to keep track of and file for reimbursement for expenses such as travel and printing; what formats to use in saving documents in various types of software, such as Word, AutoCAD, or emails; how to participate in company's Facebook groups or other internet activities. The goal is for each employee to understand what is expected of her and to know all relevant company procedures and policies.

Consistent Policies

A critical area where there must be consistent policies is in the naming, filing, and archiving of all documents relating to a particular project, including contracts, correspondence (letters, emails, faxes, etc.), drawings (from conceptual to contract documents), cost estimates, and specifications. At a minimum a record set of drawings and correspondence in hard copy should be retained in the office or in an archive for each contractual phase of a project. For example, if design development includes two phases of submissions to the client, then there should be two hard copies of all the submitted documents kept within the active project drawers and records. Usually one person, such as a principal or project manager, is responsible for organizing the files for each particular project and storing the documents. It becomes counterproductive to have two people share this responsibility as it's too likely that the firm will end up with key documents being shared between the filing systems, local drawers, and shelves of these two individuals, no matter how seamlessly they collaborate.

Social Media

Probably no innovation has had more of an impact on contemporary society than smartphones, which have enabled social media. With the release of each new smartphone upgrade, there are concomitant new applications that make social media and other applications more useful. Facebook alone has almost 2.4 billion users worldwide. Skype and Twitter have revolutionized communications. Videoconferencing with people in several locations in different countries or other locations thousands of miles apart has become routine. Texting and email communications have proliferated. Anyone competent in the use of a smartphone can take quality photos and share them instantaneously with huge number of other people.

One trend seems to be that a new software package or app will gain in popularity and usage based on its ease of use, the quickness with which it can be shared, and the instantaneousness of its appeal. All of these characteristics can be either advantages or disadvantages. Blogs, for example, which take time to prepare and post and hence involve some level of delay, are losing popularity to Snapchat and Instagram. Yet there still seems to be a good market for professional bloggers who can tell a tale and illustrate it in an engaging and robust manner, much as a landscape architect or an old-fashioned storyteller would, while Snapchat and Instagram offer snapshots of one's travels or favorite plant materials.

There is no way for me to write a detailed summary or analysis of the vast range of devices and applications available. Instead, the focus for this book needs to be on the impacts that smartphones and apps have on the professional responsibilities and duties of employees and their personal lives. The giant corporations that provide and host social media and internet services, such as Facebook, Google, and Apple, have only recently been challenged to protect the privacy of their clients, literally billions of users. Hacking of personal data is widespread. Therefore, policies established by employers must go a long way toward addressing their own privacy concerns as well as those of their employees.

Emails and texting have created an industry of instant communication, which has major advantages as well as significant disadvantages. Proceed at your own risk. It's wonderful to be able to reach people, whether clients, colleagues, or family members, in a few seconds. Long delays waiting for a letter to arrive by "snail mail" or shorter delays while the fax is being transmitted and received are no longer the norm. On the other hand, there can be a tendency to be drawn into so much emailing and texting that the result is an atmosphere of intense pressure and claustrophobia. There is less opportunity to perfect and edit the language of important communications (and less expectation that they will be perfect), so the entire process of preparing carefully thought-out letters and other documents becomes undervalued. Speedy communications have become an excuse for poor grammar, misspellings, and inaccuracies. The risk is that by accelerating the actual process of communications, the level of understanding is diminished due to misunderstandings and lack of clarity. Also, once an email or text has been sent, it's irretrievable and likely shared by many more people than one would prefer. It is still critical to look before you leap and think before you act. Use these technological tools as a way of increasing the efficiency of communications, sharing information with other key people, and soliciting opinions, but not as a way to respond to pressure-filled situations or to take shortcuts from sounder practices.

In the "good old days" it was common practice for principals or several key people to take several days or a weekend to mull over major decisions on projects or hiring of new employees. They might discuss the ramifications several times, have a few long-distance phone calls with people in different offices, and conduct a follow-up phone call with a prospective employee before a decision was made. At universities it's still common practice to have a long search for new professors led by a search committee, culminating in a two- or three-day visit by several of the highest-rated candidates. The candidates' qualifications must meet university-wide criteria for tenure, and references are carefully checked. Each finalist interviews with key faculty and usually gives a guest lecture in his or her area of specialization to students and faculty; after further review and discussion among faculty, one of the candidates is offered the job. By contrast, the other extreme might be a situation in which a draft letter to a client requesting additional fees for work that seems to be "extra" but is not yet clearly described in the text is sent by mistake because its author wrote it late at night and hit "send" by mistake.

QUESTIONS

1. When and for what purpose are the various types of social media best used? For example, what's better for ending a romantic relationship: email, texting, or neither?
2. What about inviting a date to the prom? Should one make this request in an email or text, to avoid embarrassment for either party if the prospective invitee wishes to decline?
3. When is a telephone call or a face-to-face meeting better than FaceTime or Skype?
4. Millennials seem so immersed in the culture of smartphones that they can seem uncomfortable talking to people in person compared to texting. What implications does this have for professional business and contractual relationships?

5. It used to be considered a strong social skill to be able to walk into a roomful of strangers, introduce oneself, and start to meet people one-on-one or in small groups. Part of the attraction and allure of major cities, such as New York or Chicago, is the tremendous diversity of people and culture, offering opportunities—on the subways and buses, in public plazas, in bars, in coffee shops, and at cultural venues—to mingle with strangers with whom one may have a lot of common interests. But at what point are such opportunities lost if people are afraid to actually engage in conversation and prefer to start all efforts at contact through the use of an app or texting?
6. When should there be a hard-copy backup of emails or texts?
7. If an employee is given a smartphone for ease of connecting to colleagues at work and clients, how does s/he avoid using the phone for personal purposes, with those texts or emails being subject to review by her/his employer?
8. What sort of rules governing Facebook should a company institute in order to enhance the enjoyment and skills of those employees who choose to participate in the company's Facebook site, but avoid interfering in their personal lives or act as a censor?
9. What are the limitations of videoconferencing and remote design, compared to face-to-face meetings with all key people sharing a conference room and a hard-copy set of drawings and documents?
10. Should employees be obligated to keep their company smartphones on outside of work hours and be available to respond to texts, emails, or phone calls regardless of what non-work activities they are engaged in?
11. A common use of texting is to inform a colleague or friend that one is running late and yet is on the way to a meeting or to verify its location. It's also quite useful for last-minute changes in plans—say, that the meeting is occurring in a different location than initially scheduled. When is it appropriate to engage in longer and more complex texting? When is email, a phone call, or a face-to-face meeting a better option?
12. What are the implications for employees using devices in which there is automatic backup of all texts, documents, and files by systems that many individuals as well as companies and institutions use? Backup software is now as routine as anti-virus and malware-protection software.
13. What are some considerations for sending alerts on Facebook for a groundbreaking ceremony? For a grand opening? Who should be invited? Who should be excluded?
14. How can social media be used to welcome new employees or to honor those who are retiring? What are the differences in terms of who

should be informed and who should participate in any activities associated with each?
15. What employees, if any, should be permitted to use Twitter accounts vs Instagram accounts? Why? What review process, if any, should be in place prior to someone sending a tweet or sharing an Instagram photo, and should it apply to all employees?
16. What are some advantages and disadvantages of Skype? Who should use it and who should avoid it?
17. What are the advantages of FaceTime for a large landscape architecture office? For a small one?

For all the following questions or actions, indicate whether the preferred approach for communication is (1) a phone call, (2) a text message, (3) an email, (4) a hand-delivered missive, or (5) something else (in which case, specify). Explain your reasons.

1. On March 1, you submitted an invoice for payment for the first phase of a project for which you are a subconsultant. You sent it to the representative of the firm with whom you are under contract. It's now March 2, and there has been no response.
2. On March 1, you submitted an invoice for payment for the first phase of a project for which you are a subconsultant. You sent it to the representative of the firm with whom you are under contract. It's now March 8, and there has been no response.
3. On March 1, you submitted an invoice for payment for the first phase of a project for which you are a subconsultant. You sent it to the representative of the firm with whom you are under contract. It's now March 15, and there has been no response.
4. On March 1, you submitted an invoice for payment for the first phase of a project for which you are a subconsultant. You sent it to the representative of the firm with whom you are under contract. It's now April 1. You know that the invoice was received, but there has been no further response.
5. As project manager, you have notified via email (or some other application in your computer system) ten people in the project team in different locations in New York City of an important upcoming meeting scheduled in two weeks at your office. You have received confirmations from seven people via email; one person has indicated that s/he is unable to attend. The meeting is now two days off, and you have not heard from two important participants.

HUMAN RESOURCES *(95)*

6. All but one of your office's utility bills are automatically paid via the company's checking account. The exception is the company that services your computer system monthly, including updates to virus protection programs, checks of all the networked computers in the office, and resolution of any new problems. This company sends a monthly invoice via email. As office manager, you notice that the invoice has not been received, but payment will be expected shortly.
7. Does your answer to question 6 change if the vendor is the company that provides all janitorial services, emptying all trash cans and cleaning bathrooms daily, and doing a full cleaning weekly?
8. You split your time between human resources and accounting, and have been filling in for a few days in accounting as your boss is out sick. As you are reviewing records, you find that five accounts are a month overdue—that is, invoices were sent over a month ago and no payment has been received.
9. Your office, which has twenty-five employees, is having a pre-holiday informal luncheon at a neighborhood restaurant. You want to let the restaurant know how many people to expect, but you have only heard from about half of your colleagues, with still about a week before the gathering.
10. As project manager for an infrastructure project involving your firm and several city and state agencies, you're just started to review the pre-final working drawings prior to issuing the set for bidding. You have ten days to issue final comments. You notice a major omission: a detail, its specification, and its line item in the cost estimate are not included.
11. You're expecting a routine day at your office, but you wake up at the usual time, 7:15 a.m., with flulike symptoms and a fever. There's no way that you should go to work.
12. You're scheduled to make a major presentation to a client, but you wake up with a sore throat and fever. As much as you feel loyalty to your employer and want to attend the meeting, you know that you shouldn't do so.
13. You submit corrections to the project manager at another firm for the set of drawings you've prepared as a consultant to the principal firm. The project manager notifies you that your materials have been received. You expect to review their final package to verify that what you've provided has been properly incorporated and to check one last time for corrections, but the final deadline for completing the package

is fast approaching, and you've not heard further from the project manager.

14. You're frantically working on a complex planting plan with three other employees in your office and are just starting to develop the set of specifications. On the most recent printed-out set of drawings for review, you notice a lot of careless graphic and arithmetic errors, a lack of consistency in adhering to office graphic standards, and a delay in preparing the planting schedule. Most of the problems are with the work of two of the three employees.

15. As project manager, you've just been notified by a client that your work on a local city park has won an award. You want to share the good news with the design team, located at three different firms. There may be an awards ceremony in a month.

16. You're doing an inspection of a final planting installation by a landscape contractor at a residential site a few hours' drive from your office. In one area near the front entrance, you notice that many of the plants are starting to wilt, and it appears that the irrigation system for that area may not be functioning. The irrigation subconsultant is not at the site.

17. You're putting together a proposal and have asked seven employees to forward to you their updated resumes. You've heard from five of them: four have submitted resumes, and one indicated in the morning that she would send hers by early afternoon, but it's not been received.

18. You are almost finished assembling an important proposal. You're awaiting final review comments from one of the firm's principals, and if they are received within an hour, you should have time to incorporate the comments, print out a final version, and submit by the deadline. How do you get these comments, make the submittal, and verify receipt?

19. You are a new employee, and you keep receiving "past due" notices in your email inbox for important services that were your predecessor's responsibility. What do you do? Does your response change if you have been employed at the firm for several years and the notices should have been sent to another colleague?

20. You must share preliminary review comments with a structural engineer, a civil engineer, and a lighting consultant in three different offices. The comments are clear, but there is some overlap in responsibilities and in the information or corrections each person must provide.

DOS AND DON'TS FOR SMARTPHONES, APPS, EMAILS, AND TEXTING

1. Do not give in to pressure to respond to every e-mail instantly.
2. Do not put into an e-mail, chat, or text *anything*—whether word, phrase, or image—that could somehow come back to embarrass you, a client, a friend, or an acquaintance.
3. Respond succinctly to most emails that don't require discussion: for example, acknowledge the time and place of a meeting, or provide a simple negative or affirmative answer.
4. For emails that ask for or demand thorough review, consultation with colleagues, and careful deliberations on issues that might have major impacts on a project's design or budget, proceed more cautiously and be methodical. Save such emails in another folder, or highlight them, and take adequate time to respond to them in order of importance.
5. To respond to an important email, consider preparing a draft of a response that is placed in a folder such as "Emails Waiting to Be Sent" or "Drafts." Write the basics as you have time, and save it temporarily. Review it again later in the day or the next morning before sending it.
6. Automatically empower spell-check and grammar review, but use them carefully, not automatically. (I once had a client named Mrs. B——a, and spell-check kept changing her name to "Mrs. Bulimia." I've always wondered how she might have responded to a message addressed to her by that name.)
7. Don't postpone the review or sending of emails until the very end of the day, as mistakes are more likely to occur when you're tired. The end of the workday might be more suitable for writing drafts that you'll review the following morning, when you're fresh.
8. Many email systems will complete the name of the contact and email address as you start to type the first few letters of either the person's name or email address. This is a very helpful and time-saving device. However, the risk if you're not completely alert is that you send an important email to the wrong person, whose name or email address is similar to the intended recipient's.
9. If you find yourself in a highly emotional state, such as elation or anger, this is not the time to send an important email. Wait until you are in a state of equilibrium and feel fully in control of your emotions, and therefore can act in your own and your firm's or project's best interests.
10. Do not respond to emails whose tone suggests that the sender is in a highly emotional state. Wait a day to let the feelings calm down, or, if the matter is urgent, consider calling the person directly—perhaps the next day, if appropriate.

11. Some clients or projects may require lengthy emails, which supersede the role of letters. Consider writing the text for such important communications as a Word or other word processing file, reviewing and editing it to your satisfaction, and then pasting it into the e-mail format once it's complete.
12. Save and organize emails received and sent in appropriate folders for the relevant project. There is a tendency to keep huge numbers of emails current in a single folder, so that they display as "new." The risk of this, of course, is that you can overload the capacity of the email server, and possibly lose those communications altogether if the email system crashes or experiences other errors. Therefore, each day, or certainly by the end of each week, save and/or delete all emails marked as unread. Failure to set aside a certain amount of time for this each week will result in a task that becomes overwhelming (and impossible) in scope.
13. Avoid abbreviations, acronyms, and slangy expressions in common use ("u" for "you," for example). Conciseness and brevity are admirable qualities to strive for, but not at the expense of clarity and accuracy.
14. Aim for a consistency of tone. Do not write a client using the same vocabulary and style you'd use for a friend or family member. Your purpose in writing a work-related message may vary dramatically from email to email—some emails may give detailed explanations, while others are routine approvals or acknowledgments—yet the writing style and tone should be consistent. Someone else in your office may, on occasion, have to write in your place.
15. Do not write or receive personal emails and texts on a smartphone or computer that is your employer's property unless you would not mind having your colleagues see the message.
16. If you use your smartphone to take images of work in progress at a construction site or to inventory some aspects of a site you are beginning to design or may consider writing a proposal for work on, then it's important to download the appropriate images into the office's computer system by the end of each day. This enables everyone on the design team to access the information promptly and have it organized properly for long-term use.
17. Just as you weed out extraneous emails in your desktop system on a daily or weekly basis, delete unneeded documents and files from your smartphone on a regular basis.
18. Don't try out a new app for the first time in the field *or* on a new client *or* under time pressures, when any malfunction might create significant problems.
19. Have a charger with you. Don't assume that you'll be graced with one by a friend, railroad, airplane, or fairy godmother.

20. Back up the smartphone promptly at the conclusion of any important meeting or site visit, or at a convenient opportunity soon thereafter. Organize and label the backup immediately.
21. Consider a bright-colored protective case of a durable material for the smartphone so that the phone is more likely to be easily found if you temporarily misplace it and more likely to be undamaged if you drop it.
22. Be certain that all virus protection, malware, and other software are up to date and functioning. Your office or you personally should have a protocol setting timetables for installing and verifying updates. Some software install automatically, others require your initiation.
23. Don't share your office smartphone with others in the field; maybe at lunch indoors, but not on an expansive construction site.
24. Rely on smartphones for excellent construction photos but not for publishable, aesthetic compositions meant for hard copy.
25. Learn to store the phone in the same location in the same bag, pocketbook, carrying case, briefcase, inside pocket, or other convenient and secure spot every time you are finished with it, so that this becomes a matter of habit and you don't spend valuable time trying to remember where you put the phone.
26. If you are writing a draft of an important email, do not address it fully; that is, omit the @ or .com or other key component, so that if you start to send it by mistake, it will stay put.

NOTE

The author is grateful to Dr. Howard Leifman, instructor for The SAGEWorks Bootcamp in 2016, for insights offered in job hunting, social media and networking skills.

1. Mark Twain. *A Connecticut Yankee in King Arthur's Court*. Signet Classic Paperback (Penguin New York), 2004: Chapter XXII, "The Holy Fountain," p. 156–57.

CHAPTER 5

Drawings

PART 1: SURVEYS

A topographic survey is the basic footprint on which a design will be drawn. Therefore, as a project gets started, the design team must have access to a reliable, updated survey provided by the firm's own staff or the client/owner, or be given the resources to have one done. The following guidelines and considerations apply to surveys; the specific requirements will vary from project to project:

DOS AND DON'TS FOR SURVEYS

1. A survey should be done by a surveyor licensed for that purpose. Beware the preparation of a survey by someone who is not fully experienced and licensed.
2. By definition the survey must be guaranteed accurate to half the contour interval. For example, in a survey with a contour interval of 1 foot, the elevation of any given point within it must be accurate within 6 inches. For a survey with a contour interval of 5 feet, any point must be accurate within 2½ feet.
3. Select the scale at which a survey will be drawn and the contour interval based on the type of work that will be done and the degree of accuracy required. For example, for residential landscapes a scale of 1" = 20' or 1" = 10' is common with a contour interval of 1 foot, and certain detailed areas around swimming pools or specialty gardens may be drawn and surveyed at a larger scale with a 6-inch contour interval. If a surveyor is hired for a 6-acre sloping site with a 1½-acre plateau in the center, with the most intensive design and construction likely to occur on the plateau, then the landscape architect may well prefer to have a more intensive survey done of the plateau. For

Figure 5.0 Complex pavement patterns—such as these at one of the entrances to the Memorial Arts Center in Atlanta, Georgia –are not always the clearest or the most cost effective.
Photo by Steven L. Cantor

master plans, scales of 1" = 100' or 1" = 500' with contour intervals of 5, 10, or 25 feet are common.

4. USGS maps typically are drawn at proportional scales such as 1:24,000 with contour intervals of 25 or 50 or 100 feet. Convert to 1" scale by dividing by 12; in this case, that would be 1" = 2,000'.

5. Locate all existing utilities, whether aboveground or belowground. Surveyors have tools and equipment to assist with this task. It's typical for surveyors, as they began the process of preparing a new survey, to access records from local, county, state, or federal agencies. Similarly, it's important that any relevant documents, such as old plat maps, utility diagrams, and plans of additions to existing buildings, all be provided to the surveyor prior to the beginning of his work.

6. With AutoCAD and other digital programs, it's helpful to color-code utilities and other key data onscreen as well as printing in color. If all water lines are shown as blue, for example, anyone reviewing the document will have a clear sense of their location and patterns on the site.

7. Trees and masses of vegetation must be shown. Depending on the size of the site and the scope of the work, the minimum size of trees to be recorded will vary. For example, for a park landscape where most of the work will be carried out in relatively open ground, perhaps the only key data will be the location of scattered specimen trees and the edge of masses of forest, with the location of the trees forming the transitional area being located. On residential sites where a lot of planting may be expected, it may be useful to locate all trees over 4-inch or 6-inch caliper. It is usually urgent to protect as many major existing trees in good condition as possible.

8. Another sound practice is to request of the surveyor a spot elevation on both the uphill and downhill side of a tree. In that way, if it's possible that grading, whether cut or fill, may take place near the tree, it's easier to determine how to minimize the impacts on the tree roots.

9. From the beginning, the more the design team or owner can inform the surveyor of the intent and purpose of the design, the more the surveyor can determine where and how to provide a greater degree of detail. For example, if the surveyor knows that a septic field of half an acre will be needed, she can highlight on the survey some areas that may be suitable, based on slope conditions and soil types and lack of trees.

10. The accuracy of the survey is paramount, and some information is critical. Property lines must be precisely located, along with any applicable setbacks or easements areas. Alleys, driveways, roads, curbs, and parking lots must be shown clearly. It is helpful not only to show

the existing contours occurring over such spaces but also to provide a precise spot elevation at every corner of geometrically shaped pavements and/or a grid of spot elevations for more organic shapes.

11. Since grading and drainage are often part of the scope of work for many projects, the location of catch basins, drains, manholes, and other utility structures must be shown accurately. Invert elevations and surface elevations at each catch basin or other drainage structure must be provided.

12. Whether a new or existing survey is to be used for a new project, one of the first tasks for the landscape architect is to verify the accuracy and content of the survey. A print of the preliminary survey mounted on a board can be taken to the site, and the landscape architect and/or staff member can walk the site using the survey as a guide to verify that all major buildings, structures, utilities, trees, and other major elements are shown correctly. Enough elevations should be checked to verify the general contour pattern and drainage. If any omissions are noted, then the surveyor must provide additional information. The revised survey should be verified in the field prior to conclusion of this phase of work.

13. Common errors that may occur in surveys are the mislabeling of species of trees. Surveyors are expert at recording and calculating information to scale but may not be expert at telling some species of trees apart. The landscape architect should verify all species, particularly those in critical areas that may need to be removed, transplanted or become the basis for important design elements. The time to note where critical specimen trees are found on the site and their condition is at the beginning of the design process.

14. In a survey with dozens of paved areas of many different materials, it's also common for some to be mislabeled. The landscape architect must check that all areas labeled concrete are, in fact, concrete and not asphalt or brick, and so on.

15. Sometimes important information may be missed by a surveyor due to the season in which the survey is done. Therefore, when the landscape architect is verifying the survey's accuracy, extra attention should be given to noting any areas of poor drainage or puddling of water, trees in poor condition, and so on. Although these types of conditions can overlap with the preparation of a site analysis, observations made during the survey and survey verification process may streamline the site analysis process.

16. Another critical factor to verify on site is the direction of true north. Since it is common practice during the preparation of design and construction drawings to rotate the angle of the north arrow in relation to the basic rectilinear shape of the drawing, it's essential to know that the north arrow is correct on the basic survey.

17. Verify that the scale is accurate. For example, the lengths of segments of property lines are often labeled. The landscape architect should measure several of these in the field and verify that the distances shown on the survey are correct and to scale.
18. Just as most offices have graphic standards for depicting site elements, so do surveyors. Prior to authorization of a new survey, the landscape architect should review the surveyor or engineer's graphic standards and be certain that they are the same or compatible.
19. It is more typical for designers than surveyors to have standards for layer lists and line weights. During the design process, it's often necessary to print drawings in which the relative graphic importance of the various survey elements must appear in a reasonable and logical manner, and there must also be the opportunity to freeze certain information so that it does not appear on some prints or some workstation screens. It becomes a tremendous waste of time to have to digitize or redigitize many components of a survey after its completion because the surveyor did not receive prior to beginning the work a clear list of layers, colors, and line weights to use and minimum sizes for text and labels. The other key members of the design team, such as engineers and architects, should be consulted to verify their preferences for symbols, line weights, and other graphic criteria to be used in the survey.
20. If possible, anticipate from the beginning the orientation of design drawings in relation to the survey so that the basic orientation of texts provided in the survey will not have to be rotated.
21. Legal standards usually require the surveyor's stamp and seal and the date to be incorporated in the survey document. The landscape architect may suggest the location of this information to expedite incorporating it into the design drawings that follow.
22. Once the survey is updated and certified to be accurate and complete, a digital copy is referenced as the base for future site analysis, design and construction drawings. The survey is a legal document, so printed and digital copies must be saved within the project files prior to any editing, such as adjusting line weights or text to read better in relation to whatever graphics or design elements are being drawn over it. In the future, if some potential conflict is caused by what appears to be an error on the survey, then it is a simple matter to go back to the original and verify whether the error was made by the surveyor or occurred in one of the analysis, design, or construction drawings.
23. Even if the survey has been verified and it is known to be accurate, there are some precautions that landscape architects might consider in implementing a detailed design:

a. Avoid placing a fence or other element such as a tool shed directly on a property line. If possible, it's safer to place the fence inboard of the property line by a foot or so, even though this might cede a narrow strip of property to a neighbor. This gives some protection if small inaccuracies occur in the survey that may not be detected until or unless there is a dispute.
b. Sometimes it is necessary in a design to meet the minimum standards for setbacks or other requirements. However, if space is available, try to set key elements slightly past the minimum setback thresholds.
c. Some building codes set floor area ratios and other maximum sizes for the footprints of buildings or other structures or built areas. There are some homeowners, developers, and clients who, from the beginning, want to squeeze every square foot possible from such guidelines. However, it is prudent to settle for footprints that still have a little breathing room. For example, design and living requirements are such fluid criteria. It might be that after a few years, it becomes clear that adding a tool shed near a garden, a garage, or a paved area would make a huge difference in the efficient use and enjoyment of a property. If the buildable area is at its maximum extent from the moment the design is first implemented, it becomes much more challenging and involved to exceed maximum limits. At a minimum, it may be necessary to apply for a variance or otherwise petition for exceptions to long-established zoning guidelines.
d. Criteria for septic systems are rapidly evolving. It used to be that septic fields had to be located in relatively flat to gently sloping open fields with few trees. In some settings they are now permitted in moderately sloping site with substantial trees. The criteria for percolation tests, the type of existing soil that will be acceptable to a particular health department, and whether granular soil from outside sources may be brought in for the purpose of achieving suitable results on tests all vary from county to county and state to state. If the landscape architect knows from the beginning that septic fields will be required for a project, then an excellent use of the survey is to mark on a print as soon as possible what areas are suitable for this purpose, and to determine the total area available for this purpose. It is important to have designated areas suitable for septic systems marked as off-limits from early phases of analysis and design and use these results to determine the maximum size for any buildings proposed for the site.
e. Criteria for drilling wells as sources for potable drinking water and irrigation also vary dramatically from place to place. If wells

will be proposed for a site, the landscape architect may use the survey to mark potential sites for drilling and also verify that they are uphill from and at least the required minimum distance from proposed septic fields.

24. If the metric system is going to be used for a project, then the survey is the first document to be prepared accordingly. Therefore, whatever graphic and engineering criteria are to be incorporated should be analyzed carefully prior to the beginning of the work.
25. Depending on the site and the nature of the survey work, it is sometimes helpful or necessary for the surveyor to examine or sometimes precisely locate off-site landmarks, such as the outline of a residence on an adjacent property or the mass of a forest or specific specimen trees that may cause some shade on the owner's property. The landscape architect should spell out such requirements prior to the beginning of the survey process and should be certain that he has permission in writing from adjacent landowners for such purposes. Sometimes neighbors are willing to provide digital survey information to a surveyor so that she can easily incorporate adjacent landmarks, structures, and key elements on a new survey. By having such information accurately recorded from the beginning, the landscape architect is in a much better position to screen unsightly areas and provide privacy to those living on both sides of a property line.
26. Include enough off-site features, such as edges of roads, driveways, screen plantings, and utility lines, that a clear picture emerges of the site for the new project, including important adjacent land uses and information.

(See Figure 5.1 for a typical survey).

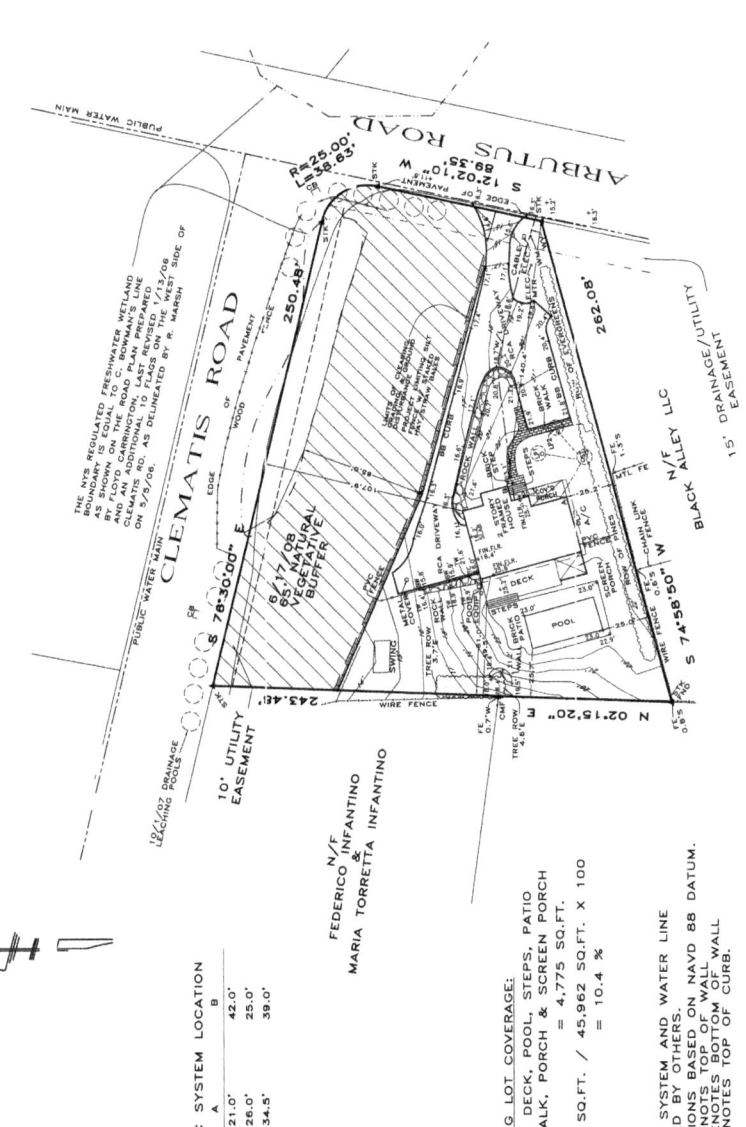

Figure 5.1 Typical topographic survey.
Courtesy of Jeff Dragan, LDGN Landscape Architects, DPC

PART 2: DRAWINGS: CONCEPTUAL, SCHEMATIC, DESIGN DEVELOPMENT AND MASTER PLAN

Many projects have contracts that require the landscape architect or firm to prepare in sequence drawings of increasing accuracy and complexity: conceptual, schematic, and design development. Sometimes a master plan drawing is also included. Upon approval of the design development drawings, or in some cases after approval of the master plan, the consultant prepares working drawings and related documents for the entire site or for its first phase of work.

Conceptual, Schematic, and Design Development Drawings

Schematic and conceptual drawings show the program anticipated for the project applied to the site for the purpose of verifying that all proposed uses of the site are being considered. The approximate location of each activity and its proximity to other uses are determined. The approximate size or required spatial area for each required activity is demonstrated as well, so these drawings can be a good method to verify what will fit on a site. Conceptual and schematic designs also are often used to test different design options. Just as there is more than one way to cook an omelet, there are many ways a program of activities may fit on a site. Spatial and functional relationships between program activities may be tested in abstract diagrams without regard to the site at first, and then in schematic sketches, before they are applied at larger more practical scales as overlays on site maps.[1] If abstract designs with strong geometric and graceful forms are developed in the schematic or conceptual phase, sometimes these can carry through into the design development and final design phases of projects, so that the built form derives directly from the concept.

By the beginning of design development, preferred options have been selected, and the landscape architect starts demonstrating the specific required dimensions of each activity, its relationship to other activities, and the materials necessary for its construction, such as the types of pavements and walls. Grading is shown to verify that the proposed activities can fit comfortably into the site. By the end of design development, the landscape architect usually has also prepared a comparable cost estimate giving the approximate cost for implementation, with a contingency of usually about 10–15% added to cover unknown factors.

Master Plan Drawings

The master plan may encompass a large or small site, but shows a design in its totality usually at the level of design development. By carefully studying the relationships between different proposed functions and areas of the site and by evaluating the costs of implementation in relation to the funds available for construction, the landscape architect may further assist the client by showing a phasing plan. It establishes the sequence and cost of each major feature of the site, and it emphasizes how all elements of the master plan design could eventually be implemented without having to damage or rebuild something that was part of an earlier phase of construction. These four phases of work—conceptual, schematic, design development, and master plan—can have a wide range of meanings, interpretations, and requirements, depending on the client and whether the client is an individual or a bureaucracy, public or private, so in the process of reaching an agreement to work on such a project, it's critical for the landscape architect to understand the scope of the work and budget the office's time carefully.

DOS AND DON'TS FOR CONCEPTUAL, SCHEMATIC, DESIGN DEVELOPMENT, AND MASTER PLAN DRAWINGS

FOR ALL DRAWINGS

1. Use as a base map a topographic survey recently developed for the site that can be printed (with appropriate layers frozen) at different scales for different drawings.
2. Incorporate north arrow.
3. Incorporate graphic and written scale.
4. Develop a title block and labeling system that will work for all drawings to be produced, from conceptual to design development.
5. If some drawings overlap, clearly develop match lines that will always be shown in the same way graphically.

DOS AND DON'TS FOR CONCEPTUAL AND SCHEMATIC DRAWINGS

1. Since early steps in the design process emphasize a variety of usually at least two or three alternative designs, consider assigning two or three people to a project, and giving them a deadline for each to independently develop a schematic plan. This may result in designs that are substantially different from one another, rather than variations on a single theme.
2. It can be argued that the most important aspect of design is testing alternatives at an early stage of the process. Have pin-ups or desk reviews of the three alternatives by several other colleagues who have an interest in the scope of work even if they are not involved with the particular project.
3. A major function or program activity may be graphically indicated by a bubble of indeterminate shape. Continuing with this exercise, the designer develops functional relationship diagrams or sophisticated bubble diagrams in totally abstract configurations, without regard to site or scale, and conceptualizes how circulation might be developed between the "ideal functional patterns of the program's major use areas."[2]
4. Gradually progress toward consideration of site elements.
5. Then respond to topography, major landforms, and site analysis, still abstractly.
6. Consider usage requirements as geometric areas, and the geometry of the site. For example, most parking functions can be thought of as rectangles; activities can be conceptualized as circles, and so on.
7. Incorporate various types of pedestrian, bicycle, and vehicular movements and connections as well as parking.
8. Conceive of planting as masses, and show specimen trees and other planting functions.
9. Respond fully to the client's and contract's requirements.

DOS AND DON'TS FOR DESIGN DEVELOPMENT DRAWINGS

1. Don't develop too much detail too soon. For example, mark different types of pavement and walls, including their approximate dimensions, but do not show joint patterns and detailed transitions of one pavement or area to another.

2. Give an approximate elevation to each feature of the proposed design, so that one can appreciate the proposed topographic relationships between major design elements.
3. Use graphic labels that coincide with major headings in the comparable cost estimate so that someone reviewing the plans can easily go back and forth from one document to another.
4. Develop outline specifications for all major proposed elements to be constructed.
5. Incorporate perhaps five details showing typical important elements of the design.
6. Include some cross sections and elevations to demonstrate relationships between proposed areas and functions in order to highlight changes in elevation between key elements.
7. Respond fully to the client's and contract's requirements.

DOS AND DON'TS FOR MASTER PLAN DRAWINGS

1. Based on the scale of the drawing, show all proposed development with the same level of detail.
2. Include some spatial sense of the transitions between one design function and another.
3. Have both a graphic scale and a labeled scale.
4. Consider rendering the entire master plan in color, or at least the area proposed for first-phase development.
5. If possible, include an enlargement with more detail of the area proposed for first-phase development.
6. Include time for review of the drawings by someone in the office unfamiliar with the particular project or at least this aspect of it.
7. Respond fully to the client's and contract's requirements.

Seun City Walk, South Korea

To illustrate a full range of conceptual, schematic and illustrative design drawings, there is Vicky Chan's proposal to restructure Seunsangga in South Korea. The future of business is about collaboration, sustainability, and customization. Seun City Walk is meant to provide a collaborative environment to connect work with play. By weaving the public area into the private buildings, the park is architecturally and structurally part of the

Seun complex (also known as Seunsangga). Pedestrians seeking pleasure can experience different programs for work, play, eating and resting. People seeking opportunities can see the park as an open platform for experimentation, collaboration, and display. The deck is typically 9 meters (29.52 feet) wide, and it is subdivided into 3-meter (9.84 feet) wide ribbons for circulation, greenery, and program. This typical relationship changes at areas designated for special programs and landscape—for example, the softscape transition from Jongmyo Park, which is more urban, to Namsan Mountain, which is more natural. The program responds to each block differently. The programs of work, play, eating, and resting are scattered through the site, but each block will have a focused theme to make the experience unique to the business nature of its corresponding building. (See Figures 5.2 to 5.7).

WeTown, Canada

Landscape architects and other designers should not forget the impact of conceptual study models. The one shown in Figure 5.8 is from WeTown, a proposal for a sustainable, walkable, and healthy city for 40,000 residents in Canada. Reducing the reliance on automobiles, the project provides a walkable loop to connect thirty-six buildings of apartments, offices, and retail. The eight-minute journey from home to work would be filled with greenery, healthy activities, and excitement. Different outdoor and indoor strategies are designed to promote an active lifestyle. A smart system is deployed to reduce water consumption, increase energy efficiency, and promote local food. (See Figure 5.8).

2nd Street Park, Perth Amboy, New Jersey

Another project to illustrate some of the themes of this chapter is the 2nd Street Park in Perth Amboy, New Jersey. The city acquired a 6-acre parcel of land by the waterfront to redevelop for community recreational use. The Rutgers University Center for Urban Environmental Sustainability (CUES) worked collaboratively with the city of Perth Amboy and the Middlesex County Improvement Authority and engaged the community with design workshops to develop a concept design for the park. The design for the park programmed educational, community, and ecological spaces and connected the park to downtown Perth Amboy and the Raritan River waterfront. (See Figures 5.9 and 5.10).

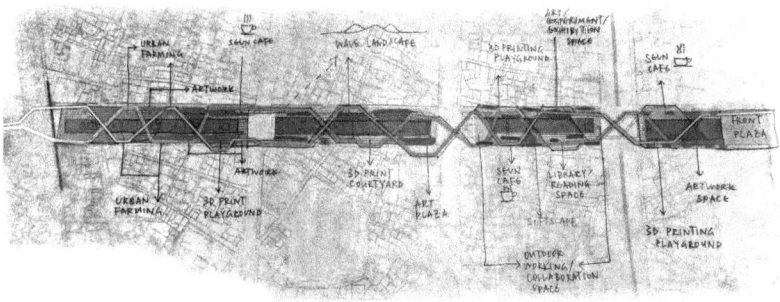

Figure 5.2 Jongmyo Park Mountain concept (Seun City Walk).
Concept Diagram of Seun City Walk showing landscape connecting programs.
Daewook Lee, Vicky Chan, and Melissa Chan

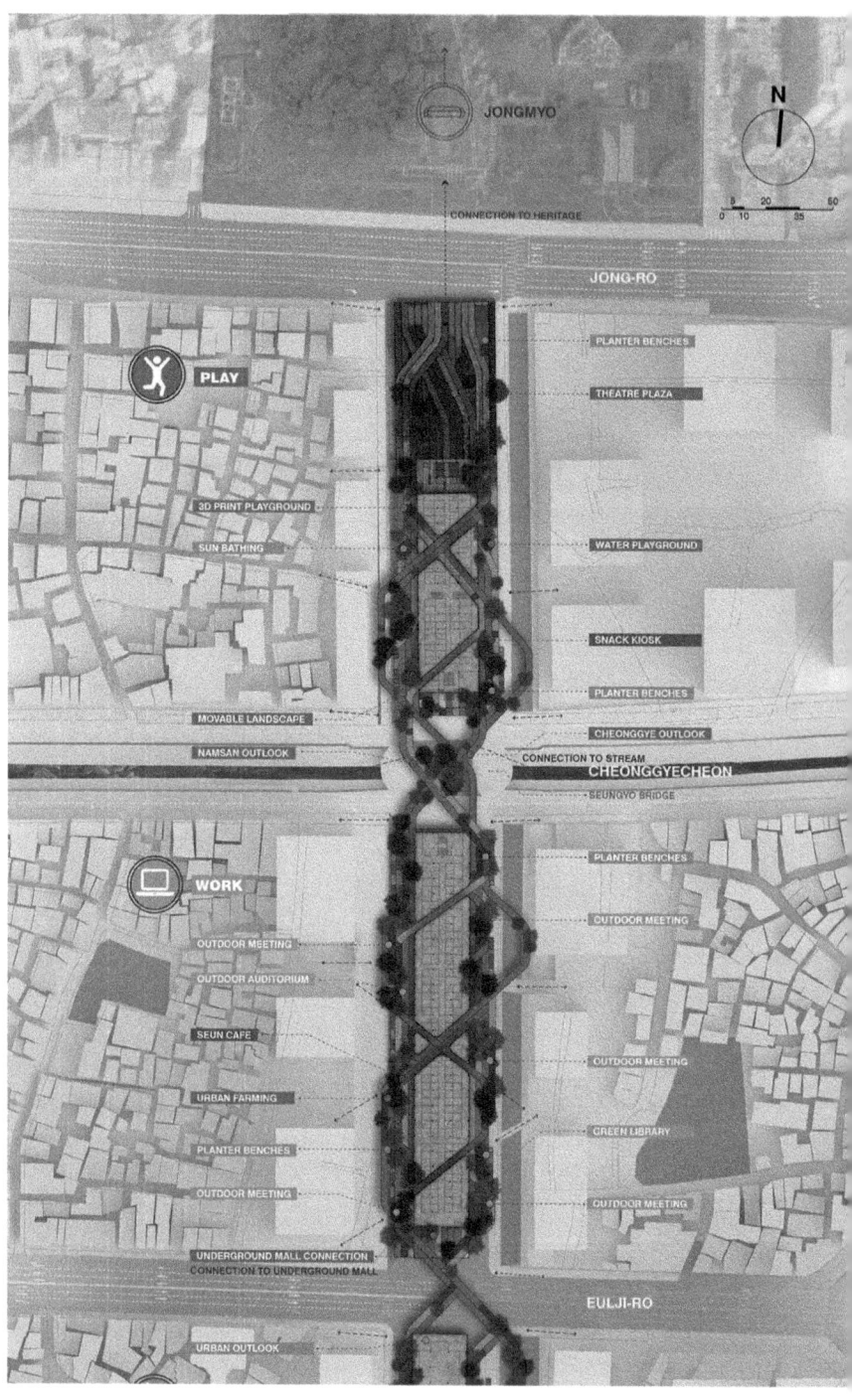

Figure 5.3 Jongmyo Park Mountain.
Concept Sketch of Seun City Walk showing landscape dividing programs into zones.
Daewook Lee, Vicky Chan, and Melissa Chan

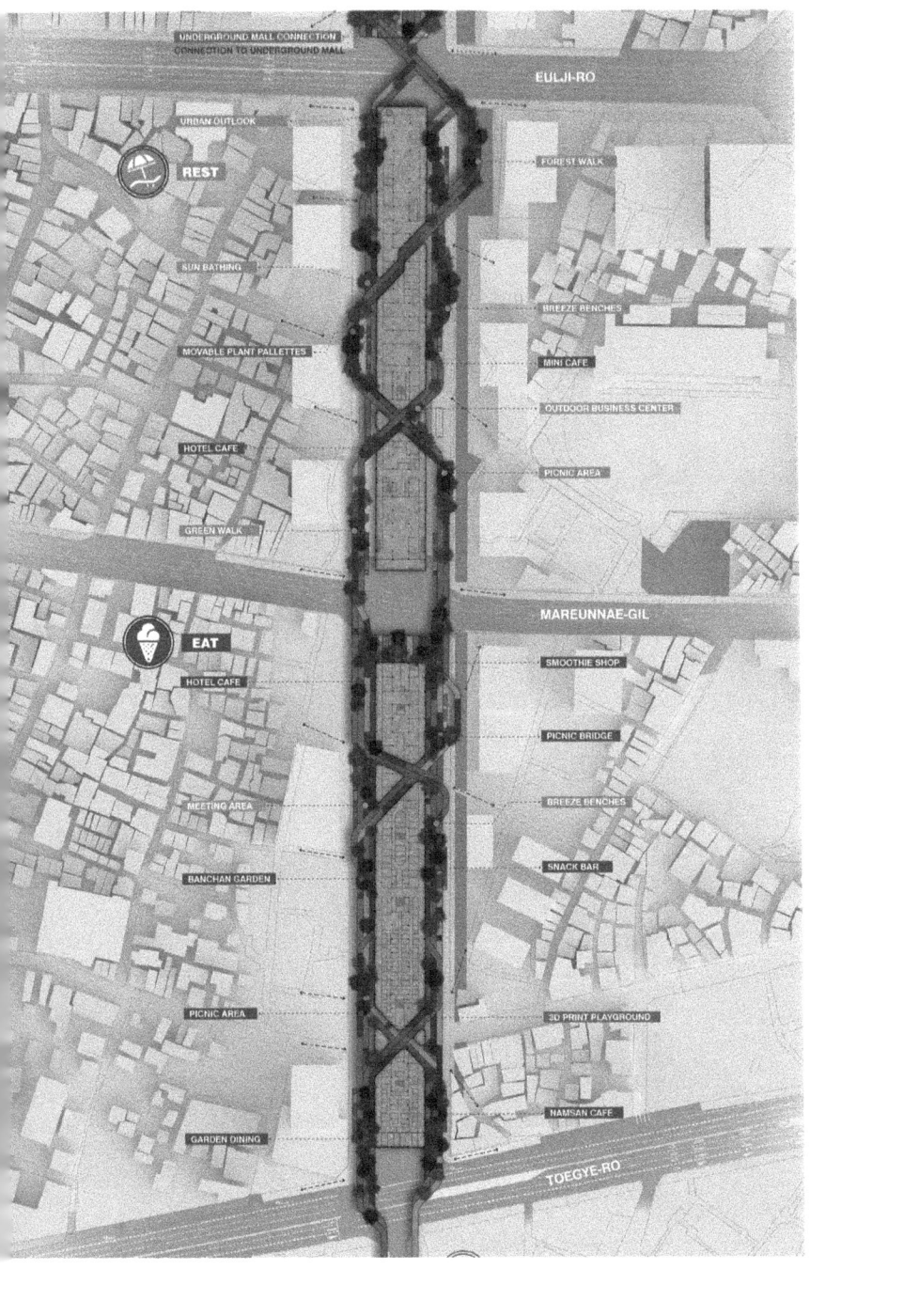

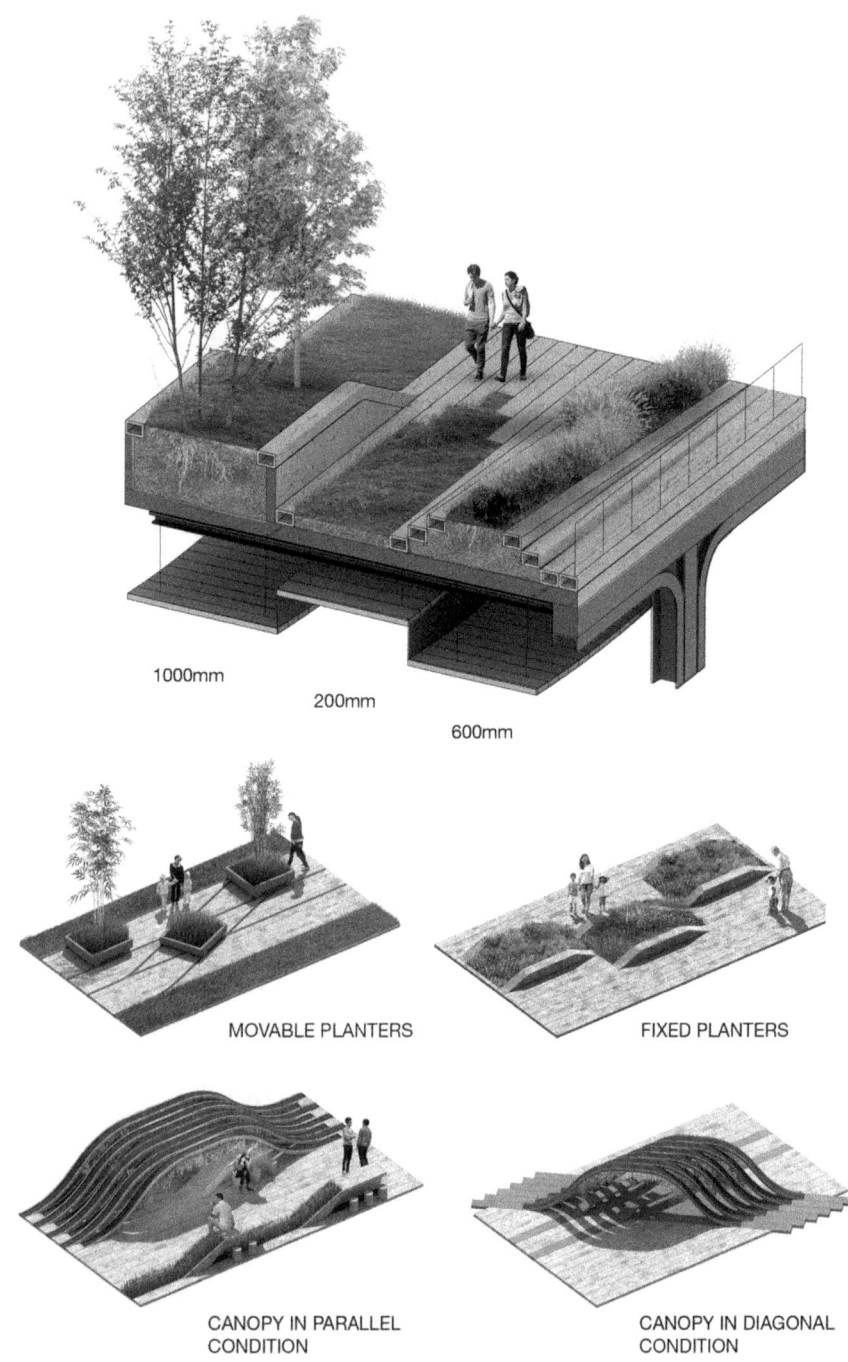

Figure 5.4 Design Taxonomy of Seun City Walk showing various types of landscape treatment above deck and below deck.
Daewook Lee, Vicky Chan, and Melissa Chan

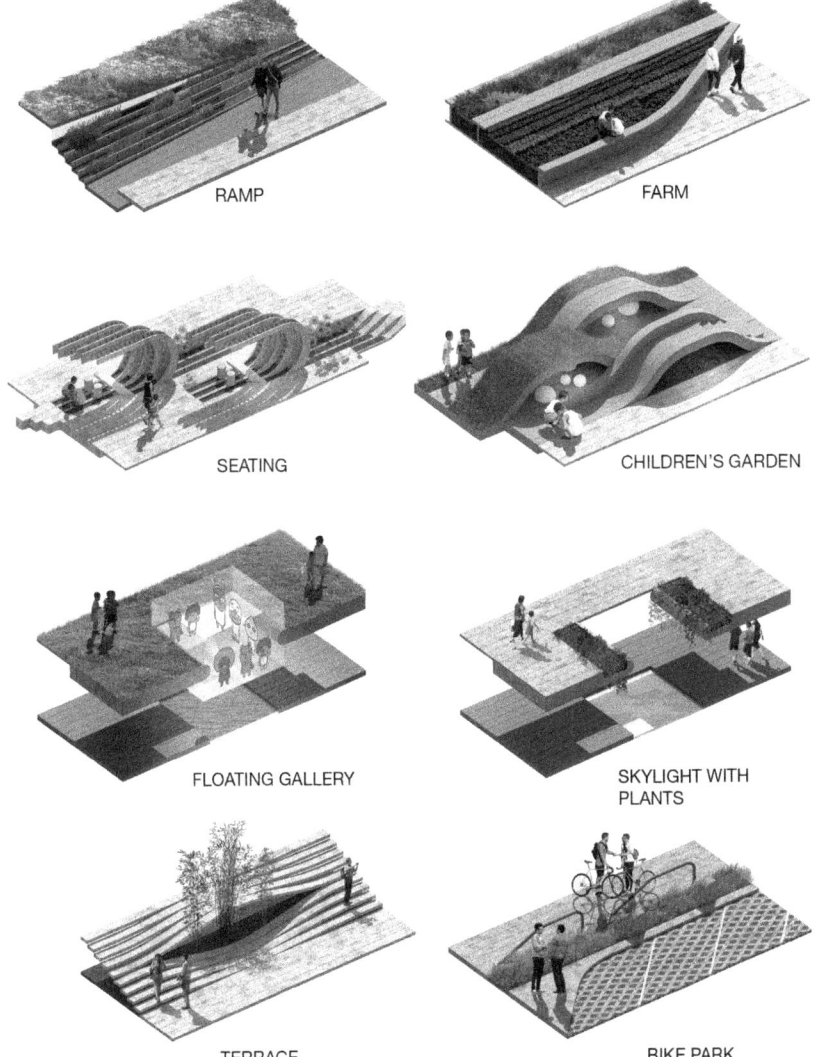

Figure 5.5 Entrance plaza, Rendered Master Plan of Seun City Walk showing programs, plants location and bridge connections to existing buildings.
Daewook Lee, Vicky Chan, and Melissa Chan

Figure 5.6 Seun City: View under bridge overpass: Below deck design of Seun City Walk with opening to above. Plants on movable carts are designed next to parking space for better mobility.
Daewook Lee, Vicky Chan, and Melissa Chan

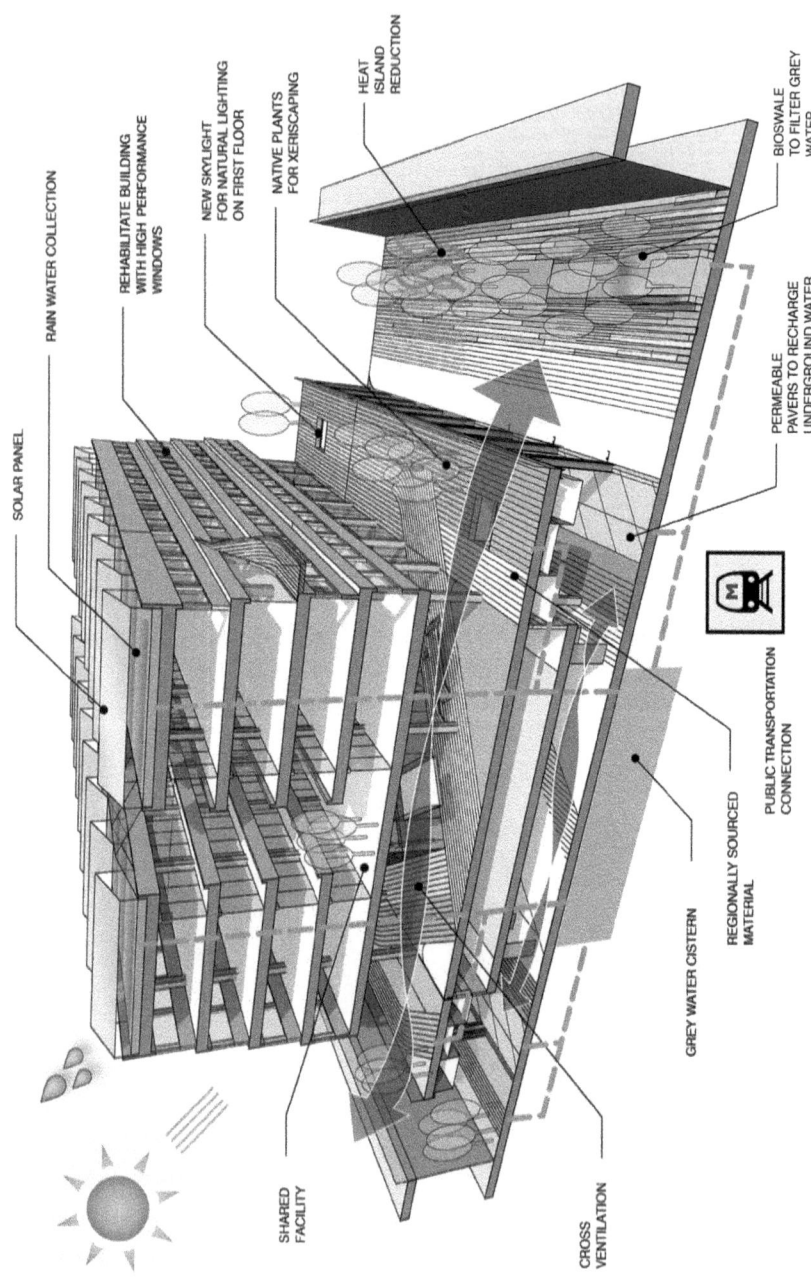

Figure 5.7 Sustainability diagram of Seun City Walk to explain air circulation, rain collection, and plants to reduce heat island effect.
Daewook Lee, Vicky Chan, and Melissa Chan

Figure 5.8 Wetown Conceptual Model, for large new urban development near Vancouver, BC, Canada.
Vicky Chan

Figure 5.9 Perth Amboy 2nd Street Park: design sketch for 6 acre park to be built on the site of a former industrial site, now a brown field opposite a public school by the Raritan River.
Richard Alomar, Rutgers University Center for Environmental Sustainability

Figure 5.10 Perth Amboy 2nd Street Park design sketch.
Richard Alomar, Rutgers University Center for Environmental Sustainability

PART 3: CONSTRUCTION DRAWINGS/ CONTRACT DRAWINGS

After the design development drawings and initial cost estimate are approved, the design team starts the process of preparing detailed *construction drawings* showing the demolition, layout, grading, drainage, planting, and details of the project, along with a final (much more detailed) cost estimate and specifications. Construction drawings are the heart and soul of a landscape architecture project. Their accuracy, consistency, and completeness help the landscape architect to execute in the field what is shown in the drawings to the satisfaction of the client, the contractor, and the designer. Although these drawings are a new product, they are in many ways based on the decisions and details shown in previous versions of the drawings. Although the whole set of drawings is a precise blueprint of how to build the project, they must encompass flexibility in how they are achieved. Not every detail or measurement can be locked into place; some must float. The graphic labels and item numbers of elements drawn in the details must match comparable items in the lists of the cost estimate and specifications. Because these serve as a set of instructions and guidelines for a primary contractor, and perhaps a series of subcontractors, each specializing in a specific trade, the design team must imagine how they would achieve the execution if they themselves were handling it, and then demonstrate with confidence to another party how to proceed.

The Floriculture Greenhouse Outdoor Living Lab at Rutgers University, New Jersey

The Floriculture Greenhouse Outdoor Living Lab at Rutgers University is charged with reprogramming, reutilizing, and reorganizing the space for horticultural teaching and research purposes. The spaces contain existing raised planters, planting areas, existing temporary and permanent structures, beehives, and rain gardens. Previous greenhouses were removed several years ago. The site has not been properly graded, and there has been poor drainage due to existing clay soils. The reprogramming includes the reorganization of the space and the introduction of new site features that will meet teaching and research needs for the landscape architecture department. There will be removal or salvaging of all the existing site features, regrading of the site to provide positive drainage away from the existing building, and a proposed shed. The grading will direct water to the proposed rain garden to encourage biofiltration and potential percolation. The rain gardens will include overflow structures to minimize flooding. At the rear of the building a new metal deck will be installed between the two greenhouses. Other new facilities such as a footbridge will facilitate circulation, access, and maintenance. Throughout this chapter are a series of individual construction drawings from this project. (See Figures 5.11 through 5.16; see also Figures 2.7a and 2.7b).

Designers often talk about integrated design, such as the High Line in New York City, in which some design elements are nested within others, often like pieces of a jigsaw puzzle. Construction drawings require a similar sense of integration. These may be drawn at a range of different scales in order to include the most appropriate part of a site or degree of detail required. Depending on the size and complexity of the site, there may be multiple drawings of each type or just one. The following are general guidelines for each type of drawing. (Please note that since planting is often a project unto itself, it is addressed in a separate section of this chapter.)

DOS AND DON'TS FOR ALL DRAWINGS

1. Plan the organization of the entire set of drawings before starting the first drawing.
2. Show the starting point of any major string of measurements. This may include demolition, layout, grading, or planting. Don't leave to chance where the contractor would originate a complex operation unless you, the designer, are completely convinced it doesn't matter.
3. Show bold north arrows. The convention of having north point up across the length of the drawing is still worthwhile.
4. Show a graphic scale and a precisely labeled one. If the drawing is enlarged or reduced, at least one scale will always be correct.
5. Show match lines and be able to imagine how an entire series of drawings could be pasted together to create a map of the entire project area.
6. Demolition, layout, and grading drawings may be on the same topographic base, with different layers turned on or frozen depending on the purpose of each drawing.
7. For complex designs, use the same scale for layout drawings and grading drawings, if possible, so that it is easier for the contractor to coordinate between the two.
8. Verify the legibility of all the drawings, particularly if they must be printed at half-size for ease of carrying around the site during construction. Crisp, clear lines and images are useless if they are not accompanied by legible dimensioning and other necessary labeling. Many clients, particularly government agencies, now require that

a half-size set of drawings must be made available to all key personnel prior to the start of the work. The quickest way to have such drawings rejected is if they are not fully legible.

9. Plan the set of drawings so that the work for each major trade—demolition, planting, masonry, metalwork, woodwork, and so on—is shown on separate drawings, so that the prime contractor would retain an entire set of drawings for coordination but be able to hand out to each subcontractor a set of drawings that encompasses only that sub's work.

10. Provide a set of general notes that apply to the entire set of drawings. These notes may be on a stand-alone drawing. Every subcontractor can be given one set of these notes, to encourage coordination and so that each contractor has a sense of what other work is going on at the site or will be scheduled.

11. Provide a set of specific notes applicable to each trade, which appear on the relevant drawing. Only the landscape contractor needs to know the requirements for topsoil and mulch. Only the mason needs to know exactly what type of brick or stone is specified, and so on. The landscape architect must sort through a considerable amount of information and choose how it is provided to each contractor for each area of the design.

12. Coordinate with all utility companies so that directions on the drawings for capping, removal, additions, or installations of utility systems are absolutely clear to everyone. This is so important to utility companies themselves that they often provide toll-free numbers to call for them to come and stake out, at no cost, their existing utility systems. These may include stormwater, sanitary sewer, electric (both high- and low-voltage), gas, septic systems, and fiber-optic cables.

13. Include language on daily maintenance and long-term upkeep of the site, no matter what the stage of construction, so that the contractor (and subs, if required) is responsible for keeping the site clean, safe, and in excellent condition for the duration of the work, and so that after completion, information about long-term maintenance requirements is still available.

14. Provide a list of those items of work requiring shop drawings, so that it's clear to the contractor from the beginning what will be required. Similarly, in the specifications, spell out clearly the requirements for typical shop drawings—for example, to scale, at a minimum size, with clear labeling, the review process, and so on.

15. Indicate how and where samples will be stored on-site. Often the landscape architect or owner or some consultant or contractor may find it necessary to review a particular product to coordinate its color, texture, or character with something in another scope of work. The more readily available an organized system of storage and labeling on-site, the better the likelihood that all contractors will coordinate and have their work dovetail properly.

16. Don't repeat the same information in more than one drawing (or within more than one technical specification). Redundant information often becomes misleading or inaccurate. It's too easy to make a slight mistake in language and create a major problem in misinterpretation.
17. For areas of layout in which there is a degree of uncertainty, provide for staking out and trying out a particular area. For example, if a parking space for the owner's SUV or sports car is paramount, and yet it must fit within a limited amount of space, show it on the drawings but allow for adjustment of the measurements based on having the owner drive the car into the staked area. The importance and usefulness of engaging the owner in decision making about her own project cannot be underestimated.
18. Toward the end of the development of the construction drawings, schedule at least one meeting with the owner to review the drawings. Be certain that there is a sufficient comfort level with final selections of materials, finishes, major plants, and so on. Even if the landscape architect has a design contract that allows him to charge more if changes are made after previously approved design development drawings, it's still better to make adjustments on the drawings before materials, products, and plants are brought to the site and installed than afterward.
19. Include a north arrow and a graphic scale for every drawing in plan view to scale.
20. Comply with all office or design team standards as to line weights, text font and size, layers (both colors and symbols and line weights) in AutoCAD, and other requirements. There may be a comprehensive legend that applies to all drawings or each type of drawing may be provided with a legend.
21. Anticipate the sequence of drawings carefully. Sometimes a client agency will require a certain order of drawings to be in compliance with their standards. In general, a standard sequence is (1) cover sheet with list of drawings, (2) topographic survey, (3) demolition, (4) layout, (5) grading and drainage, (6) planting, (7) details, and (8) special elements.
22. Verify the maximum (and minimum) sheet sizes. Often government agencies will have required sheet sizes.
23. If multiple drawings are required to show the complete site for the scale that is suitable for a particular drawing, be certain that all match lines are clearly identified and there is enough overlap that the intent of each drawing remains clear.
24. Date the drawings and allow for revision stamps. Some clients have requirements about where this information must appear on the title block of each sheet. Most design offices develop their own standard so that they know where to look for the latest revision date on a particular drawing.

25. When the drawings are nearly complete, have someone unfamiliar with the project or at least someone unfamiliar with the drawings themselves review them for completeness, accuracy, and clarity.
26. Be prepared to sign and seal the set of drawings (sometimes this is every sheet in the set, sometimes this is the cover sheet on a bound set) with the design principal's name who is registered as a landscape architect in the state in which the work is being done.
27. Create a set of PDFs or JPEGs, as per client requirements, for distribution of the drawings to others in the design team, contractors for bidding or review, and any other required participants.

Demolition Drawings

Demolition drawings range from very simple to complex, so anticipate the most important elements to show the contractor. For example, even a small amount of asbestos abatement requires a contractor trained in this process and a step-by-step review and approval process, and usually a separate set of demolition drawings, whereas most demolition operations involving removals of pavements, trees, and walls involve fairly standard procedures. Before you start the drawing have a sense of what order of operations may likely be the most efficient, and what permits the contractor may be required to obtain. Familiarize yourself with whatever codes apply in terms of hauling materials off-site to landfills, and what hours of operation may be permitted.

DOS AND DON'TS FOR DEMOLITION DRAWINGS

1. Use as a base sheet a topographic survey in which all the information identified has been verified in the field by the landscape architect or other staff member or consultant such as an engineer.
2. Freeze any layers that are not relevant so that the graphics may focus on demolition items without confusing extraneous information. The complete topographic survey may be included as a separate sheet for reference for the design team and contractors so that all existing information is identified. Just as one example, it's useful to see the length and bearing of each property line on a topographic survey for the purpose of calculating areas and verifying directions, but on a drawing focusing on demolition, this information may be a distraction.

3. Develop a legend of different types of shading, crosshatching, or other patterns, each of which represents a specific type of demolition—for example, trees to be removed, concrete pavement to be removed, asphalt pavement to be removed, sidewalks to be removed, specific types of walls to be removed, utilities to be removed, and so on.
4. Call out any demolition that requires special permits or standards of performance, as per specifications or general practice. For example, removal of asbestos or other contaminated material may be so restricted as to be shown on its own sheet with specific notes about the demolition process. A specific state or federal agency may be required to review the work in the field.
5. Develop a cost estimate in which the quantities and volumes of each item to be demolished are carefully calculated. However, it's better *not* to call out the quantities (e.g., linear feet of fence, square yards of 4-inch-thick concrete, cubic yards of gravel) because if any errors in these numbers are found on the drawing, it can become the responsibility of the landscape architect or whoever is contractually obligated as the principal in charge of that drawing to absorb the cost of the error. It's much better to label types of demolition carefully, leave it to the contractor to calculate quantities, and then check this work.
6. Verify the order of sequence of operations. Demolition is often a "performance" type of operation—that is, as long as everything is disposed of legally and as shown on the drawings, it does not necessarily matter what equipment is used or what order of operations is followed. However, if specific restrictions or requirements apply, these must be emphasized on the demolition drawing. For example, if a lot of heavy debris, such as broken concrete or masonry, is to be hauled off the site, there may be limits on the size and capacity of dump trucks permitted on the adjacent road system in order to prevent damage to the pavement, or certain items may have to be removed first in order to allow for the proper sequence of construction to be implemented.
7. Know the local building codes and be aware of any restrictions placed on demolition operations. Be certain that the demolition operations proposed are in compliance with code requirements. Such requirements may range from the complex, such as erosion and sedimentation control measurements, to the simple, such as misting or spraying some areas to prevent the accumulation of dust.
8. Where possible, show details for major operations such as erosion and sedimentation control directly on the demolition plan, or at a minimum, include clear notes referring to other sheets that contain this information.
9. Include on the drawing any listing of permits that must be obtained by the contractor prior to the beginning of demolition operations. Also, note any permits that the owner (or operator of the site) may need to obtain.

10. Reference on the drawing any necessary contact information, such as names and phone numbers of utility companies, environmental compliance agencies, or other entities that may have jurisdiction over key aspects of demolition. For example, many utility companies will come and stake out for free the location of existing utility lines to minimize the potential for exposing or damaging them during demolition operations. Sometimes contractors are required to notify government representatives or utility companies if certain conditions or elements are discovered—for example, the exposure of a live electrical line not properly insulated, or any sort of potentially toxic substance.
11. Demolition operations may depend on weather conditions but often require steady, fully staffed and supervised crews at work. Sometimes it's practical to include a liquidated damages clause in this aspect of the contract to ensure that this phase of work occurs in a timely manner and therefore there is no delay in the beginning of the subsequent phases of the work. There may be a higher initial cost of demolition itself, but a more timely progression to the overall work on the site.
12. Demolition plans must show not only what is to be removed but also what existing elements on the site must be protected in place, such as major trees, and which will tie into the new design. Sometimes existing elements, such as bluestone pavers, might be salvaged from the site, stockpiled, and reused in the new design.

Perhaps the single most bizarre unexpected demolition delay I've ever experienced was at horse farm estate in Virginia. Plans called for the construction of an equestrian field on a gently sloping site, formerly farmed but parts of which had some dense young forest and undergrowth. The contractor started doing standard clearing and grubbing operations, so the site was gradually cleared to allow for the beginning of grading operations, but one of the backhoe operators noticed something shiny. It turned out to be part of a plane that had crashed. This being a small community, where most people in the equestrian and farming communities knew each other, one of the local people working on the site recalled an incident from thirty or forty years earlier when a plane on a training mission had disappeared amid a sudden, strong thunderstorm. A little more excavation was done to verify that the crash was a military plane and that there might be human remains. The contractor called the Army Corps of Engineers, who sent an archaeological team to the site, which carefully excavated the plane, verified the plane's identity, and recovered the human remains for proper burial. It was only after this work was finished that the demolition and grading could be completed. One must learn to expect the unexpected. (See Figure 5.11).

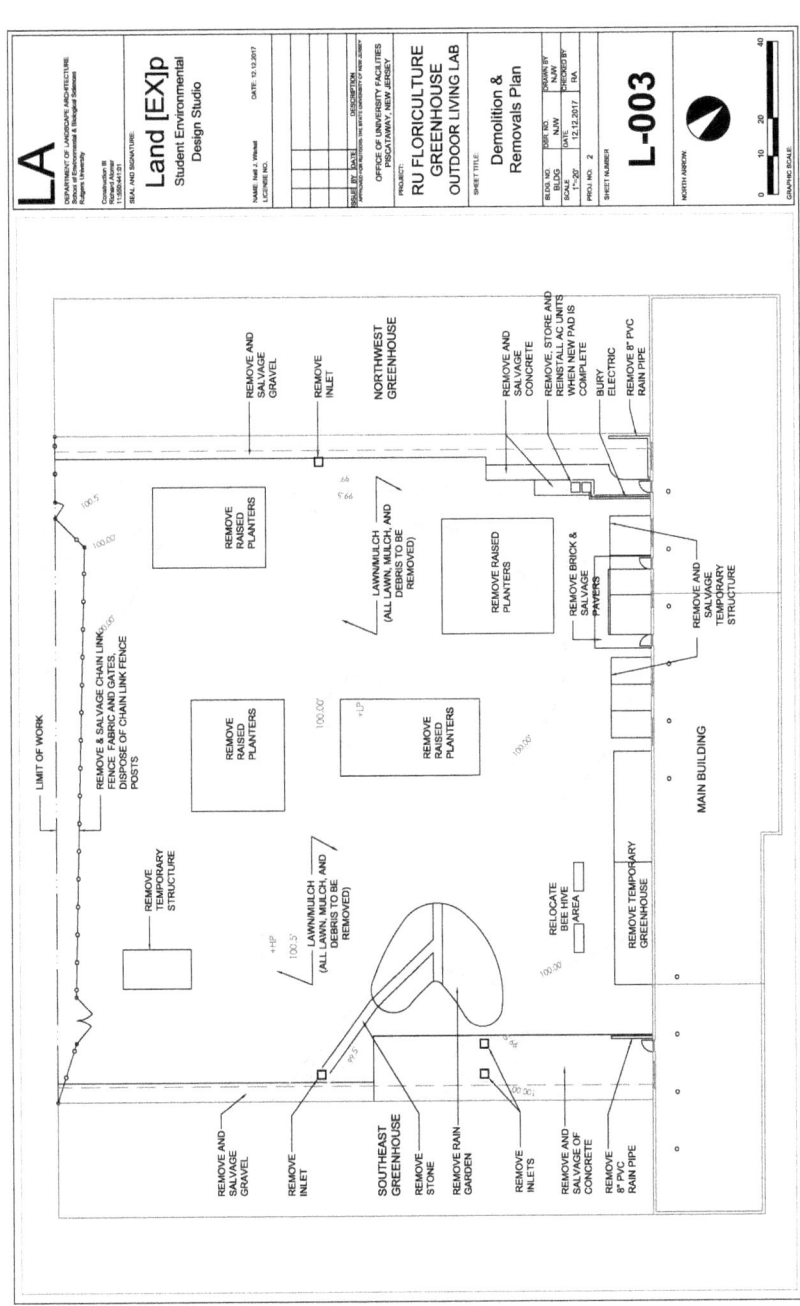

Figure 5.11 Floriculture Greenhouse Outdoor Living Lab at Rutgers University. Demolition and Removals Plan

Layout Drawings

The layout drawing is the most important of the construction drawings because the other drawings follow from it and emerge from the information provided on it. You should be able to check gradients on the grading plan by verifying distances on the layout plan. Key details are located on the layout plan. Often what is demolished is in relation to what is shown to remain on the layout plan, so this is the first plan to show an integrated sense of the overall design. It must show both an accurate sense of the design, with a hierarchy of line weights revealing the strong and light points of its structure, but also how to achieve it with enough flexibility to overcome any discrepancies within site conditions and inherent inaccuracies of the drawings themselves.

DOS AND DON'TS FOR LAYOUT DRAWINGS

1. Imagine the layout in your head, and think of key starting points on the site. For small and simple designs, there may be only one; for larger and more complex projects, there could be several. Identify these starting points or baselines in strong graphics so that they stand out boldly on the drawing.
2. Verify that starting points and baselines are easily determined and measured on the site, without substantial potential for error, for example, a point that is 20 feet from the midpoint of and perpendicular to an existing wall of a residence or a baseline that is parallel to an existing wall at a set distance from it.
3. Where possible, locate starting points and baselines in relation to existing elements that will remain, not something that will be demolished as the work progresses. Until the layout work is finished, there should always be available a simple method for verifying their basic accuracy.
4. Allow some *dimensions* to float. This may be as simple as using a ± symbol so that a leftover space at the end of a long string of dimensions can be labeled as an approximate distance rather than a precise one. This tells the contractor executing the work where to allow for adjustments without undermining the consistent aspects of the design.
5. Let some *elements* float to absorb error and allow for graceful adjustments in the field. For example, imagine that you are designing a bilaterally symmetrical design of a garden in a large rectilinear site between two parallel walls. The centerline should be

precisely shown, and the layout of each symmetrical element must be shown. However, the distance between the *outermost elements* on each side of the centerline to the faces of the existing walls should not be marked exactly. Either provide a ± dimension, indicating a range that is acceptable, or use = marks, or none at all, if there is certainty that as long as the elements and dimensions shown are achieved, then the distance to the walls will not matter substantially.

6. Know your margin for error and the degree of accuracy required. For some applications, it's important to show dimensions to the nearest inch or centimeter; in others, more or less accuracy is required. Don't provide more or less precise information than is necessary to develop the correct layout and dimensions of the design. For example, a layout of a fence with posts 6 feet on center set 1 foot off the property line is clear and within the standard degree of accuracy of this element. By contrast, the layout of a sidewalk with brick in a basketweave pattern in which the designer wishes to eliminate fractions of brick may require more precise dimensions.
7. Comply with office standards for graphic clarity of labels. The size of dimensions may be increased or decreased depending on their relative importance to the layout concept and detail.
8. If you provide a long string of dimensions that appear graphically to add up to a total length that is called out, be certain the sum of these dimensions equals the total length shown.
9. Provide a legend.
10. Emphasize in a set of notes the key operating principles for the drawing.

(For layout drawing, see Figure 5.12).

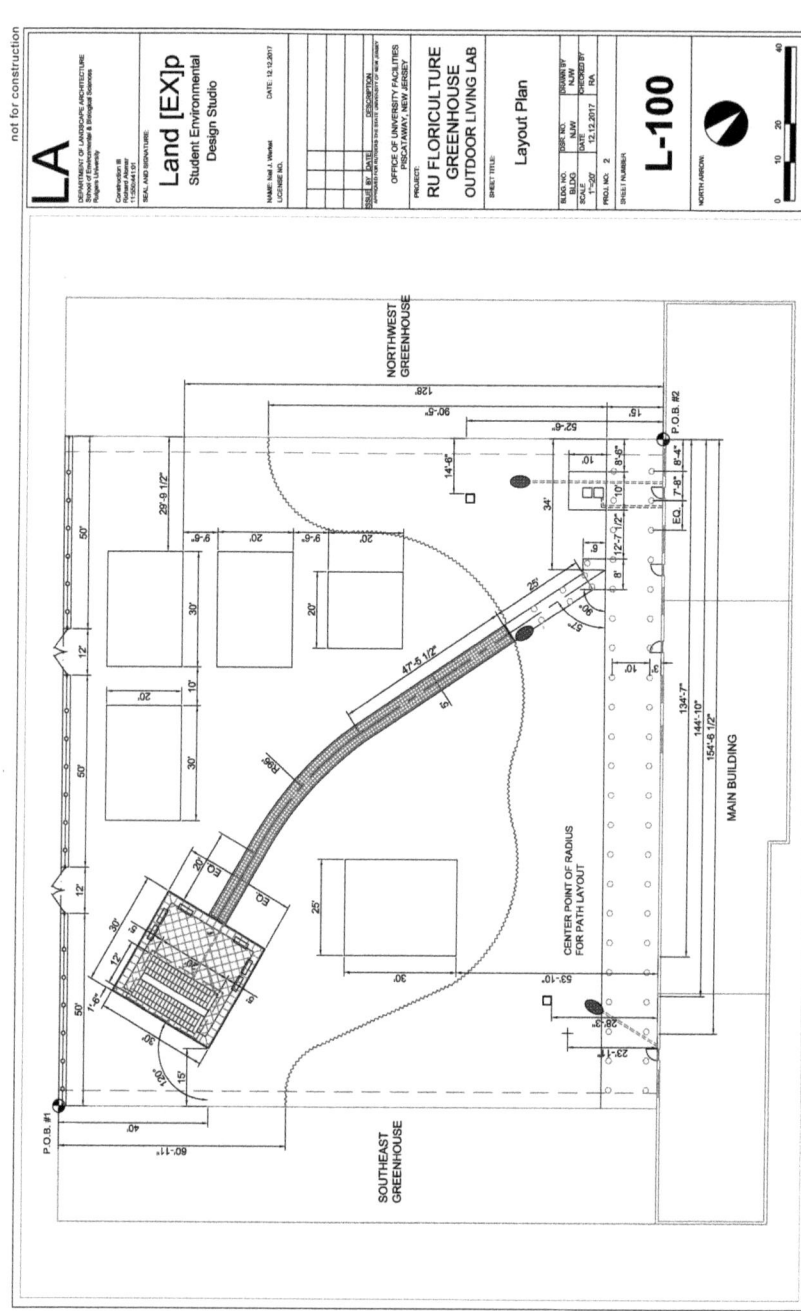

Figure 5.12 Floriculture Greenhouse Outdoor Living Lab at Rutgers. Layout Plan

Grading and Storm Drainage Drawings

Whether you are developing a simple grading and drainage plan showing discrete flows away from the house on all sides with appropriate drainage structures to intercept any stormwater before it leaves the site or you're preparing a complex undulating landscape, consultation with a civil engineer on grading and drainage is often advisable. The engineer can verify that the drainage design complies with all code requirements, that the underlying soil types don't create hazards for grading at the slopes proposed by the design, and that storm drains and sewers are connected correctly to existing sewer systems or released to the existing topography without risk of erosion.

DOS AND DON'TS FOR GRADING AND STORM DRAINAGE DRAWINGS

1. Develop a labeling system for contours so that it's easy to recognize swales and ridges. The simplest way is to consistently label each contour on its high side.
2. For critical areas, include occasional notes identifying the percentage of slope or gradient of specific locations.
3. Identify high points and low points.
4. Identify centerlines of valleys and ridges, which are often the low points and high points of a topographic system or physiography.
5. Use a contour interval appropriate for the scale of the drawing and the intended application. The grading should be no more or less accurate or precise than the layout. Include a set of notes identifying crucial operations, such as sedimentation and erosion control (sometimes shown on the grading plan, sometimes shown on other independent drawings), how and where to stockpile topsoil or other materials for grading operations, maximum rate of cutting and filling permitted, and so on.
6. Identify the size and length of each pipe, but, similar to layout drawings, allow some flexibility in the length of either the first or last pipe in each system.
7. Call out and use a different symbol for each type of drainage structure, such as catch basin, inlet, manhole, or pipe of a certain dimension.
8. Study carefully the location of each outfall and be certain that it will be stable and not at a site that will set off erosion problems.
9. Be aware of code requirements. For example, it's usually the case that drainage across property lines may not exceed what existed prior to disturbance of the site as a result of construction operations.

10. Balance cut and fill. This implies that all soil excavated on-site is appropriate for filling elsewhere. This is not always the case. If excavated material must be hauled off (it's too silty, too high a percentage of clay, has some level of contamination, etc.), factor in the cost of hauling and permits required, if any, before finalizing such a grading concept. If fill material must be brought in, determine the likely source and hauling cost and create a landform that achieves the design goals with the minimum volume of material. This does not mean that you avoid a bold design. If a berm is needed for screening, for example, establish the minimum height that will achieve the desired result in combination with planting to be incorporated on it.
11. Maintain sheet flows if possible, and avoid concentrating runoff. If an existing site has considerable areas of sheet flow, try to leave such a system intact. At the same time, if swales or other grading methods must result in concentrating and collecting water, locate the drainage structures, such as catch basins, area drains, and so on, in locations that are not key visual elements within the landscape.
12. Strip drains are effective methods of intercepting and collecting stormwater at the base of steps, the edge of a pavement, or within an expanse of pavement. In their contrast to pavement materials, they are usually highly visible elements in the landscape, so select and specify attractive ones. There are many available in historical styles and a range of materials.
13. Allow for an increase in future runoff; that is, don't design and size pipes and drainage to such tight standards that if another room is added to a residence or another garden is developed with some pavements, that you don't need to upgrade the storm drainage system. Another way of thinking of this: the cost between installing an 8-inch diameter storm drainage line and a 6-inch diameter one is not that significant compared to the considerable cost of removing a 6-inch system and upgrading it to an 8-inch system several years later when additional facilities increase runoff to be collected.
14. Find ways to retard the peak flow of stormwater that collects immediately after a heavy rain. Consider bioswales, detention and retention basins, weirs, and check dams (which collect stormwater and allow oils, silts, and other elements to settle out before flowing through the storm drainage system).
15. Avoid minimum acceptable gradients—for example, < 1% for pavements—although pavement with a gradient of even 0.5% will still drain but will very slowly, with a lot of sediment settling out and potentially staining the pavement. It's usually better to find a way to engineer a steeper gradient. At the same time, conditions may dictate occasionally having a very flat slope, for example, to avoid scraping the surface roots of a major existing shade tree.

16. Be careful in large-scale grading not to disturb a larger area at one time than can be carefully protected against erosion and sedimentation damage. Depending on the site, its gradients, and types of soil, determine a maximum area to be graded at any one time and all of the silt fences and other devices needed to prevent erosion and sedimentation from undermining what is being created. Hydroseeding can be effective over large areas, but if heavy rains strike such an installation at the wrong time, the stormwater can wash away a great deal of soil.
17. Minimize lawn areas. Of course, many clients want graceful and gracious lawn areas, but they require vast quantities of water, fertilizers, herbicides, and pesticides, whose impact on the soil profile may be quite damaging. Fewer microorganisms in the soil and more chemicals means fewer insects and nutrients for birds, for example. Find a way to provide an appropriate expanse of lawn without incorporating acres for mowing and maintenance.
18. Lay out and anticipate the grading at a macro scale before executing it at a micro scale. Set and mark grade stakes at the elevations shown on your drawings and walk the distances between them, visualizing the finish grades to be achieved. If strong and fluid connections appear achievable, then proceed with detailed grading. If there appear to be conflicts between a few high and low points too near one another, then imagine a resolution in the field and update the drawings.
19. Cut slopes are more stable than fill slopes, so the rule of thumb is not to cut at a gradient steeper than 2:1 nor fill at a gradient steeper than 3:1. Exceptions can occasionally be made, but consult with an engineer.
20. At the same time, finesse or improve the design with sinuous shapes and with convex and concave slopes, which should be carefully drawn and labeled in the grading plan and emphasized to the contractors who will execute the work as an important feature of the design.

(For grading drawing, see Figure 5.13).

Detail Drawings

The information shown on details drawings must be carefully drawn to coordinate with the particular drawing—demolition, layout, grading, drainage, et cetera—on which each detail is called out. The graphic vocabulary of details has many conventions, which should generally be followed. Some details work best as a simple cross-section view; others require a combination of plan, section, and/or elevation, while others are best done as perspectives or isometrics. Use whatever method is the simplest and clearest. Think

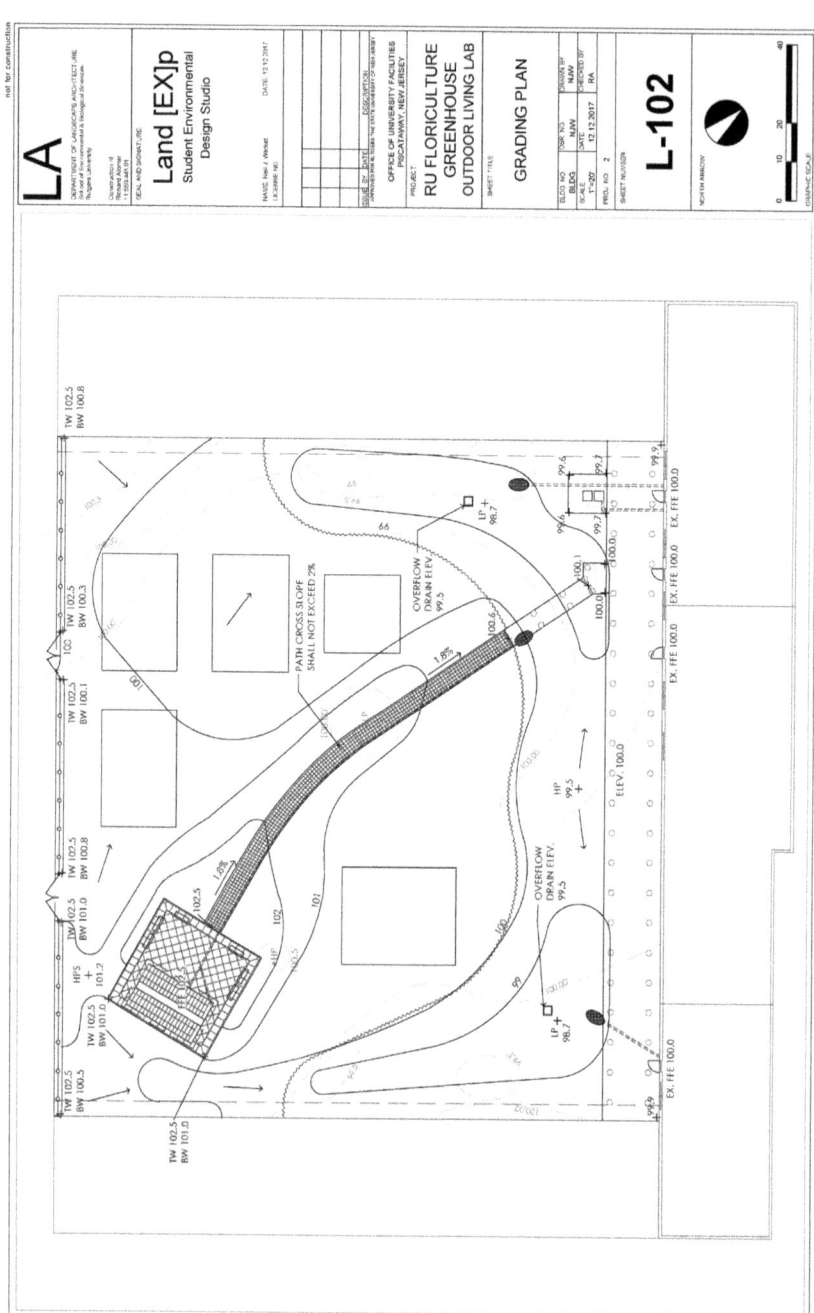

Figure 5.13 Floriculture Greenhouse Outdoor Living Lab at Rutgers. Grading plan.

carefully about what scale is best to draw each detail. In some cases, there may be a series of details that have some common features or connections, so it's best to draw them at the same scale for clarity. At other times, a complex detail may need to be drawn at a much larger scale than simpler ones.

DOS AND DON'TS FOR DETAIL DRAWINGS

1. Organize the details by type and trade—for example, all pavements on one sheet and all walls on another, and/or work to be done by masons versus work to be done by paving contractors.
2. If possible, use the same scale for details of the same type and element. With the exception of close-ups of specific parts of details, it's confusing when a section is drawn at a different scale as an elevation for the same detail or style of construction.
3. Where applicable, include in the detail drawing the specification or item number appropriate for each detail.
4. Apply uniform standards for terms and items such as "compacted subgrade," "undisturbed grade," "fill," "cut," and so on. Be certain that throughout the set of details these elements are the same and consistent with one another.
5. Achieve graphic clarity with the symbols, images, labels, and dimensions.
6. Reference the details in the plan documents in the order that you feel is best to show them on the detail sheets.
7. Graphic clarity is essential. Pay particular attention to line weights and to the size and clearness of the text, including all quantities and dimensions.
8. Don't overlabel details. Coordinate them with the specifications. Information highlighted in the specification need not be repeated in the graphic detail and vice versa. The two together should give complete clarity as to what is to be installed or built. Repetitions in labeling always run the risk that some correction—particularly one done late in a project, when there may be a time crunch, even if the change is a very necessary one—may be shown on the detail drawing but not erased and revised in the specification. Or a revised specification might incorporate dimensions and quantities that have not been corrected in the detail drawings. Confusion will result when there is more than one set of information about the same item.
9. Because details are so complex, always have a colleague unfamiliar with the project review the preliminary and final set for completeness and accuracy.
10. Similarly, review the cost estimate to be certain that all details are incorporated as item numbers within it, with a specification assigned to it.

(For detail drawings, see Figures 5.14 and 5.15).

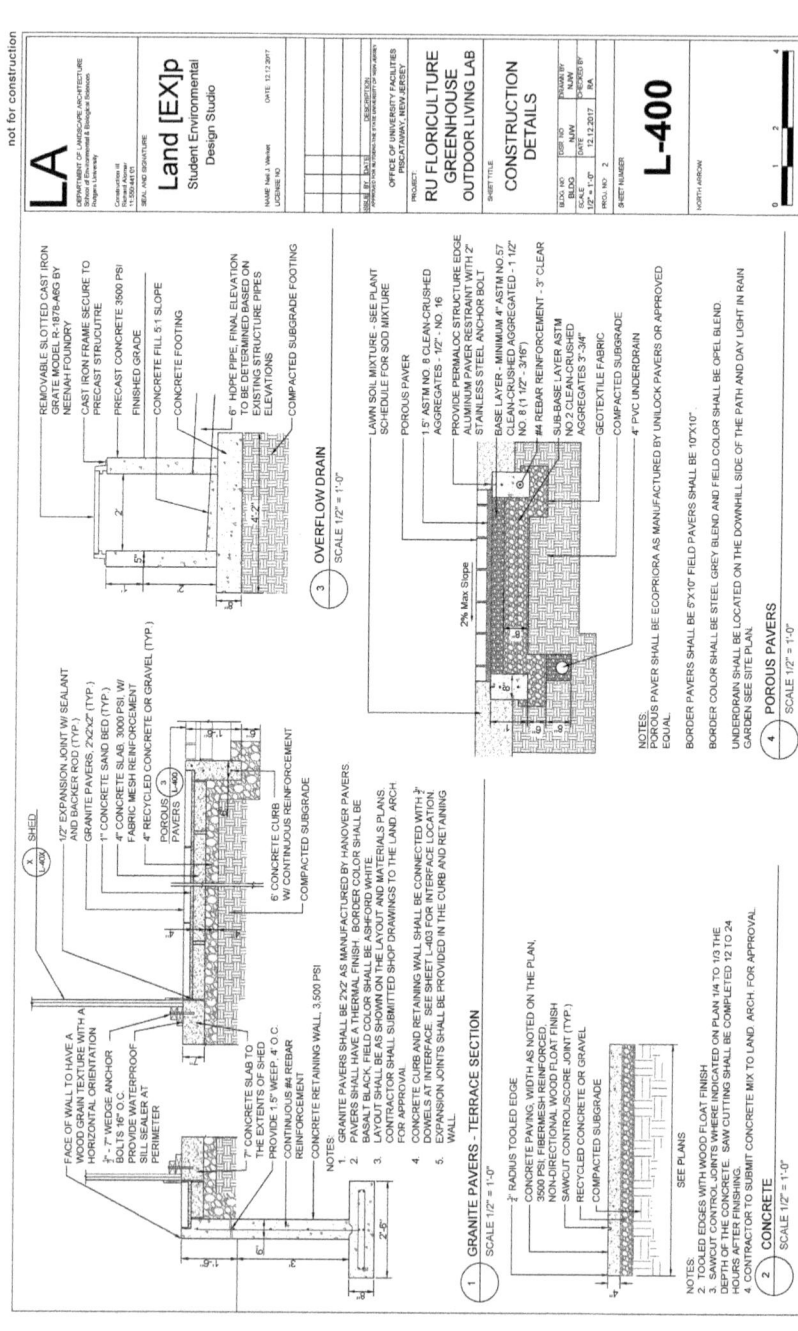

Figure 5.14 Floriculture Greenhouse Outdoor Living Lab at Rutgers. Details 1.

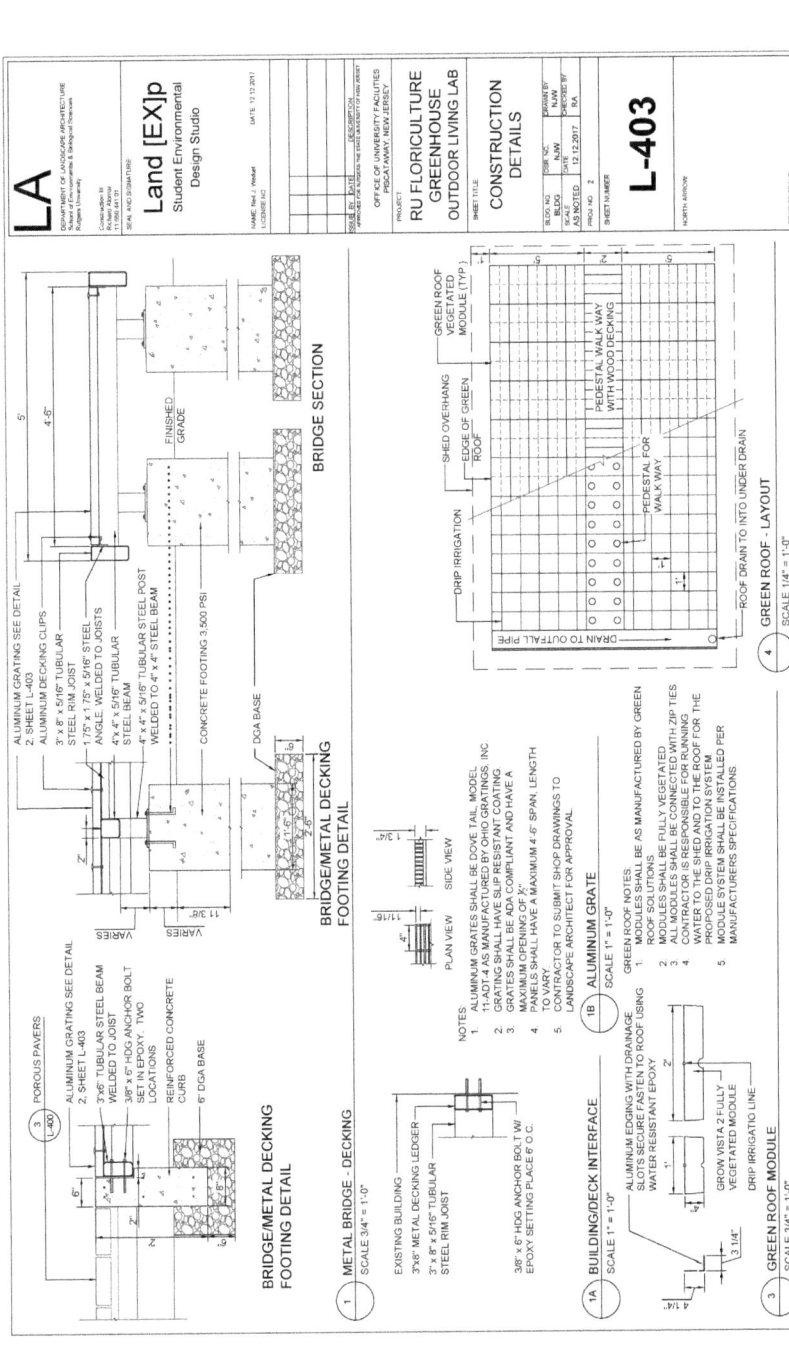

Figure 5.15 Floriculture Greenhouse Outdoor Living Lab at Rutgers Details 2.

Problem 1

DETAILS

Develop one comprehensive detail through a set of five coordinated and interconnected drawings for pavements, curbs, or related details for an urban design setting *(imagine one that challenges your capabilities and interests)*:

1. One plan view (to scale).
2. Two sections through the plans (to scale).
3. Two enlargements of the sections (to scale).
4. Focus on the use of one primary material, such as stone, pavement, metal, and so on. You may use several materials, but one should take precedence.
5. Provide enough detail to show how to completely construct the element.
6. Challenge yourself.

PART 4: PLANTING DESIGN

The Olive is a tree the least known in America, and yet the most worthy of being known. Of all the gifts of heaven to man, it is next to the most precious, if it be not the most precious. Perhaps it may claim a preference even to bread; because there is such an infinitude of vegetables which it renders a proper and comfortable nourishment.
    —Thomas Jefferson, letter to William Drayton, Paris, July 30, 1787[3]

Planting design is traditionally the most important function for landscape architects. As the profession has grown, there are many landscape designers and horticulturalists who may not have technical skills in construction, grading, and drainage design and who therefore choose to develop and focus their expertise on creating gardens of unusual variety and beauty while using a diverse array of plant materials. There are vast resources about plant materials, planting design, styles, and methods of planting. This discussion will focus on professional and practical considerations.

Sometimes planting is the icing on the cake: all of the other goals of the design have been resolved—such as siting a residence and ancillary facilities, laying out the main buildings of a campus, doing the layout and grading for a complex of athletic facilities—and the planting is the last phase of work. It must tie everything together. In other situations, the planting design is the entire scope of work. With such a broad and diverse array of possibilities, what are some criteria to apply?

Structural Planting Elements

Start with the structural planting elements, which define the spaces, and which also align and harmonize with, or contrast with, architecture or other built elements if they are present. Such plants may be specimen trees that form an allee, or parade, down both sides of a driveway and connect to other spaces. These plants can also be hedges, the anticipated height for which is taller than 5 to 5½ (1.5 to 1.7 m), the effective eye level of most people, so that these plants may block views and form the boundary to a space.

Sequence of Spaces

Think in terms of the sequence of spaces being created, and how the edges overlap. Transitions are critical to any cohesive design. Sometimes sudden and jarring changes from one space or another create excitement and tension prior to some moment of relaxation. Most commonly, however, an overlapping connection is more desirable, because each of the adjacent spaces gains in character from the edge they borrow from another.

Scale

Always be aware of the scale of the spaces. Even if you have years of experience, it's common to imagine spaces drawn on paper or at the computer to be larger than they are. From early phases of design, imagine people within or walking by the spaces, and the emotions you wish the spaces to help generate as the people continue to explore. If possible, stake out the overall length and width of spaces to be defined by planting, as a way of testing your perceptions. Once you have tried out some possibilities and have confidence in some concepts, return to the studio to finalize the dimensions.

Gradually fill in the blanks. Add understory layers, plants for variety and contrast, masses of shrubs and perennials, and perhaps some unique details. Use your graphic style to show groups of shrubs and perennials, rather than individual ones. Progress from the general to the specific, and keep exploring until you are confident with the results. A planting plan may be an illustrative rendering to show the client the design intent, so that one can see the sequence of spaces created or reinforced by the proposed planting and distinguish easily between existing plant materials, particularly trees, and proposed planting.

Plan Graphics

When the design is at a level of detail sufficient to convert to a planting plan drawing, there are important steps to take to mold it into a clear construction drawing. As with any plan drawing, think in terms of a hierarchy of line weights, so that the boldest lines refer to the most important and dominant elements proposed for the landscape. Specimen trees and hedges can be shown with the darkest and thickest line weights and/or the largest circles, so that they pop up within the graphic. The hierarchy steps down in boldness as the plant materials depicted step down in size and level of importance. A planting plan as part of a set of construction drawings must be organized in the clearest possible way in order for the landscape architect and landscape contractor to determine quickly what plants are proposed for which locations, the total quantities of plant materials of each species, and the size of these materials.

Labeling

The labeling of each and every plant material shown on the drawing is a challenge not just as a graphic exercise but also as a set of directions to the landscape contractor, whose goal is to be able to understand the concept of the design and how it is to be achieved, and to have a quick way to review the drawing and come up with plant quantities. The labeling must be accurate and clear.

The following method of labeling materials in the plant list with two-letter symbols is quite helpful:

> *Trees:* Two upper-case letters using the first letters of each word in the tree's botanical name, such as "QR" for *Quercus rubra*.
>
> *Shrubs:* One upper-case letter followed by one lower-case letter, again using the first letters of each word in the shrub's botanical name, such as "Bt" for *Berberis thunbergii*.
>
> *Perennials:* 2 lower-case letters, using the first letters of each word in the perennial's botanical name, such as "ho" for *Helleborus orientalis*.
>
> Third letters are added occasionally to avoid repetition, such as if you're specifying a number of varieties of a particular species or if the first two letters are the same for two species.
>
> These two- or three-digit labels are entered alphabetically as part of the plant list, and the quantity of any mass of plants is labeled clearly on the plan itself.

This methodology offers a number of advantages for use in contract documents. At a glance, anyone reviewing the drawing can immediately distinguish between trees, shrubs, and groundcovers, and also see where groupings of the same plant materials occur. There is little confusion between different types of material. The drawing itself is not nearly as cluttered as if all labels had the same graphic style. When there is an extensive plant list with a large number of each category of plant species (trees, shrubs, perennials), it becomes much easier for anyone reviewing the drawing and plant list to distinguish between them and not get overwhelmed by the repetitiveness of both the graphics and the labeling.

This methodology assumes that the person creating the drawing and the person reviewing or using it are knowledgeable in the nomenclature of botanical names. For example, someone familiar with trees might guess quickly that "AR" stands for *Acer rubrum*, red maple. For someone ignorant of the jargon, it will take a while for the organization and content to be clear. It is conceivable that the key symbols could be based on common names instead of botanical names, such as "RM" for red maple, but as most nurseries, catalogs, and specifications are organized according to the botanical names, the first method is preferred. The following Plant List is a representative example.

SAMPLE PLANT LIST (CHAPTER 5)

KEY	QUANTITY	BOTANICAL NAME	COMMON NAME	SIZE	NOTES
	TREES				
AC	5	*Abies concolor*	White fir	10′–12′, B&B	Full to ground
*ARO	1	*Acer rubrum* 'October Glory'	October Glory red maple	3″–3½″ cal., 14′–16′ ht., B&B	Specimen
*ARR	2	*Acer rubrum* 'Red Sunset'	Red Sunset red maple	2″–2½″ cal., 12′–14′ ht., B&B	Matched pair
BN	3	*Betula nigra*	River birch	2″–2½″ cal., 10′–12′ ht., B&B	Multi-trunk
MS	4	*Magnolia stellata*	Star magnolia	7′–8′ ht.	Multi-trunk
*PP	4	*Picea pungens*	Colorado blue spruce	10′–12′ ht.	Matched quartet
PS	5	*Pinus strobus*	White pine	7′–8′, B&B	Full to ground
*QR	6	*Quercus bicolor*	Swamp white oak	5″–5½″ cal., 16′–18′ ht., B&B	Matched to line both sides of driveway
*TA	5	*Tilia americana*	American linden	4½″–5″ cal.	
	SHRUBS				
Bt	15	*Berberis thunbergii*	Barberry	5 gal.	Red foliage
Bs	12	*Buxus sempervirens*	Common box	3½′–4′	
Fs	27	*Forsythia suspensa*	Weeping forsythia	30″–36″, B&B	
Hv	7	*Hamamelis virginiana*	Native witch hazel	4′–5′	Yellow flower
Hq	12	*Hydrangea quercifolia*	Oakleaf hydrangea	3 gal.	

KEY	QUANTITY	BOTANICAL NAME	COMMON NAME	SIZE	NOTES
Ig	19	*Ilex glabra* 'Shamrock'	Inkberry	5 gal., 24"–30"	Hedge
Pj	5	*Pieris japonica*	Japanese pieris	5 gal., 24"–30"	
Rs	6	*Rosa* sp.	Shrub roses	3 gal.	2 each of 3 varieties, each a different color
Sv	3	*Syringa vulgaris*	Common lilac	4'–5'	Violet or pink
	PERENNIALS				
ar	45	*Ajuga reptans*	Carpet bugle	1 gal., 8" o.c.	Green foliage
fg	7	*Festuca glauca*	Blue fescue	1 gal., 12" o.c.	Bluish foliage
hn	12	*Helleborus niger*	Lenten rose	2 gal., 24" o.c.	White variety, evergreen foliage
ho	7	*Helleborus orientalis*	Christmas rose	1 gal., 18" o.c.	Purple variety, evergreen foliage
hs	25	*Hemerocallis* sp.	Day lily	2 gal., 24" o.c.	Yellow bloom
hp	17	*Heuchera* 'Palace Purple'	Heuchera	1 gal., 15" o.c.	Pink bloom
pa	7	*Pennesetum alopecurides*	Fountain grass	2 gal.	Brown or red tassels/plumes
pq	5	*Parthenoissus quinquefolia*	Virginia creeper	3 gal., min. 3 runners	Specimens, train on arbor
sa	7	*Sedum* 'Autumn joy'	Stonecrop	2 gal.	Red or pink foliage/bloom

* = Tree tagged at XYZ Nursery on 10/3/20__.

The planting plan should be drawn to scale. The on-center spacing of major trees and shrubs need not be labeled on the drawing, as with the use of an architectural or engineering scale, as per the legend on the drawing, anyone reviewing it can determine the approximate distances. With perennials, groundcovers, grasses, and bulbs, however, it's sensible to indicate the spacing, for example, 6 inches on center (6" o.c.; 15 cm or 150 mm) in the plant list. Often one of the first ways to cut costs if the total bid is too high is to increase the spacing of major groundcovers, so having these dimensions clearly evident on the initial drawing is helpful in this regard. Plant materials are most commonly shown on plan drawings as a mass with a particular texture, but unlike some shrubs and trees, which may be rendered to suggest the shape or character of what they are, perennial masses most typically are just abstract graphics. Therefore, incorporating additional labeling to make them more tangible is always helpful.

Planting Details

The planting plan should show the method of planting each type of plant material—at a minimum, one detail for trees, one for shrubs, and one for groundcovers or perennials. More complex plans sometimes distinguish between methods of planting deciduous versus evergreen trees, but this is not always required. Although some people argue that the sheer weight of the root ball will anchor a major tree in place, since severe winds can suddenly occur anywhere, and the shifting of the root ball even by small amounts may result in damage to the tree or new rootlets just emerging, it is prudent to show staking details, whether one, two, or three stakes per tree. Each method may have certain advantages and disadvantages. For example, a single stake is certainly less costly and less labor intensive than two or three stakes per tree. The labor cost per tree is also significantly less if there is one stake to remove per tree after a year or so when the tree is well established, than two or three stakes. On the other hand multiple stakes create a natural barrier giving some visual protection to the tree, plus the added protection of ropes, twines or wires used to anchor the tree to the stakes.

Tree Planting

More research is still needed as to the best way to plant a tree. For example, over many years some have changed a typical tree planting detail from one in which the pit is excavated deeply enough to place the bottom of the root ball on a platform of loosened excavated material to a detail in which the depth of the tree pit matches the depth of the root ball. The latter method is less expensive in terms of excavation costs and may prevent the root ball from settling too much, which sometimes results in a tree not growing well because it was planted too deeply.

Street Trees

Learn from experience the best way to plant in specialized settings. Many cities have parks departments or arborists in charge of street trees. For example, New York City's Department of Parks and Recreation developed standardized details for tree pits, including required dimensions and paving materials, for placement along city streets and a list of approved medium and tall tree species for locations in the five boroughs of the city. During my more than three decades as a landscape architect in New York City, the department has regularly updated its approved tree list in light of new hazards and challenges, such as the Asian longhorn beetle, as some tree species are more susceptible to it than others and some neighborhoods are much more prone to infestations.[4]

Planting of trees is more complicated when they are placed within pavements, particularly in the sidewalk in the right-of-way of a street. City or other local codes may specify both a minimum and a maximum on center spacing for trees, and further regulations or simple common sense will dictate to the designer how far from a utility line (storm drainage, sanitary sewer, electrical conduit, gas line, etc.) a new tree must be placed. Since each tree is surrounded by pavement on three sides and a street curb or pavement on the fourth side, the planting pits are isolated from one another. One innovation developed in several cities, including New York City, is to use a compacted structural fill underneath sidewalks or unit pavers. New York City's product, as developed with Cornell University, is called CU Structural Soil, and it consists of a specified mixture of 83% #4 crushed angular stone, 17% clay loam, and 1 ounce copolymer hydrogel per 200 pounds of stone.[5] Used as a subbase underneath tree pits or pavements, this unique material provides adequate strength for support for pavements and vehicles, but also provides aeration for tree roots and allows them to spread continuously in long, connected tree pits, thereby enhancing the potential for vigorous growth, even when some pavement is set adjacent to the tree pits.[6]

Staking

Since plant materials may be thought of as the heart and soul of landscape designs, it is not surprising how much difference of opinion there is on every aspect of planting. How many stakes should be used for a tree? How deep and wide should a tree pit be excavated? Should a prepared soil mixture be used in planting or should the soil native to the site and excavated from the planting pit be used? Should the trunks of trees be wrapped or not? Should the burlap wrapping around a shrub or tree be removed prior to planting? How should drainage be handled when trees or shrubs are planted in areas with wet subsoils or perched water tables?

Mulching

Mulching is essential to hold moisture in the soil, prevent evaporation, and prevent weeds from germinating and competing with the tree for nutrients. The depth and type of mulch depend on what is available in a particular locality, the cost of the material, and climate conditions. One consideration is the pH of the bark material, with pine bark mulch being more acidic than shredded hardwood bark mulch.

Odd or Even Numbers of Plants

There has been considerable discussion and debate about how to lay out a mass of planting. Generally, in groups of under a dozen plants, a better aesthetic arrangement is achieved with staggered rows and an odd number, say seven instead of six or eight, because the plants reinforce one another as an organically shaped mass, and the staggered rows prevent one's view from looking down the aligned openings, which would occur if the spacings were uniform front to back. It's somewhat similar to the alignment of rows of seats in a theater: with staggered rows, everyone's view tends to be better, but with all seats aligned with those in front and in back, a number of theatergoers will have their views blocked by someone sitting directly in front of them. The exception to this asymmetrical spacing may occur in very formal settings of a limited number of shrubs or trees planted as a clearly geometric group, say four or eight. Yet some might claim that even in a formal setting, a trio is better than a quartet.

Tagging or Pre-growing

For planting projects with significant budgets, and with nurseries with which the landscape architect has a good history of solid work, it can be useful to pre-tag specimen trees to be used on a project, and incorporate these species into the plant list with a clear note identifying the source. All potential bidders would then become aware of the intent to use tagged material for particular species in particular settings. Another pre-planting procedure is pre-growing, in which the landscape architect directs the nursery to start growing in pots of a specified size a particular species of perennial or other plant in relatively short supply for which a considerable quantity will be needed for the project being bid. Again, in this case, there is a note on the plant list indicating that the contractors bidding the work must purchase this particular species and quantity in the size noted for use in the project. There may still be considerable variation in what each contractor bids, because there can be differences in what each charges for excavation, planting, mulching, watering, and so on.

Herbicides and Pesticides

There is a long history of use of herbicides and pesticides for control of insects, other pests, and plant diseases. The challenge is to find the right balance between protecting the plants being planted or established on the site with the strength of the chemicals involved, which often have side effects. Systemic applications, in which the chemical is worked into the soil and watered in so that absorption into the affected plants is through their roots, may be preferable to sprays, the application of which may include a lot of airborne mist that could drift to locations where it's unwanted or even dangerous.

Fertilizers should be used judiciously. Overfertilizing can create excessive growth that lacks hardiness or prevents the plants from naturally adapting to their new environment. Particularly in roof gardens and planters where there may be limited room for root growth, fertilization should occur with restraint so that the plants acclimate to their setting. Slow-acting, organic fertilizers are preferred to fast-acting, chemical ones.

Soil Testing

Most localities in the United States have county agriculture extension services or comparable bureaucracies that will test soil samples for free or for a modest cost. This can be invaluable to determine if the soil in different areas of a site is deficient in key nutrients or may contain something hazardous. For example, it's not uncommon to find soil in residential sites with a history of agricultural use to be contaminated with arsenic or other dangerous chemicals. Often testing can verify the pH of soils, an important factor in determining its suitability for different types of plantings. For example, if soil is not sufficiently acidic for shrub plantings of azaleas, rhododendrons, and other ericaceous materials, a particular fertilizer or chemical may be added that could lower the pH to a suitable range. Similarly, if a soil is too acidic for a lawn area, agricultural lime may be added to raise the pH to a more neutral level. In urban settings for community gardens, testing for lead from vehicular exhaust and flaking paint is an important criterion, particular if any food crops are planned.

Drainage

Drainage of the planting beds for any proposed installations is critical. Even with site analysis and testing that has occurred prior to bidding and the beginning of the work on the property, it's possible that drainage problems may be missed. It's prudent to include in the specifications and notes on the drawings a direction of what to do should poor drainage be found in a critical area designated for planting. Typical solutions may be a continuous French drain running the length of the planting bed with linkages to the base of tree pits, in which a layer of gravel may be placed, and a release point into the storm drainage system or appropriate outfall. The bid form might require the contractor to indicate a price per linear foot for a specified construction of a French drain so that if this issue occurs on the site, there will be some way of controlling for its cost.

Guarantee

With few exceptions, the landscape architect must insist on a guarantee of all new plantings for at least a full growing season—or, better, for a full calendar year from the date of acceptance of the planting being complete. Sometimes trees or large shrubs transplanted from off-site locations or being brought in from other locations within the site may not be guaranteed. However, root pruning of plants to be transplanted a year ahead of time often leads to a better chance of success. The root pruning (and compensatory pruning of the tree or shrub) ensures the development of a denser root ball so that the transplanted material has a better chance of acclimating.

Maintenance

Similarly, maintenance is critical. There is no point in installing a landscape that appears beautiful upon completion only to allow it to deteriorate quickly once the contractor has left. It's best to incorporate the development of a maintenance manual to be provided to the client upon completion of the work, and to find (whether through negotiation or bid) a company to do maintenance, taking over from the contractor who has installed the work. Some landscape contracting companies have a maintenance capability and often agree to maintain their own work, since they feel vested in it from the beginning and know where potential problems might occur.

(For planting drawing, see Figure 5.16).

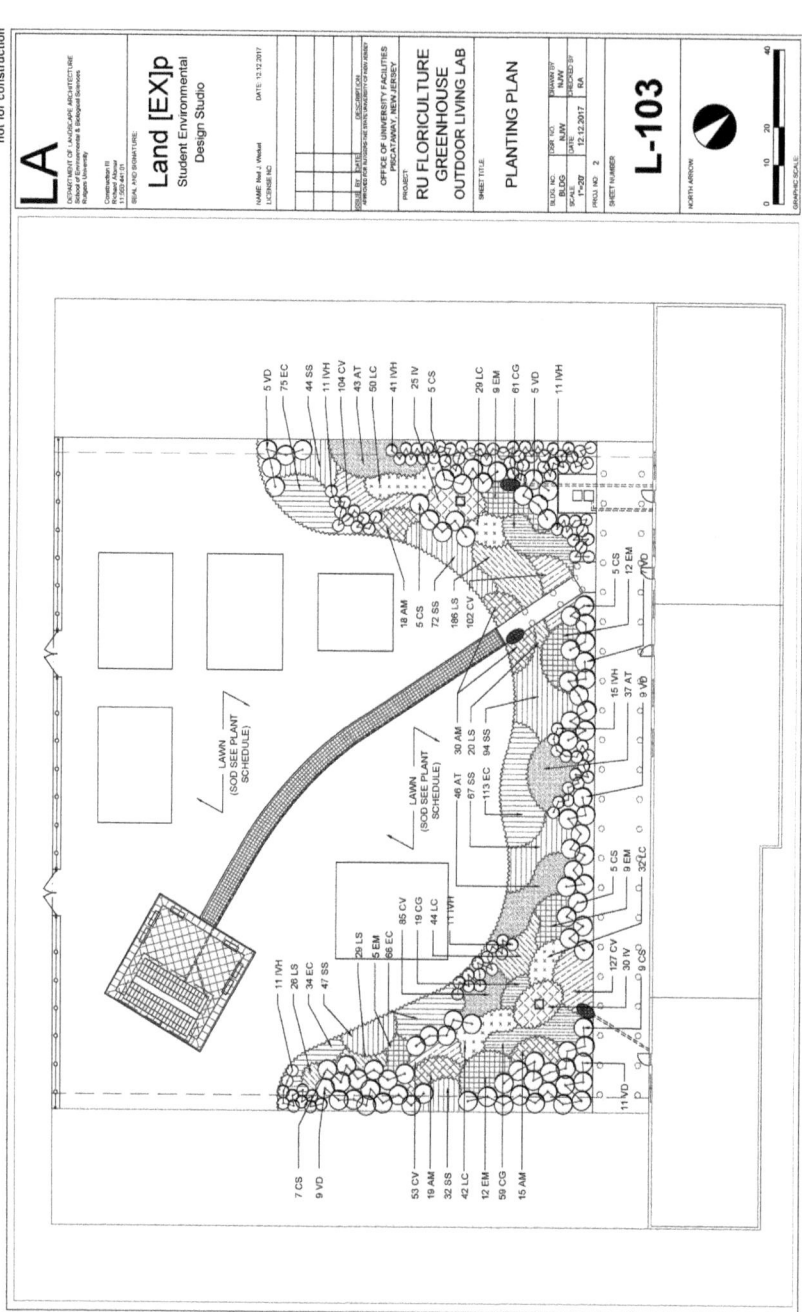

Figure 5.16 Floriculture Greenhouse Outdoor Living Lab at Rutgers Planting Plan. (Fig. 5-11 to 5-16) Neil J. Werket and Richard Alomar

PLANTING DESIGN DOS AND DON'TS

Planting design is perhaps the most variable service that landscape architects provide. The plant vocabulary changes drastically across the vastly different regions of North America, from deserts to wetlands to mountains, with widely diverse climates and temperature extremes. There are obvious differences in the hardiness and adaptability of plants across regions of the United States and Canada and tropical locations, but even within the same city or area, local site conditions have a major impact on what will grow. The horticultural vocabulary needs to be adapted to fit a sunny site or a shady site, and both could be much more restricted if they are buffeted by winds. The landscape architect must be expert about general requirements for plant materials, but also knowledgeable about unique requirements for the particular area and site in which the planting design will be executed. All of the following considerations may contribute dramatically to the success of a good planting design:

1. Insist on a guarantee.
    a. No matter how well the plantings are installed—so that the client, landscape architect, and contractor are pleased with the results upon the completion of the work—if there is no guarantee spelled out in the specifications, there is little motivation for the contractor to follow through with maintenance until the plants are well established and the client/owner or gardener can take over.
    b. One-year guarantees used to be typical, but two-year guarantees are common now.
    c. The specifications should require that any plants that die before the expiration of the guarantee period must be replaced with the same species, unless there are clear reasons an alternative species is a better choice. There should be a description of the smallest size of bare area that must be reseeded or resodded with grass or replanted with groundcover.
2. Insist on maintenance.
    a. Maintenance goes hand in hand with a guarantee.
    b. The client must be encouraged to hire the contractor who installed the plantings, or in some situations a different contractor with specific expertise, to maintain the plantings after the guarantee has expired. In some designs, the size of the completed planted areas is small enough that the client/owner can manage maintenance without much assistance, but in most situations, having a maintenance consultant who visits the site several times a month during the growing season and less often

during the winter is a way to ensure the continued success of the plantings.
   c. Requirements for ongoing maintenance may typically be spelled out in the specifications.
   d. It is prudent to require in the specifications that the landscape contractor provide a maintenance manual to the owner or client in which all maintenance operations, including a list of materials, equipment, and chemicals, such as fertilizers, mowers, and herbicides, respectively, are spelled out thoroughly, and to require one meeting between the contractor and whoever will take over maintenance at the conclusion of the work (whether the owner or another contractor) to ensure smooth and clear transition at the conclusion of the project.
3. Include irrigation or at least a source for providing water to the contractor responsible for maintenance and also to whoever will take over maintenance when the contractor's work is finished.
   a. Lack of water is one of the principal reasons that even well-installed planting designs may die. Even drought-tolerant plants require watering until they are established. Although average precipitation may be approximately the same for a site from year to year, it is common to have periods of heavy rains and periods of drought. Even when they are well established, during long periods of drought the plants can still be under a great deal of stress, so watering keeps them in a healthy state.
   b. A spray or drip irrigation system is a major expenditure. By comparison, installing hose bibs and quick coupler valves in a pattern so that the entire site can be reached by 20-foot hoses is less expensive if the people are available to use them. Either method is far better than trusting to chance that the plantings will be blessed with just the right amount of precipitation each week over the first few years of growth.
4. Consider phasing if there is a significant chance that the total cost of executing the entire planting design may strain the available budget.
   a. Planting is often the last phase of a landscape architecture project to be implemented, so it often occurs at a time when the budget may be tight as a result of previously unanticipated expenditures, such as the cost of the imported stone the client insists on, or cost overruns from adding another section to a seating area or pergola. Therefore, it is prudent to anticipate sections of the planting design that could be deferred for a year if necessary.

    b. Phasing should emphasize not only the most important areas, which must be planted first, but also recognize those areas that can be planted in the future *without causing a serious impact on the previous phases of work.*
5. Stress inspections for completion.
    a. The contract and specifications must spell out a means for the landscape architect to walk the site with representatives of the owner/client and landscape contractor in order to verify that the work is completed, and also develop a checklist of what work remains.
    b. A second follow-up inspection should occur after an agreed-upon interval to allow the contractor to correct any problems or omissions noted on the checklists.
    c. Payment of the last portion of the landscape contractor's fee should not be made until the second inspection occurs and there is verification that all planting and corrections have been implemented to the satisfaction of the landscape architect.
    d. Without such inspections, the final completion of the work may drag on and many little details may not be implemented.
6. Include shop drawing requirements for difficult detail areas.
    a. Occasionally there may be complex areas of design—for example, where pavements in particular patterns must align with different plantings, or an accent of a curving mass of plant materials is inserted within rectilinear planting beds. Until the pavement has been installed or the rectilinear planting beds laid out in the field, it is not possible to know the exact layout of these last plantings. There might also be some areas within a planting design where a great density of perennials is called for, and the general planting plan is not at an adequate scale to show them precisely. Therefore, the specifications may require the landscape contractor to provide shop drawings, based on measured dimensions of as-built conditions, to finalize the dimensions and requirements of these last planting details.
    b. Shop drawings are a safety and security blanket for the landscape architect for difficult areas.
    c. There should be clear guidelines for shop drawings within the specifications: how many sets of shop drawings are provided, who receives a set, how they are to be marked up and stamped, and how revisions must be agreed upon and distributed.
    d. Shop drawings should be drawn to scale, and based on—or have a starting point that dovetails precisely with—some measurable aspect of the design.

7. Consider the local expertise of landscape contractors and nurserymen.
    a. Despite the expertise of the landscape architect, he or she may not have detailed knowledge of a particular site compared to landscape contractors and nurserymen who have worked in the local area for many years.
    b. If, in reviewing the plant list for a site, the landscape contractor and/or nurserymen recommend *not* using a particular plant material or recommend restricting its use to a limited part of the site, they are usually correct. As long as the landscape architect verifies that such recommendations are being made to improve the planting design and not because the landscape contractor or nurserymen are merely short of the plant material in question, following their recommendations usually makes sense. In other words, trust the experts who are part of your design team.
8. Be aware of the geographical range of sources for plant materials, and use local sources whenever possible.
    a. The shorter the distance that plant materials must be shipped prior to planting, the more likely they are to be in good condition upon arrival at the site. (An added benefit is points earned toward LEED certification.)
    b. Since local climate and growing conditions can vary dramatically from region to region, the landscape architect should be very careful about accepting even healthy-looking plant materials that come from some distance. For example, a sweetbay magnolia (*Magnolia virginiana*) grown and dug from a nursery in the South and planted at a site in the North may not adapt well. In the South, these trees are often evergreen; in the North, they tend to be semi-evergreen or deciduous. It might take several years for a tree grown in the South to adapt to the harsher conditions in the North, or it may not survive at all.
9. Know when to be flexible and when to hold the line on substitutions. When plant materials are critical to the structural elements of the design, it is important to use what is specified. If the plant materials take on a more supportive and less crucial role, then substitutions may be quite practical, just so long as the landscape architect verifies that the substitute is hardy, and will, in fact, provide the same horticultural design elements as the originally specified (and unavailable) plant material.
10. Anticipate how to cut costs if necessary without limiting the design's impact. At first without eliminating any planting on the site, significant savings, as much as 10% to 15%, may be reached just by any or all of the following:

a. Adjusting the spacing of groundcovers
   b. Adjusting the spacing of shrubs
   c. Adjusting the specified size of groundcovers
   d. Adjusting the specified size of shrubs
   e. Adjusting the specified size of trees
   f. Adjusting the depth of mulch
11. Don't create potential conflicts of interpretation.
    a. If the drawings are highly detailed and call out in the plant list and labels the precise size of all anticipated plant materials, then the specifications should be general and refer to the plant list and drawings.
    b. Sometimes in public projects, the specifications use item numbers for each class of plant materials, and enough detail must be provided in the specifications so that these distinctions are clear.
12. Don't test new plant materials or recently introduced species, varieties, or cultivars in important locations on the site.
    a. Although it is an admirable goal and sound practice for the landscape architect to keep expanding the plant palette, to include new varieties or species previously unfamiliar, these plants should be used with caution. Incorporate them in areas where they are not too prominently displayed, so if they die or in some way do not adapt to the site to the degree expected, the significance of the loss is diminished.
    b. Use the most reliable plant materials in the most important locations: allees of trees, architectural elements, accent plantings.
13. Build into the project's budget for implementation enough time for ample meetings between the landscape architect and landscape contractor installing a planting to allow adequate review and approval of the work. Depending on the nature of the work and the client, at least some of these meetings should include the client.
    a. It's often helpful to have several meetings toward the beginning of implementation so that the landscape architect and landscape contractor can develop rapport, learn to work together, and know what to expect from each other.
    b. A useful goal is to agree on those circumstances or situations when the landscape architect must be contacted prior to any further work proceeding, in contrast to those situations when the contractor can simply add to a list of questions for review in the future. Writing an email or making a phone call may sometimes quickly determine which category a situation falls into.
    c. Some clients are hands-on and want to see the work progressing and have it explained to them, whereas others are less involved

and just need occasional meetings for reassurance that the work is proceeding well and on schedule. However, it is critical to develop trust with the client and have the client's approval and understanding at certain important points.

14. Keep careful records of all actions taken while inspecting the progress of the planting installation or in otherwise meeting with the landscape contractor and/or client.
    a. The client and contractor should review such documents and the landscape architect should revise or update them as necessary in order to keep an adequate record of what happened and what was agreed on to solve any particular problem.
    b. Such records can include checklists of items to be replaced after each inspection, reminders about future work, or anything relevant to the ongoing activities of the project.
15. Review invoices promptly for accuracy and completeness and authorize payment in a timely matter.
    a. There is no quicker way to engender ill will and hurt feelings than to delay payment unnecessarily. The specifications should spell out milestones at the completion of which the contractor may bill for a portion of the fee. At the same time payment of the last invoice for payment in full should occur only after an inspection in which all parties agree that the work is complete.
    b. Supporting documentation to accompany an invoice, such as itemized lists from nurseries, bills from suppliers, and equipment fees, should all be organized clearly in a format agreed to prior to the first submittal.
16. Include prequalification review of landscape contractors.
    a. For specialized projects such as roof gardens, not every landscape contractor has sufficient expertise. The success of the implementation may depend on specialized skills. Try to incorporate into the project the process of having contractors interested in bidding submit a list of comparable projects they've implemented. Reject from the bid process any contractors you find unsuitable or without adequate experience.
    b. Sometimes prequalification is not permitted. In that case, try, to the extent possible, to spell out in the request-for-bids package or the specifications the minimum skills necessary for implementation so that inexperienced contractors will be wary of bidding.
17. For projects which include a considerable amount of planting and when it is known that there is potential for difficult planting conditions, it is prudent to anticipate what happens if the soil is heavily compacted or a perched water table is present. In the

event that either is necessary, such information can help expedite the work.
a. Standard directions for aeration of existing planting beds may be provided.
b. Another option may be installation of a French drain system.
c. The specifications may spell out how to calculate the maximum allowed cost per unit of measurement for either option.

Carl Schurz Park, Manhattan, New York City

Carl Schurz Park in Manhattan is a historic park that surrounds Gracie Mansion, the official residence of the mayor of New York City. Calvert Vaux designed the park in the early twentieth century. (He was Frederick Law Olmsted's somewhat less famous but equally remarkable and multifaceted collaborator on Central Park, Prospect Park, and other public works projects). The Carl Schurz Park Conservancy commissioned Banford Weissmann to design, renovate, and maintain several garden areas within the park. Her scope of work has included everything from formal displays to woodlands. Using carefully selected plants for each site, the arrays of color, form, and texture bring beauty, ecologic strength, and vitality to the park.[7] (See Figures 5.17 and 5.18).

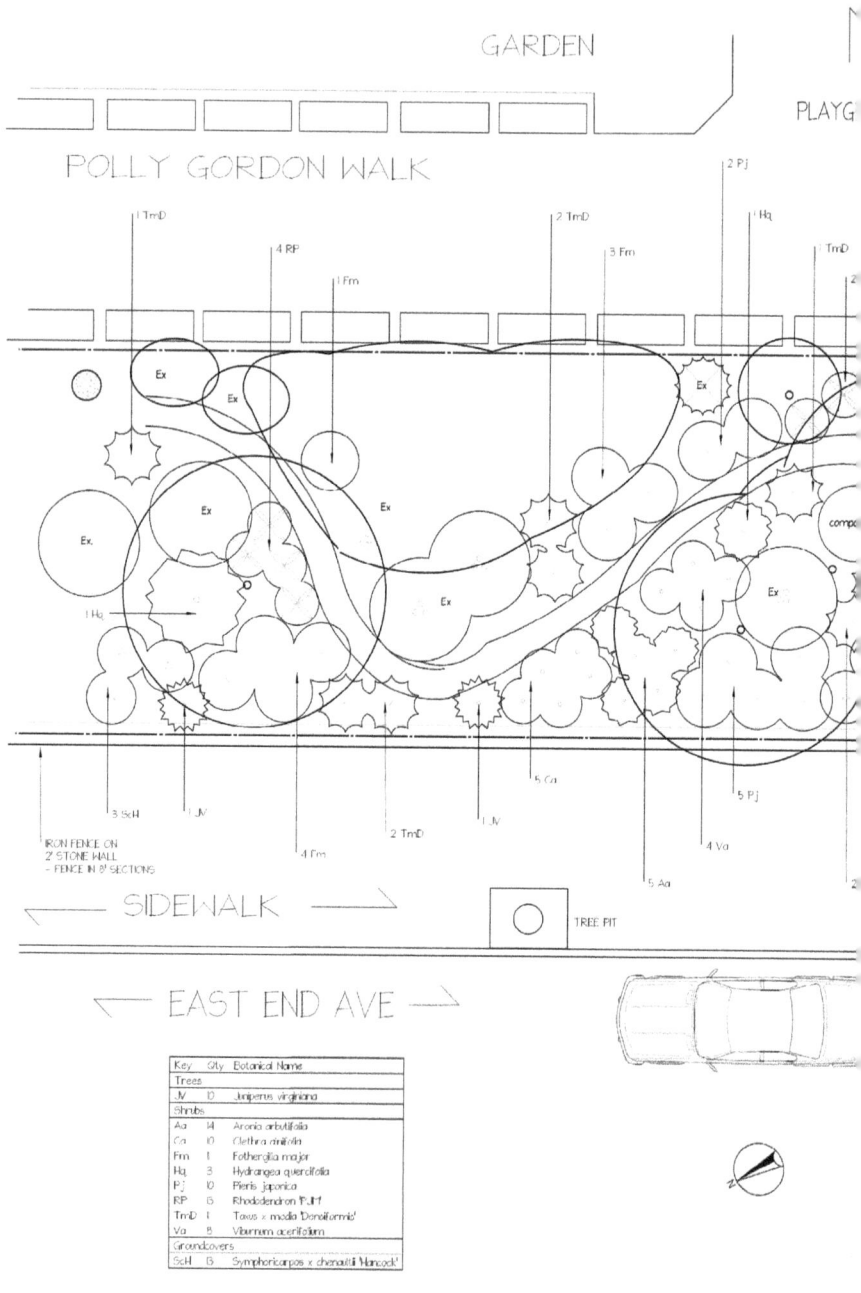

Figure 5.17 Planting Plan for South End of Long Berm in Carl Schurz Park, 15 acre park in Manhattan.
Courtesy of Banford Landscapes LLC

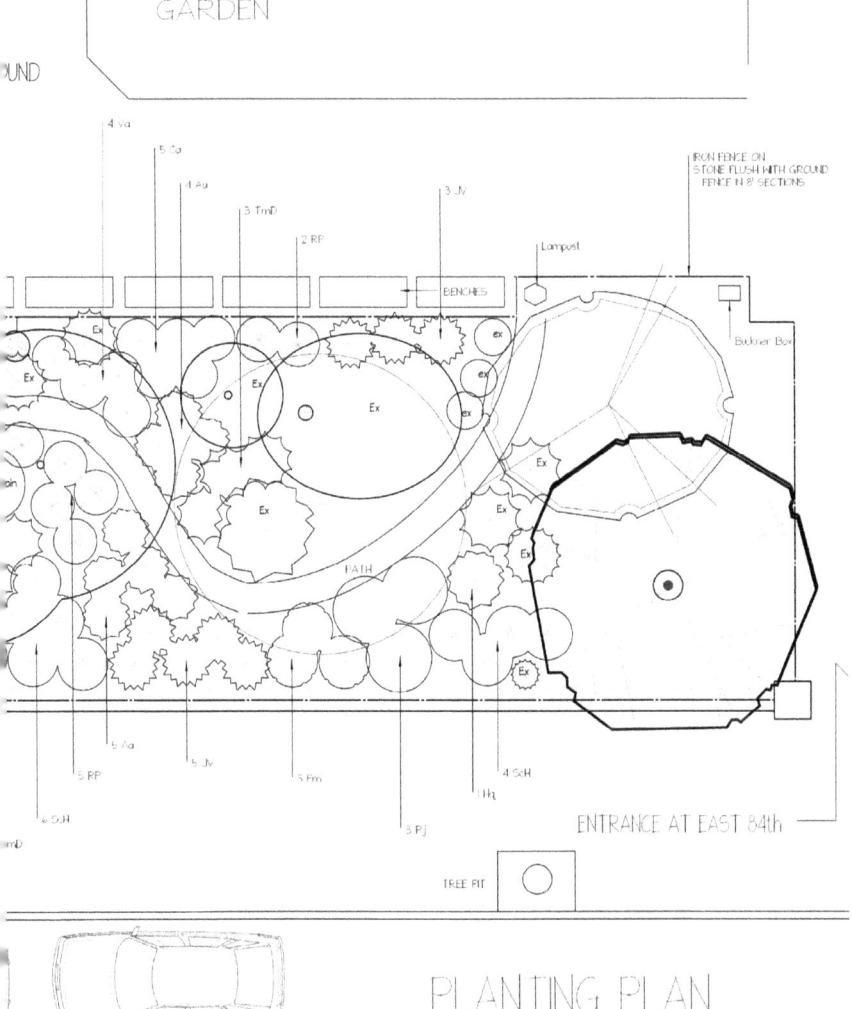

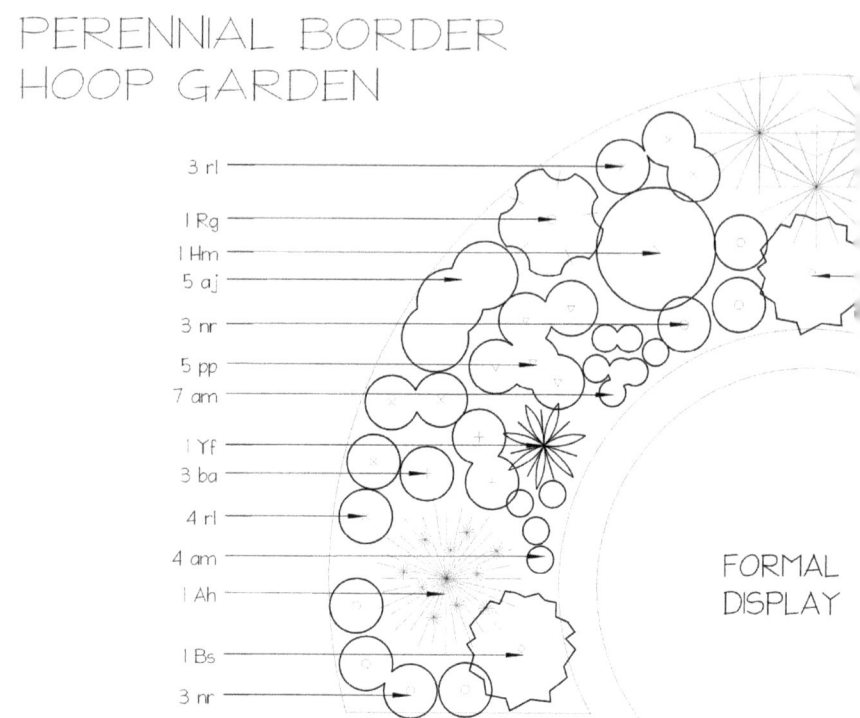

TYPE	KEY	QTY	BOTANICAL NAME	COMMON NAME	SIZE
SHRUBS					
	Ah	2	Amsonia hubrichtii	Bluestar	3 gal
	Bs	3	Buxus simpervirens		3 gal
	Hm	2	Hydrangea macrophylla 'Penny Mac'	Penny Mac hydrangea	3 gal
	Rg	2	Rosa glauca	Redleaf rose	5 gal
	Yf	2	Yucca filamentosa 'Variegata'	Variegated yucca	2 gal

TYPE	KEY	QTY	BOTANICAL NAME	COMMON NAME	SIZE
GRASSES					
	ms	3	Miscanthus sinensis 'Morning Light'	Eulalia grass	3 gal

TYPE	KEY	QTY	BOTANICAL NAME	COMMON NAME	SIZE	NOTES
PERENNIALS						
	aj	10	Anemone japonica	Japanese anemone	2 gal	white flowers
	am	22	Alchemilla mollis	Lady's mantle	1 gal	
	ba	6	Boltonia asteroides	False aster	2 gal	lilac flowers
	nr	12	Nepeta racemosa 'Walker's Low'	Walker's Low catmint	2 gal	
	pp	10	Phlox paniculata 'Fireworks'	Fireworks garden phlox	1 gal	pink flowers with red
	rl	14	Rudbeckia laciniata	Cutleaf coneflower	2 gal	yellow flowers

Figure 5.18 Banford Weismann: Planting Plan for Perennial Border Hoop Garden in Carl Schur Park.
Courtesy of Banford Landscapes LLC

Not for construction

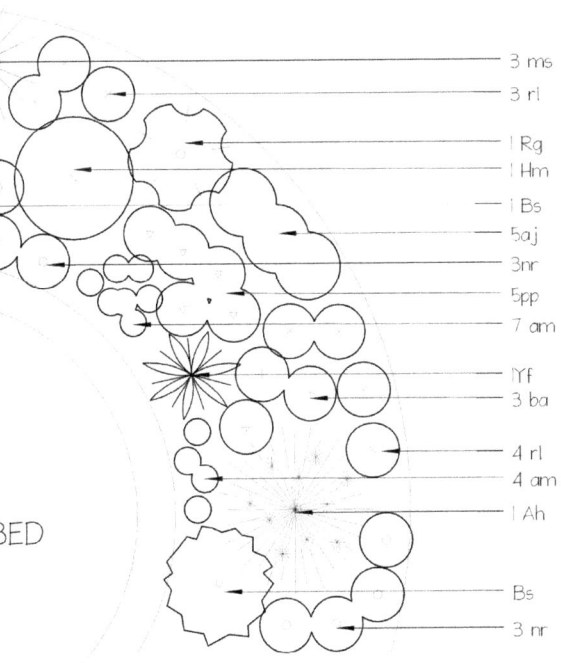

- 3 ms
- 3 rl
- 1 Rg
- 1 Hm
- 1 Bs
- 5 aj
- 3 nr
- 5 pp
- 7 am
- 1 Yf
- 3 ba
- 4 rl
- 4 am
- 1 Ah
- Bs
- 3 nr

Problem 1

PLANT LIST AND COST ESTIMATE

(In the following problem, any *terms in italics* may be changed to suit the particular instructor or team leader, or your own personal, work or studio setting. Typical changes might be specific people, such as clients or contractors, plant materials, or the locations of sites, people, institutions or companies.)

1. You have prepared a planting plan for a client for a residence in *Zone 5*. Review the plant list included below. (*You may explore other hardiness zones and appropriate nurseries and locations.*)
2. The client has accepted your recommendation to negotiate with a landscape contractor, who in turn recommends *Moon Nurseries, with locations in Maryland and Pennsylvania*, as the source for the plants.
3. You review the *Moon Nurseries catalog* (www.moonnurseries.com or http:moonnurseries.com) and find that there are some plant materials you specified in the plant list that are not available or are in different sizes.
4. There are also some materials, such as crabapples and roses, for which you need to find available varieties.
5. Use the American Standard for Nursery Stock, available at http://americanhort.org/documents/ANSI_Nursery_Stock_Standards_AmericanHort_2014.pdf, and compare what is listed for five plant material species in this reference to what the Moon Nurseries catalog lists.
6. Prepare a cost estimate for planting. Discuss the content and format in class; use Excel or another approved format.
7. For any plant materials in the plant list that are not included in the Moon Nurseries catalog, propose substitutions, and use the size and price listed for that substituted item.
8. Include a subtotal for the cost of trees, a separate subtotal for shrubs, and another for perennials. Then have a grand total of the combined cost of these three categories.
9. On a separate sheet list five examples of comparison with the American Standards for Nursery Stock.

PLANT LIST (CHAPTER 5, PROBLEM 1)

KEY*	QUAN.	BOTANICAL NAME	COMMON NAME	SIZE	NOTES
	TREES				
AC	5	*Abies concolor*	White fir	10'–12'	
AR	1	*Acer rubrum*	Red maple	3"–3½" cal., 14'–16' ht., B&B	Specimen
BN	3	*Betula papyrifera*	Paper bark or canoe birch	2"–2½" cal., 10'–12' ht., B&B	Multi-trunk
MS	4	*Magnolia stellata*	Star magnolia	7'–8' ht.	Multi-trunk
	12				4 each of 3 varieties of crabapple
PP	4	*Picea pungens*	Colorado blue spruce	10'–12' ht.	Matched
PS	5	*Pinus strobus*	White pine	7'–8', B&B	Full to ground
QR	6	*Quercus bicolor*	Swamp white oak	5"–5½" cal., 16'–18' ht., B&B	Matched to line both sides of driveway
TA	5	*Tilia americana*	American linden	4½"–5" cal.	
	SHRUBS				
Bt	15	*Berberis thunbergii*	Barberry	5 gal.	Red foliage
Bs	12	*Buxus sempervirens*	Common box	3½'–4'	
Fs	27	*Forsythia suspensa*	Weeping forsythia	30"–36", B&B	
Hv	7	*Hamamelis virginiana*	Native witch hazel	4'–5'	Yellow flower
Hq	12	*Hydrangea quercifolia*	Oakleaf hydrangea	3 gal.	

KEY*	QUAN.	BOTANICAL NAME	COMMON NAME	SIZE	NOTES
Ig	19	*Ilex glabra* 'Shamrock'	Inkberry	5 gal., 24"–30"	Hedge
Pj	5	*Pieris japonica*	Japanese pieris	5 gal., 24"–30"	
Rs	6	*Rosa* sp.	Shrub roses	3 gal.	2 each of 3 varieties, each a different color
Sv	3	*Syringa vulgaris*	Common lilac	4'–5'	Violet or PINK
	PERENN.				
Ar	25	*Ajuga reptans*	Carpet bugle	1 gal.	
Fg	7	*Festuca glauca*	Blue fescue	1 gal.	Bluish foliage
Hn	12	*Helleborus niger*	Lenten rose	2 gal.	White variety, evergreen foliage
Ho	7	*Helleborus orientalis*	Christmas rose	1 gal.	Purple variety, evergreen foliage
Hs	15	*Hemerocallis* sp.	Day lily	2 gal.	Yellow
Hp	17	*Heuchera* 'Palace Purple'	Heuchera	1 gal.	Pink
Pa	7	*Pennesetum alopecurides*	Fountain grass	2 gal.	Brown or red tassels/ plumes
Pq	5	*Parthenoissus quinquefolia*	Virginia creeper	3 gal., min. 3 runners	Specimens, train on arbor
Sa	7	*Sedum* 'Autumn Joy'	Stonecrop	2 gal.	Red or pink foliage/bloom

* Two-letter symbols are used to indicate the type of plant materials, as follows:
Trees: Two upper-case letters, such as "QR."
Shrubs: One upper-case letter followed by one lower-case letter, such as "Bt."
Perennials: Two lower-case letters, such as "ho."
Third letters are added occasionally to avoid confusion. This methodology, along with its advantages for use in contract documents, is discussed in the text.

NOTES

1. Albert Rutledge and Donald J. Molnar, *Anatomy of a Park* (New York: McGraw-Hill, 1971), 98–101. See especially the section on Relationship Diagrams in chap. 6., which, despite its age, remains an antique gem describing the survey, analysis, and synthesis of the design process.
2. Ibid., 99–100 and Figure 6.8.
3. Thomas Jefferson letter to William Drayton, Paris, July 30, 1787 in The Papers of Thomas Jefferson, Julian P. Boyd, editor, collection within *The Garden and Farms Books of Thomas Jefferson*. edited by Robert C. Baron, Fulcrum Press, Golden, CO: 1987, p.181.
4. https://www.nycgovparks.org/trees/street-tree-planting/species-list (also includes information about Asian longhorn beetle); www.C:/Users/Yamaha/Downloads/Tree-Planting-Standards.pdf, *Street Tree Planting Standards for New York City*, 2016, 6, 8, 9, 16–19; See also (compare) Michael Bloomberg, Mayor and Adrian Benepe, Commissioner, City of New York Department of Parks and Recreation, *Tree Planting Standards*, September 2009. This document shows evolution of tree species and tree planting approach over a decade.
5. http://www.hort.cornell.edu/uhi/outreach/pdfs/custructuralsoilwebpdf.pdf; See also Michael Bloomberg, Mayor and Adrian Benepe, Commissioner, City of New York Department of Parks and Recreation, *Tree Planting Standards*, September 2009, 18–21, 23–24.
6. Ibid.
7. "Carl Schurz Park," Banford Landscapes, http://banfordlandscapes.com/carl-schurz-park.html.

CHAPTER 6

Building Code, Specifications, and Cost Estimates

PART 1: BUILDING CODES AND RELATED PLANNING ISSUES

When any design project begins, it's essential for the landscape architect to check and verify building code requirements and any other relevant codes. This discussion will be an introduction to some of the most important areas where local building codes have a major impact on the design and construction drawing process. It is urgent to review major code requirements for a property and neighborhood prior to entering any sort of contractual arrangement to design anything within that property because it's possible that code requirements may seriously limit what it is possible to implement on the site. It is a helpful exercise to do a sketch drawing to scale in which all the zoning requirements, code restrictions, and other legal limitations of a property are illustrated.

When reviewing and applying a building code to a new project, it is prudent to remember that the basic purpose of the code is to ensure the health and safety of those using the relevant property and protect passers-by or stray wanderers from being harmed. Codes are not etched in stone never to be revised or challenged, but before a landscape architect requests a *variance* (the legal permission to implement something that at face value is not in compliance with or is a violation of the code), it is prudent to develop a rationale for how the proposed change will further protect the public, and not to explain why a particular client deserves a taller fence, a larger pool,

Figure 6.0 Water feature, Queens Botanical Garden. The design and management of water features within an urban environment are special challenges in landscape architecture. Here it is successfully achieved and demonstrated, as all water is recycled. See also http://www.greenroofs.com/projects/queens-botanical-garden-visitor-and-administrative-center.
Photo by Steven L. Cantor

or a taller pergola than a neighbor's. Variance applications vary in their requirements; usually there is enough paperwork, time-consuming review, public hearings, and related expenses to discourage an applicant unless it's felt to be urgent.

Zoning

The codes will include maps designating different geographic categories within the municipality, county, city, or other government area. For example, some areas may be zoned for light industry, others for multifamily housing, still others for single-family residences, and so on. Some zoning codes will have a uniform use and set of requirements for an entire neighborhood; others will allow or even encourage a mixed use, the specific requirements for which vary street to street, lot to lot, or building to building.

Setbacks

The codes will define what the applicable setback requirements are for a particular site within its jurisdiction. Front yards, side yards, and rear yards are clearly defined, with some restrictions being placed on the uses that may occur within each, such as lawns in the front yard, swimming pools in the rear, limited uses in the side yards. It is common for swimming pools to be forbidden in the front yard, and sometimes in the side yards due to concerns about the potential impact on neighbors and the public. (See Figure 6.1).

Sanitary Sewer or Septic, Water Supply, and Other Utilities

Restrictions and requirements for utilities systems vary widely, and often there are significant changes from one locality to another, and more substantial differences from one state to another. As desirable neighborhoods become substantially full, with few available building lots or buildable areas left, restrictions on sanitary sewers or septic requirements are becoming more restrictive due to fears and environmental damage caused by poorly designed or poorly built septic systems and settings where the demand on the septic systems is greater than their capacity. Even when lots are served by public sanitary sewer systems, it is important to determine that there is available capacity to receive the sanitary load for a new building or

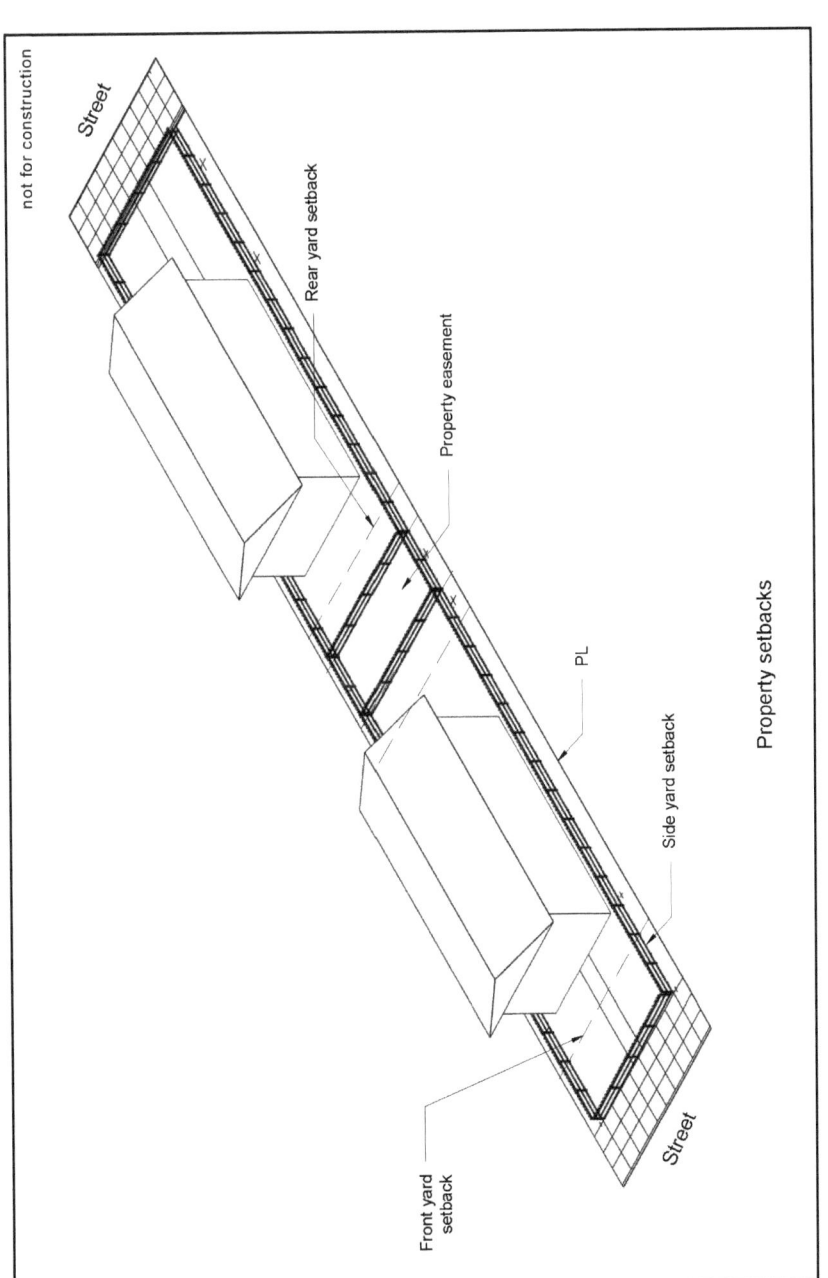

Figure 6.1 Typical front yard, rear and side yard zoning setbacks for residences. It is prudent not to place a fence precisely on a property line.
Richard Alomar and Vicky Chan

addition being proposed. Therefore, no matter how attractive aesthetically a buildable lot appears to be, it is urgent fairly early in the design process to study the soil types and septic requirements and carry out percolation tests to verify that adequate capacity exists within the site to support the septic requirements of a proposed building, such as a residence. Often the code requires a certain reserve capacity as well, so that after, say, twenty-five years, when the installed septic system has reached the end of its life span, there is space available for a new system.

Alternative Sustainable Design Systems

Building codes are often glacially slow in adapting and catching up to modern advances, in part because the failure of something new that has been widely approved could have disastrous consequences. On the other hand, the need is dire. Septic systems are a good example, as some areas in the northeastern United States have many restrictions on new systems because most of the sites that have soils with acceptable percolation rates have already been developed. Therefore, domestic water use strategies that reduce, reuse, and recycle are gaining traction. Gray water, which comes from baths, sinks, and showers, is often separated from sewage and reused for irrigation of gardens or related purposes. Wastewater can be treated and recycled as fertilizer. Rainwater on roofs can be harvested for reuse. Building codes are gradually accommodating such sensible and practical alternatives.[1]

Stormwater Runoff and Wells

Usually it is not permitted for any proposed construction or activity within a site to result in an increase in the volume of stormwater flowing off the site or a change in its discharge location into other adjacent properties. Water supply is another area where restrictions may apply because a neighborhood or region may have reached its capacity for wells or water supply from public water mains. Where wells are permitted, there are requirements setting a minimum distance between the well and the nearest septic line or sewer line, as contamination of the water supply from leaking sewage lines or the natural flow from septic fields over time cannot be permitted. Just as grading operations may be restricted within the drip line of a large specimen tree, it is common to set a minimum radius from the center point of a proposed well in which no disturbance or construction of any kind is permitted within the defined circle as a way of protecting the potential water supply from contamination.

Floor Area, Height, and Number of Stories

Building codes frequently specify a maximum height and/or maximum number of stories for buildings of a given type. A floor area ratio formula is often included, which allows some flexibility in the total area of a building in relation to its height and the size of the lot.

Easements

There are different requirements and categories for *easements*, which permit a utility company or other approved entity to maintain and control an access route of a set width through otherwise private property. Easements are common for overhead or underground power lines, storm drains, sewers, and other utilities. Near rivers and within watersheds subject to seasonal flooding, easements are often created to absorb overflow storm drainage and allow for drainage flow during periods of intense rainfall. It is often forbidden to fill in such easements without compensatory cuts and to prove that any proposed grading will not impede the flow of water.

Right-of-Way and Property Lines

Right-of-way is defined as the legal width of ownership of a street that passes by or through another property. *Setback requirements* are defined as the minimum distance a building must be from the right-of-way. Sidewalks, plantings, and some utilities may be permitted within the right-of-way of a road, but usually the owner, such as a municipality, county, or state, has the legal right to widen the road, and thereby encroach on plantings or sidewalks within the right-of-way. For this reason, it's vital in every project to carefully evaluate what, if anything, to place within the right-of-way. It's a sad situation for a homeowner to set screen plantings within the right-of-way, not realize that he or she has encroached on public land, and find that just as the plantings reach maturity, they are removed for a widening of the road. It's a standard precaution to place screening elements, such as fences, hedges, and rows of trees, fully within the private-property side of the right-of-way line.

A corollary to this caution is for property owners to verify the precise locations of their property lines and have them staked out so that it is verifiable that any construction the owner authorizes is taking place entirely inside the property. For example, a fence or hedge might be centered a foot (or more) inside the owner's property. Why take a chance that the staking is slightly off and years later you are found to be encroaching on a neighbor's property who is then free to cut down what you have lovingly nurtured for a decade or longer?

Driveways, Garages, and Carports

Even though architects have a major role to play in buildings, landscape architects are often involved in addressing how the building meets the street, how a driveway arrives at the entrance or garage, how parking is handled, or whether a porte cochere or some type of outdoor shade structure is placed over outdoor parking areas. The building code often has restrictions on and requirements for the size and orientation of garages, the width and overall length of driveways, the materials for pavement, and how close to side or rear property lines the driveway and parking areas can be placed. Many families and businesses have multiple cars, but it's not unusual to find restrictions on the number of parking spaces permitted within a lot, particularly for a residential or commercial use. The code may also permit with restrictions or forbid off-street parking.

Specialized Requirements

Depending on the location of a property and its nature, building codes may restrict or regulate many other functions or uses. For example, if a district is within a landmarked zone, many restrictions and requirements may be in place to ensure that any new construction or renovations comply with detailed requirements as to materials, finishes, styles, heights and colors of walls, fences, lighting standards and fixtures, gates, and even pavements. Seemingly innocuous requirements may sometimes have major impacts on design. For example, it's not uncommon for landmarks restrictions to require that nothing built on the roof of a new building be visible from the streets below. At a minimum, this might require the landscape architect to set back pergolas, other built elements on the roof, and even plantings a considerable distance from the parapet of the building, but the result might be a space on the roof too small for use. The design team would likely have to do studies from every direction in which there is a view of the roof to verify that what's proposed will stay out of sight.

Tree Protection

Some cities or communities pride themselves on protecting existing trees, so if any trees over a certain size are proposed for removal, a review process is initiated and it must be proven that no practical alternative for the design remains that would allow for the existing tree to remain. Sometimes, when it is permitted to remove existing trees, the code will require compensatory plantings of trees according to a mathematical formula based on an equivalent number of caliber-inches of new trees equal to the total caliber-inches of the trees removed. Still, the best solution is almost always to remove as few major existing trees as possible, as it may take a lifetime for new trees to grow large enough to compensate for the canopy that was removed.

Swimming Pools

There are many variable requirements for swimming pools. Codes usually forbid placing them in the front yard, open to public view. Various setback requirements indicate minimum distances that pools must be set back from side and rear property lines, and sometimes from the nearest wall of the on-site residence or other building. Further requirements sometimes stipulate a width of paved terrace adjacent to the pool edge or both the maximum and minimum sizes of pavements for entertainment adjacent to a pool. Other requirements may govern the extent and number of speakers for an outdoor stereo or grills for outdoor use, particularly if they are supplied by gas lines or other utilities.

Fences and Gates

Numerous requirements apply to fences and gates in general, and there are more specific ones for fences and gates at swimming pools, since it's within the public interest to require that every swimming pool be entirely fenced with a lockable gate so that an unsupervised child or other person—for example, someone intoxicated—does not wander in and fall into the pool, with dire results. The heights and designs of fences sometimes include restrictions about the grid of horizontal and vertical members to prevent a crawling baby or even an inquisitive dog from inadvertently putting its head into the fence and becoming stuck. Fences around tennis courts often stipulate a maximum height and also dimensional requirements of the grid of wood or chain link mesh, small enough to prevent balls from getting lodged into it yet large enough to be economical. Some codes may require special treatments for fencing—for example, that it be wood or vinyl instead of chain link. Galvanized or painted finishes may be required as well.

Miscellaneous Recreational Activities

Although most recreational activities require primarily adequate space, some activities are intensive in requirements for water supply or fertilizer and could have impacts on adjacent properties. There can be code requirements restricting activities based on the amount of noise they generate (say, from a basketball court) or the odors that might result (for example, from a competitive equestrian ring or horse barn). There are usually very different restrictions on the use of animals, and facilities for them, in rural areas, where they are essential and welcome for farming, compared to urban areas, where they may be considered a nuisance. Some municipalities may consider the tall goal with basket for basketball to be unsightly and forbid placing it in the parking court by a garage if it's visible from the street. My own experience includes a wealthy community that permitted the construction of a full-size ice hockey rink, complete with Zamboni, in the side yard of a large residential property. In the off-season it doubled as a large basketball court. The building to house the Zamboni had a dual function as a garden and tool supply shed.

Invasive Species

With the emergence of much scientific research on endangered and invasive species, it's not unusual to find that some communities ban plants that other communities celebrate. Again, the landscape architect should review all relevant codes and ordinances. There is an increasing emphasis on the use of native plants, but many non-native plants are commonly used and appreciated. *Lagerstroemia indica*, crape myrtle, is native to Asia but is a common ornamental shrub or tree throughout the South and as far north as New York City. In the family Lythraceae, it's appreciated for the beauty of its flowers, its reddish fall color, and the exfoliating bark that gives it a sculptural quality in the winter. However, another plant in the same family, *Lythrum salicaria*, purple loosestrife, is considered highly invasive and is banned in many states. One characteristic the two plants have in common is abundant pollen. Hay fever sufferers can be made miserable during peak periods of seasonal pollen from either or both plants.

Irrigation and Lawns

Many codes now regulate irrigation, in response to water use restrictions due to droughts or depleted wells and reservoirs. There may be restrictions on the extent of water use, both as to which days water for irrigation may be used and as to how much. With the development and improvements in many types of drip irrigation systems, it's increasingly common to find code requirements banning spray systems, which are comparatively wasteful of water. It behooves the landscape architect specifying irrigation systems or components to design with water conservation in mind.

Many Americans seem to still appreciate large areas of lawn. Some codes restrict owners from *not* having a lawn, and forbid having a meadow or more naturalistic landscape instead. There is evolution of this concept but a long way to go before lawns will not be ubiquitous in the public landscape. As more water restrictions are put in place during droughts or as routine requirements in neighborhoods in arid settings with limited water supply even in the best of weather conditions, it is likely that codes will trend toward encouraging xeric landscapes and other plantings that conserve water or do not require nearly as much daily or weekly use.

Planning codes come into play when a landscape architect is designing an entire subdivision, studying a large property with multiple uses, or analyzing existing properties or neighborhoods for proposed long-term improvements or development. A prosperous city may be anticipating restrictions or guides on future growth; another city might be studying how to encourage and attract new development. The focus tends to be more on understanding the complexity of the tapestry rather than its individual components.

Building Codes

In their book *Building Codes Illustrated: A Guide to Understanding the International Building Code*, the authors Francis D. K. Ching and Steven R. Winkel, FAIA, summarize the development of the International Building

Code (IBC). Although the book is intended for architects, it has much useful information for landscape architects. For example, the authors illustrate the requirements of the Americans with Disabilities Act of 1990, many of which relate to access issues. For the complete discussion, see the ADA guidelines and chapters 10 and 11 of the Ching and Winkel book.[2]

The International Building Code, first introduced in 2000, provides one modern building code to supplant the three previous building codes governing the United States: the BOCA National Building Code, developed by the Building Officials and Code Administrators; the Uniform Building Code, developed by the International Conference of Building Officials (ICBO); and the Southern Building Code (SBC), developed by the Southern Building Code Congress. When I worked in the southeastern United States as a resident of Atlanta, Georgia, for example, there was a long period when the SBC took precedence. Even though the IBC has established traction and it's very helpful to have uniform standards that apply across the country, the landscape architect must still be aware that federal, state, and local code requirements may apply for particular applications, such as swimming pools or guardrails on rooftops. Most of the IBC applies to multistory buildings, often of nonresidential use. When working on residential buildings, such as single-family residences or apartments, it's all the more important for the landscape architect to carefully review local and state building codes for residential applications. However, there is also an International Residential Code, meant to "regulate construction of detached one- and two-family dwellings and townhouses that are not more than three stories in height with a separate means of egress." The landscape architect should verify which code applies to the residential sites which he or she is studying. Particularly rigorous requirements may discourage a designer from taking on a design commission, as it may be too difficult to meet the client's needs while also complying with zoning standards.[3]

Americans with Disabilities Act of 1990

The ADA, the Americans with Disabilities Act of 1990, a federal law, was a life-changer for many people with disabilities, as the force of federal law applied to all buildings, not just new buildings being planned, and required that all buildings be accessible. The law was enacted to protect Americans with all sorts of disabilities, not simply the limited number of people confined to wheelchairs. Because its impact is so large and there are sometimes disagreements about interpretation of the law, landscape architects should review its criteria carefully when evaluating a scope of work for a new or existing building. Over the years the scope of the law has expanded, in part depending on the role and budgets for enforcement of each federal administration since its enactment. This is not so different as the varying degree of emphasis the federal government places on the enforcement of civil rights laws throughout the United States. Some people chafe at what they think is an overly aggressive policy; others feel there is not nearly enough enforcement. The application of the law has been expanded as a result of court challenges in which plaintiffs claim that failure to cover their disability is a violation of their civil rights, and therefore demand coverage and enforcement; such challenges have met with varying degrees of success.[4]

This law applies to all new construction, and also requires improvements to existing buildings when such modifications are deemed practical. There is always the risk that a federal reviewer or administrator might deem some change to a building to be readily achievable when the owner does not. What is deemed readily achievable and practical becomes a matter of cost, which may generate considerable discussion and review. Historic buildings, most of which were never designed to accommodate people with disabilities, are still not exempted from the law. Therefore, when a landscape architect undertakes a project encompassing a historic building or site, the costs and requirements for access and use by people with disabilities should be carefully assessed at the beginning of the project. If any public funds are to be used in the project, then a careful review of how the ADA might be implemented should be undertaken, because it is likely that some government official will eventually require its application. For example, it is far better to have a modest plan ready to implement and integrated into the proposed design that meets the criteria for an accessible route than to have one imposed.[5]

Access

For landscape architects, most of the requirements of the law relate to access issues. The requirements are generally the following:[6]

1. There must be at least one clear, accessible access to the main entrances of buildings. The intent of an accessible route is to allow persons with disabilities to enter a building, get into and out of spaces where desired functions occur, and then exit the building.
2. The basic requirement is that accessible spaces in a building must have an accessible means of egress.
3. A ramp may not exceed 30 feet in length without a landing.
4. Cross slopes in a ramp may not exceed 1 in 48 (2%).

5. Ramps in a means of egress may not exceed a 1-in-12 (8%) slope.
6. Ramps are limited to a vertical rise of 30 inches (762 mm) between intermediate landings.
7. Landings are to be provided at the top and bottom of each ramp and at changes in direction.
8. Landings that accommodate a change of direction must be at least 60 inches by 60 inches (1,525 mm by 1,525 mm).
9. A sloped surface with a pitch shallower than 1 in 20 is not considered a ramp. It is acceptable to use such slopes in accessible routes if level landings are provided at doors or changes in direction.
10. Other ramps may not exceed a 1 in 8 (12.5%) slope. It is recommended that the designer never use ramps with a pitch steeper than 1 in 12, even in non-accessible paths of travel. The use of 1-in-12 ramps makes those paths of travel accessible and safer for all building users.

Parking

1. The basic criterion is that at least 2% of parking spaces must be accessible.
2. Requirements for accessible parking at rehabilitation and outpatient facilities are higher, being 20% of the total.
3. For every eight accessible spaces, or fraction thereof, one space is to be a van-accessible space.
4. Parking is to be located such that the accessible route of travel is the shortest possible path from the parking area to the nearest accessible building entrance.
5. The international symbol of accessibility is to be located at accessible parking spaces.[7]

Detectable Warnings

Detectable warnings are a series of truncated dome shapes in a pattern of specified size and spacing that is detectable by visually impaired people as a way of indicating that a change in grade follows. Some detectable warning systems are required by local codes to be of a color and texture that contrast with adjacent sidewalks or other pavements. In many locations, detectable warnings are also required at street intersections as a warning to visually impaired pedestrians of the edge of a vehicular zone—that is, that the person is entering a crosswalk and must be aware of vehicles.[8]

Other Factors

The general requirements of the ADA act as a check on architecture for those buildings subject to it, and the landscape architect must ensure a good fit between the indoors and outdoors. For example, seating, signage, lighting, and drinking fountains that occur outdoors must first be situated along the primary accessible route.

Interpretations of and new guidelines for the ADA are developed by the Department of Justice and were most recently issued in 2010. See https://www.ada.gov/2010ADAstandards_index.htm. At this same website is additional information on the ADA, such as technical assistance, specific design standards (including graphic diagrams), and other relevant resources.

Although a great deal of progress has been made in standardizing code requirements in some areas, such as the rules and regulations of the ADA, the requirements for the design of some elements remain variable. A good example is the height and dimensions of railings and guardrails over steps and at roof decks. (See Chapter 7, Part 4, "Roof Gardens and Green Roofs," for more information.) Therefore, even when information in the International Building Code and other relevant codes seems clear, the designer should still review applicable local codes. (See Figure 6.2).

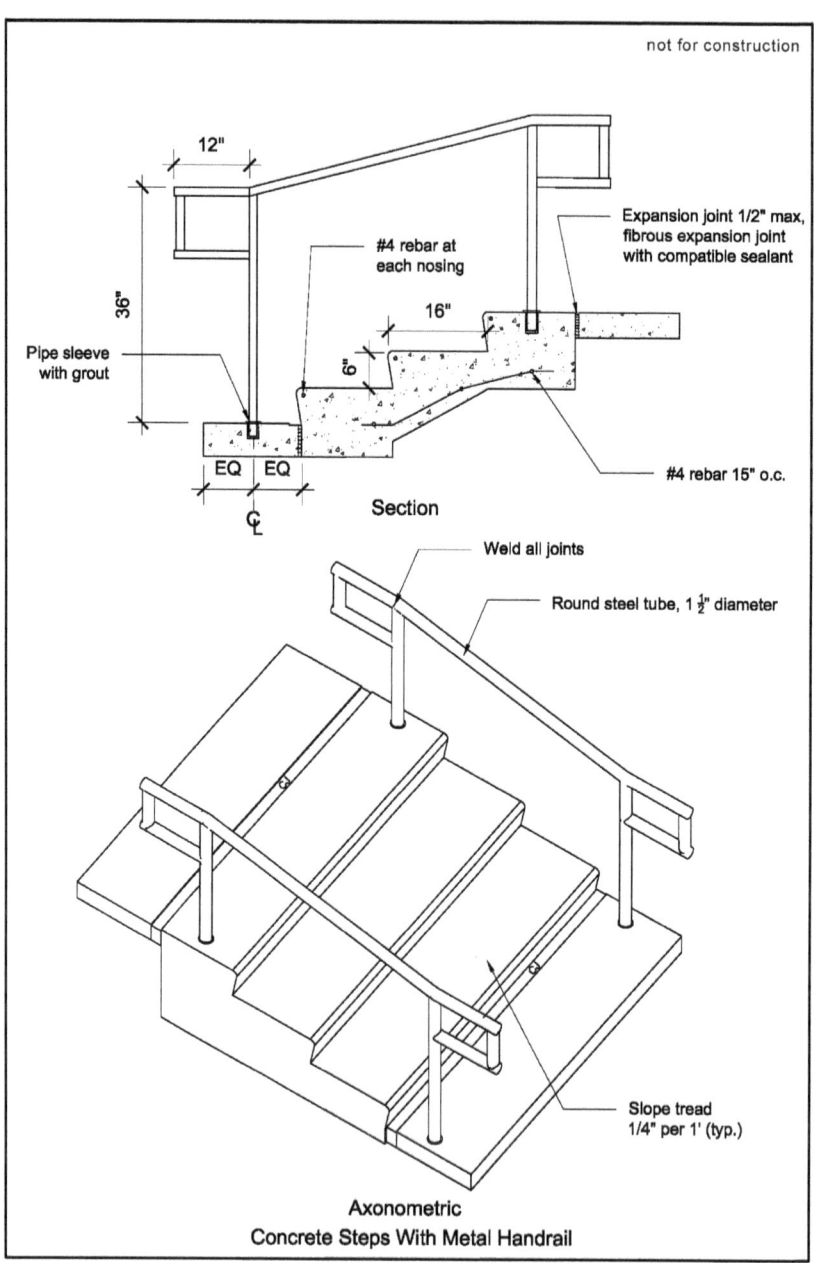

Figure 6.2 Landscape architects must verify that handrails in specialized settings comply with local building codes.
Richard Alomar and Vicky Chan

Problem 1
ZONING VARIANCE

(In the following problem, any *terms in italics* may be changed to suit the particular instructor or team leader, or your own personal, work or studio setting. Typical changes might be specific people, such as clients or contractors, plant materials, or the locations of sites, people, institutions or companies.)

You have been working as a landscape architect on a residential site plan for an existing residence *at a site and location* in which the client wants to add a swimming pool to a half-acre site. The residence is approximately in the middle of the lot. There are standard setbacks for the front, rear, and side yards. However, about a third of the rear of the property includes a wetland, *town* regulations for which do not permit any construction within fifteen feet of its mapped boundary. Upon studying the buildable area in the rear yard, you see there is not sufficient room for a swimming pool. You therefore study the side yards and find that one of them is wide enough and long enough to be able to incorporate a pool, except that the edge of the pool and walk closest to the neighbor's property would necessarily encroach by 5 feet past the setback limit. You see no other alternative in terms of the placement of a swimming pool. The client is enthusiastic about this proposed solution, but when she discusses it with the neighbors, a couple with whom the client has had good rapport, they nevertheless object.

Therefore, you prepare further studies of options for the swimming pool and its immediate surroundings. You prepare a variance application requesting that the town grant you permission to build the pool in the side yard and come within x feet of the side property line. Sketch the plan that you've developed. Write a letter, no more than one page, in which you summarize to the town's zoning review department your rationale for allowing the variance. What is your rationale for applying for a variance on behalf of your client? What design elements might you propose to mitigate the negative impact on the neighbor's property? Is there any way that the installation of the pool and related amenities associated with it could improve the neighbor's prospects? You will need to make a public request and presentation to the town zoning board at its next scheduled meeting.

Prepare two sketches showing alternative layouts for the swimming pool and side yard of the residence.

Prepare a draft of a letter to the town zoning board in which you request a variance. *Address the letter to the town zoning board where you are working or which is relevant to your area of practice.*

Problem 2
HISTORIC BUILDING AND AMERICANS WITH DISABILITIES ACT OF 1990

(In the following problem, any *terms in italics* may be changed to suit the particular instructor or team leader, or your own personal, work or studio setting. Typical changes might be specific people, such as clients or contractors, plant materials, or the locations of sites, people, institutions or companies.)

You have agreed to study a historic building on a 3-acre site in *Arlington, Virginia*. It dates from the 1850s and features *Victorian architecture*, including some beautiful *Tiffany windows and lamps*. The building is 40 feet long by 25 feet wide, with the entrance centered on the longer side. The building sits on a relatively flat site; perhaps for this reason, the main floor of the building is raised about 6½ feet above existing grade, and access is via formal steps, 15 feet wide, with three ornate handrails (one in the center and one at each edge). From inside the grand entrance, there is a nice view of an agricultural landscape. There are two sets of 8 steps, with each set separated by a 5-foot-wide landing; each step has a riser of 5 inches, and the treads are 15 inches wide.

The State of Virginia Department of Historic Preservation has asked you to develop a plan that will sensitively provide an accessible entrance to the building via a series of ramps. You must comply with the requirements of the Americans with Disabilities Act of 1990. The state's main concerns involve accessibility for people in wheelchairs, people with disabilities who cannot negotiate steps, elderly or other citizens who might prefer a less strenuous option than climbing the steps, and visually impaired people. Also, it wishes to add a perimeter fence in a historic style compatible with the architecture of the building.

1. How would you propose to bill for this work? Write a one-page letter giving your proposal and scope of services. Address it to the appropriate *agency*.
2. Show a design, with an illustrative plan, a layout, and a grading plan. Work at the scale of 1" = 10' or 1" = 20'.

PART 2: SPECIFICATIONS

Make thee an ark of gopher wood; rooms shalt thou make in the ark, and shalt pitch it within and without with pitch.

And this is the fashion which thou shalt make it of: The length of the ark shall be 300 cubits, the breadth of it fifty 50 cubits, and the height of it 30 cubits.

A window shalt thou make to the ark, and in a cubit shalt thou finish it above; and the door of the ark shalt thou set in the side thereof; with lower, second, and third stories shalt thou make it.

—Genesis 6:14–16

Introduction

The specifications are the third part of the trio, along with the working drawings and the contract that form the contract documents. The most important function of specifications is to tell the contractor how to execute and manage construction. They also provide key information to cost estimators, manufacturers, inspectors, contractors, and the owner about the amount and quality of work to be performed. This section gives an introduction to specifications: what they are, why they are needed, and how they are written. After reading this section, one should have an overview of the subject and a clear sense of how to begin. Since the topic is so broad and is covered by many references, the focus here is on general concerns relevant to landscape architecture, practical applications, definitions of basic terms, specific examples of specifications and other documents.[9]

When the landscape architect is hired to execute a design, he or she must prepare a set of contract or construction drawings, which consist of three components: the contract, the working drawings, and the specifications. The contract specifies the services the landscape architect will provide for a specified fee with a set completion date. The drawings, called working drawings or construction drawings, show graphically—as precisely as drafting allows—what work will be done for the project. The specifications include (1) the *general conditions*, the legal, contractual, and administrative requirements and responsibilities of the contractor, owner, and landscape architect, and (2) the *technical specifications*, a description of the work to be done. Since specific legal requirements for a project may vary substantially, a section of *supplementary conditions* is often included too, comprising additions, subtractions, and modifications of sections of the general conditions.

General Conditions

The general conditions include legal requirements and responsibilities of those under contract. For most small- to moderate-sized projects, the general conditions include the following information on all nontechnical aspects of the work, such as:

1. A summary of the work to be included in the contract
2. A calendar of implementation
3. Administration (permits, licenses, billing, and daily operations)
4. Insurance requirements
5. Safety and security requirements
6. Changes in the work
7. Corrections to the work
8. Guarantees and terminations of the work

For large and complex work, three separate sections may be used: general conditions, supplementary conditions, and general requirements. The general conditions section then is strictly limited to specifying legal responsibilities and requirements of the parties under contract. The supplementary conditions, still limited to legal requirements, include modifications, additions, and deletions from the standardized general conditions. The general requirements, consisting of all nontechnical requirements and information the contractor must have to plan and construct the project, become part of the technical specifications. For substantial projects, the general conditions are often incorporated into a set of specifications as a preprinted edition in a standardized format prepared by the American Institute of Architects or some other professional association. The subjects included are in Division I of the CSI standards, discussed later in this section.

Bidding Requirements

To plan for the execution of the job, bidding requirements are also essential. Technically, they are neither specifications nor contract documents because the bidding requirements apply prior to the signing of a contract for the execution of the work. However, the bidding requirements are often printed separately, but included in and distributed within the specifications. There are three components: the *invitation to bid*, the *instructions to the bidders*, and the *bid form*. The invitation to bid notifies all parties who might be interested in submitting a bid. The content of this document is limited to information that will help a potential bidder decide if he/she should give the project serious attention. An interested potential bidder reviews the instructions to bidders, which provide all the information necessary to submit a bid. The bid form or proposal form is a document on which the bidder lists a breakdown of the bid in a standardized format convenient to the purposes of the landscape architect and owner. All bidders use the same form, to ensure a uniform basis for comparison.

Bonds

In order to protect against dishonest, incompetent, and irresponsible contractors, the owner may decide to include bonds in the contract. The general conditions and instructions to bidders often will include specific surety bond requirements. A *surety bond* is a financial guarantee by a surety company of the faithful execution of the contract and the payment of bills. Four types of bonds are most common: the bid bond, the performance bond, the payment bond, and the labor and materials bond. The *bid bond*, submitted with the contractor's proposal, guarantees that the contractor will agree to the contract if the proposal is accepted. If the contractor withdraws upon acceptance of his proposal, the surety company forfeits the amount of the bond, usually 5% to 10% of the bid amount. The surety company may then sue the contractor, but the owner is awarded the bond amount, which can cover the expenses of locating another contractor to execute the work.

The *performance bond* guarantees that the contractor will execute the work. It usually includes protection to the owner against defective work by the contractor. If the contractor stops work, the surety company has several options. It may act as a general contractor to finish the work already begun; it may pay the owner the amount of the bond less a proportional amount reflecting the percentage of the work completed; or it may locate another contractor to finish the work.

The *payment bond* guarantees that the contractor will pay all bills required by the contract. This protects the owner from claims filed against him after the owner has made all payments to the contractor. A *labor and materials payment bond*, a special type of payment bond, guarantees that the contractor will pay all bills for labor and materials. This bond gives protection to workers, subcontractors, and material suppliers. The performance bond and the labor and materials payment bond are usually issued together. In all cases, if the contractor defaults or fails to meet his obligations under the terms of the contract, the surety company pays the amount of the bond to the owner.

A bid bond of 5% to 10% of the contract price is usually required. A performance bond of the full contract price is standard. For payment bonds and for labor and materials payment bonds, most specifications will set the amount between 50% and 100% of the contract price. Some bonds are required by law, and specific requirements vary from state to state. Federal contracts will often require performance bonds, labor and materials payment bonds, and bid bonds.

A *liquidated damages* clause in a contract is not a bond, but it does provide additional protection to the owner. This clause in a contract sets a date for completion, beyond which the contractor must pay a penalty per day (or some other time period) until the work under contract is completed. As with bonds, it is customary to notify the contractor in the instructions to bidders that such a requirement will be included. Initially, it may cause an increase in the bid price, compared to a project without this clause. However, considerable savings accrue to the owner if the work is completed in a timely manner, without significant delays. Absent such a clause, a contractor who is quite busy on a number of projects may delay completion at considerable inconvenience to the owner, which could result in costly delays of other projects planned in sequence or simply a delay in having activities on site return to normal. Of particular concern are housing projects, schools or other settings in which many people could be adversely affected if they no longer had full access to their usual facilities and activities.

Four Types of Technical Specifications

The heart of a set of specifications is the technical specifications section, in which the "nuts and bolts" of the work are described. This section follows the general conditions. Technical specifications generally follow one of two formats. One approach is to describe in detail a method of execution, much like a recipe, beginning with a list of the materials needed and proceeding through a step-by-step procedure. The alternative approach is to require certain results or performance, leaving to the contractor the choice of materials and methods as long as the performance standard is met.

There are four different types of technical specifications: (1) descriptive, (2) performance, (3) proprietary, and (4) reference. A *descriptive specification* describes in a thorough, logical manner the materials needed and the methods and standards of workmanship to be used to install the specified item. This type of specification gives the designer the most control in the step-by-step process of construction, and is often used for the most important items of a project. A *performance specification*, by contrast, traditionally has been used when the landscape architect requires the contractor to match an existing condition. The result required of the contractor and the standards for verification of those results will be indicated, but the method of execution is left to the discretion of the contractor. A *proprietary specification* indicates a specific product or manufactured item to be installed, such as a particular type of bench or light post. Under some contracts, the designer is not permitted to specify a single choice, so the term "or equal" will be added to the text of the specification. This clause may increase the competition among bidders for a project, but the "or equal" clause also removes the landscape architect's control over design quality because there are often conflicts in determining whether a substituted item is actually equal to the one specified. If contract conditions forbid specifying only one brand of product to be installed, a fairer approach is for the landscape architect to specify several options, such as three suitable bench types, but not include the "or equal" clause. The last type, the *reference specification*, specifies a standard established for a material, execution technique, or testing procedure. There are many reference standards. Some of the most common are ASTM (American Society for Testing and Materials), ANSI (American National Standards Institute), AASHTO (American Association of State Highway and Transportation Officials), and ASNS (American Standards for Nursery Stock). The consistent use of reference specifications considerably shortens the length of a set of specifications. Examples follow for each type of specification.

1. *Descriptive specification example.* Topsoil shall be fertile, friable, natural topsoil, brown in color, and reasonably free of weeds and foreign matter. It shall be broken up and free of clods. It shall not be handled in a frozen or muddy condition. Every effort shall be made to preserve organic matter. The pH measurement shall be 6.7 to 7.3, as an average of at least three test samples taken from what has been delivered to the site. [It is often typical to specify other materials to be mixed with the topsoil in particular proportions, such as sand, lime, and pine bark to define a "planting soil mixture" used for planting operations for trees, shrubs, and groundcovers.]
2. *Performance specification example.* Black chain-link fence shall match existing vinyl-clad 6-foot-high chain-link fence along the north property line. Submit shop drawing showing post spacing, top and side rails, and mesh size.
3. *Proprietary specification example.* Brick pavers shall be 1-5/8" × 3-½" × 7-5/8" "Best Brushed Buff" solid brick paver from Bubba's Brick Company, Albany, Georgia. See www.bubbasbrick.com. Pavement pattern shall be running bond, as shown on the drawings.

4. *Reference specification example.* Welded smooth wire fabric shall conform to "Specifications for Welded Steel Wire Fabric for Concrete Reinforcement," ASTM 185-73, and shall be fabricated from plain wire conforming to "Specifications for Cold-Drawn Steel Wire for Concrete Reinforcement," ASTM A-82-76. Maximum wire spacing shall be 12 inches. [Rather than cite the entire reference specification, it is helpful to cite the specific relevant sections.]

Interrelationship Between the Drawings and Specifications

One must distinguish between the role of the drawings and that of the specifications. An analogy may be helpful. Think of a wall of mortar and brick. Specifications and drawings must be bonded just as integrally as these two components. If one is weak, the effort fails. The drawings are graphic, the specifications are written; the two must reinforce each other. The drawings show what cannot be easily written, and the specifications show what cannot be easily drawn. The drawings show precise form, size, shape, dimensioning, layout, detail, and interrelationships between important design features and materials. The specifications indicate the type and quality of materials and workmanship, methods of installation and construction, and test and inspection requirements. The drawings and specifications must complement each other but should never repeat each other. Any attempt to describe in words what is drawn or draw what is written most often will result in inaccuracies or inconsistencies that a contractor may misinterpret. At a minimum this could lead to confusion until the contractor clarifies the situation; worse, it might result in an improperly constructed element of the design, the correction of which may be quite costly and delaying.

In the event of conflicts that must be resolved in the courts between the drawings and specifications, the specifications, since they are written, usually are given priority over the drawings. But even in the most carefully planned projects, there are bound to be some discrepancies. To avoid protracted conflicts among the parties to the contract, means for resolving such problems (often arbitration) should be included in the general conditions.

Organization

Specifications are organized in many different ways, depending on the size of the project, the nature of the work, and the style of the office. Methods of organization have evolved to ensure accuracy of estimating and ease of reference. Generally specifications are organized into a series of sections, each of which deals with all the work of a particular trade to be executed by one subcontractor. Traditionally, these sections have followed a chronological order: demolition, followed by excavation and backfill, then concrete and other pavements, then masonry, then planting. For projects of limited scope, this method still works well.

For larger, complex projects involving work by many different subcontractors, a more standardized format is appropriate. The Construction Specifications Institute's (CSI) Format for Construction Specifications is now widely used. The American Institute of Architects, the Associated General Contractors, the Sweets catalog, and many building manufacturers now all follow the CSI system. There are sixteen major divisions, each with subheadings. In the list that follows, subheadings are provided for Division 2 as an example. It is typical to use subheadings for all divisions.

    Division 0: Bidding Requirements, Contract Forms, and Conditions of the Contract
    Division 1: General Requirements
    Division 2: Site Work
      02010 Subsurface Investigation
      02050 Demolition
      02100 Site Preparation
      02140 Dewatering
      02150 Shoring and Underpinning
      02160 Excavation Support Systems
      02170 Cofferdams
      02200 Earthwork
      02300 Tunneling
      02350 Piles and Caissons
      02450 Railroad Work
      02480 Marine Work
      02500 Paving and Surfacing
      02600 Piped Utility Materials
      02660 Water Distribution
      02680 Fuel Distribution
      02700 Sewerage and Drainage
      02760 Restoration of Underground Pipelines
      02770 Ponds and Reservoirs
      02780 Power and Communications

02800 Site Improvements
02900 Landscaping
02910 Shrub and Tree Transplanting
02920 Soil Preparation
02930 Lawns and Grasses
02950 Trees, Plants, and Groundcovers
02970 Landscape Maintenance

Division 3: Concrete
Division 4: Masonry
Division 5: Metals
Division 6: Wood and Plastics
Division 7: Thermal and Moisture Protection
Division 8: Doors and Windows
Division 9: Finishes
Division 10: Specialties
Division 11: Equipment
Division 12: Furnishings
Division 13: Special Considerations
Division 14: Conveying Systems
Division 15: Mechanical
Division 16: Electrical

Different Ways to Organize a Set of Specifications

The majority of the work done by most landscape architects could be included in Divisions 0–6. For example, Division 2, Site Work, includes all clearing and grubbing, storm drainage, paving, grading, and plantings, which is the great bulk of many small landscape architecture projects. However, many specification writers, rather than fit all the work into the many numbered paragraphs and subparagraphs of CSI Division 2, prefer to use their own system, organized section by section. Each subsection of the CSI Division 2 is then written as a separate section, permitting a simpler organization and numbering system.

On the other hand, in a project involving many different disciplines, such as landscape architecture, architecture, electrical engineering, and civil engineering, the CSI format is often the most appropriate. Then each designer is responsible for the preparation of a particular division of the specifications, which includes all of his/her areas of expertise and responsibility. A consistent organization is achieved even with many authors and contributors. Under these circumstances, the landscape architect may be a consultant to an architect and would prepare the specifications for Division

2. Even though there are distinct CSI division numbers for concrete, masonry, metals, and wood and plastics, the landscape architect might include them all under Paragraph 02800, Site Improvements. In this way, all of the work for which she is responsible is included in only one division of the specifications, and the content of Divisions 3–6 will refer only to the architectural application of these materials.

It is quite common for the chain of command to be reversed, that is, for the landscape architect to be the principal and the architect to be the subconsultant. Depending on the nature and scope of each designer's responsibilities, the organization of the specifications could be similar to or different from the above example. The operative principle is for the landscape architect and other key members of the design team to organize, format, and assign specifications in a manner that will achieve clear results.

There are situations in which certain types of work that the landscape architect is responsible for should be specified in other divisions. Swimming pools, for example, are included in Division 13, and lighting in Division 16. The installation of lighting and swimming pools is generally carried out by specialized contractors, whom the general contractor usually finds. These items do not adapt themselves to inclusion in Division 2.

Many landscape designers focus only on planting design. Even so, it is a mistake for them not to develop a set of specifications governing their landscape projects. Problems may easily occur with procurement and delivery of plant materials, the quality of the plant materials, minor grading and drainage issues, all aspects of the planting operations, and the inspection and guarantee of the work. No matter how experienced one is, how simple the project is, and how smoothly one expects it to be executed, a well-written and time-tested set of specifications would provide clear directions on procedures to follow and what to do should difficult eventualities occur.

The CSI expanded its format from sixteen to fifty divisions in 2004, in order to encompass the complexity and specialization of modern construction and new areas of technology.[10] Some of the paragraphs within the newer system are still not yet used, but landscape architects should review them, particularly for complex projects. Still, a standard landscape architecture project may well fit into the older system. Another newer system is GreenFormat, in which the basis of organization is according to materials, products, systems, and technology associated with sustainable design; this is useful in the LEED process.[11] As with specifications themselves, the landscape architect must choose the system of organization that is most practical and effective for the project at hand. Just as a pianist would not perform on a Steinway concert grand with the lid fully open in an intimate den for an audience of five, a landscape architect should not use a fifty-part

CSI system of specifications for a small residential design when a sixteen-part organization or a well-placed series of notes will suffice.

Unique Specifications for Government Agencies

Many government bureaucracies, such as school systems, city parks departments, and state and federal government agencies, have their own specifications. Although these systems may be somewhat similar to the CSI format, the landscape architect must review and understand them from the beginning of a project. The labeling of details on the drawings and the preparation of cost estimates may have to include the precise specified language, such as an item number for a particular drainage detail or bench, from these required specifications. One challenge of doing work for government agencies is that their specifications may include unfamiliar requirements or organization for specifications, which will take the landscape architect considerable time to master. These bureaucracies, however, will have their own specification departments, which generally will be quite cooperative in issuing master sets of specifications to consultants, including the assigning of unique specification numbers to unique details developed by the landscape architects as part of the scope of work for a new project under contract.

Content and Writing Style

Writing specifications requires mastery of both content and style. The specification writer must use good grammar and syntax in effectively communicating technical requirements. In some ways one should write no differently than if one were planning a novel or nonfiction article: there are universal attributes of effective prose. Writing in clear English rather than legalistic jargon will be more effective. On the other hand, the technical nature of specifications requires certain rules of both style and content not common to other types of writing.

Content

1. *Accuracy.* Give complete and correct information.
2. *Appropriateness.* Give appropriate information for either a descriptive or performance specification, but no reasons, justifications, suggestions, or explanations.

3. *Reasonableness.* Specify tolerances and degrees of precision no more accurate than necessary. Use measurable, standard tests and methods.
4. *Practicality.* Use standard sizes and patterns. Work within the limitations imposed by the materials and craftsmanship available.
5. *Fairness.* Avoid ambiguous language, conflicts between workmanship and performance, and conflicts between the technical requirements and the resulting characteristics.
6. *Consistency.* Develop similar formats and use similar language for specifications within the same division.

Style

1. Use either the active voice or the imperative voice, with key words and phrases at the beginning or end of sentences.
2. Use one tense consistently.
3. Be concise. To shorten a sentence is to strengthen it.
4. Use parallel construction, writing expressions of similar content and significance in the same form.
5. Use nouns and verbs, rather than adjectives.
6. Be definite, specific, and concrete. Write the particular details that matter.
7. Write statements in positive form.
8. Use negative statements only as warnings or for specific purposes. Negative statements may be perceived as creating too many obstacles, which would discourage contractors from bidding on or executing the work.

Writing Methods

Just as the landscape architect prepares preliminary and final drawings for a project, the specification writer packages both a preliminary and a final set of specifications. The preliminary set, in outline form, is often used in preliminary estimates for a project, to assure the landscape architect that the work proposed is within the budget. The final set, reflecting changes to the working drawings, is comprehensive and reinforces the whole concept of the project.

As a project is concluding the design development phase, the specification writer assembles outline specifications. These briefly describe the materials essential for construction. Key design elements of the plan are labeled and matched to an outline specification that describes that type of element. Standards of workmanship or special techniques of construction are not necessary at this stage, since the purpose is to secure a preliminary cost estimate. A set of preliminary drawings and specifications is given to a contractor or office estimator to verify anticipated costs.

Once the final design development plan is approved and working drawings are well under way, the task of writing the final specifications begins. The writer matches every item detailed in the working drawings with the specification explaining it. A landscape architecture office will accumulate a file on different types of specifications. For a particular project, the specification writer, based on the design intent and conference with the designers, will revise an older version (using a cut-and-paste approach) until a satisfactory and effective specification is created. Work proceeds not at a steady pace, as may be the case with working drawings, but in fits and starts corresponding to the completion of a particular detail or drawing and the resolution of discussions about certain elements being detailed. Some sections of the final specifications will require research, while others will be straightforward adaptations of earlier efforts. It is prudent to have a member of the office unfamiliar with the project review the final working drawings and specifications for consistency, effectiveness, and completeness.

Large offices and those that deal in specialized types of projects are increasingly using computer-aided specifications or those linked to computer word processors. The CSI format is most often used. Standard specification data is stored on disks or on the hard drive for a variety of projects. Since as much as 80% of the specification from an old project can be used for a new project, the computer can be used to store great amounts of information and assemble new sets of specifications from previous examples.

Prewritten Software Packages

There is now available a whole range of software packages that organize information about specifications into easily retrievable packages that can be edited quickly and accurately. There are several advantages to the computer methodology. First, rather than working from an old set of specifications that might not include important information needed for a particular new project, the specification writer starts with the master specification, which includes all possibilities. S/he needs only to delete irrelevant material and occasionally add specific paragraphs covering special conditions. Second, the master can be updated continuously to reflect current practice, thus avoiding the inclusion of obsolete materials, a problem that may occur when an old set of specifications is being cut and pasted to prepare a new set. Third, word processors and computer printers increase office efficiency and the accuracy of the product. Printers commonly produce pages at a very rapid speed with limited supervision, thus freeing employees to work on other tasks. As the master set is constantly updated and corrected, the resulting specifications become closer and closer to a specifier's dream: a perfectly accurate document.

DOS AND DON'TS FOR SPECIFICATIONS

1. Avoid citing reference specifications for simple specifications. Few landscape designers, no matter how experienced, have a clear memory of the content of an ASTM standard. It's better to at least call out the main point of a particular ASTM reference in a specification than to just list it with no clear explanation.
2. Be aware of when to use or develop a descriptive specification, a proprietary specification, a performance specification, or a reference specification.
3. Keep current with changes in standards.
4. Tailor the specifications to the level of complexity of the proposed project.
5. Don't repeat in the specifications any dimensions that are shown in the detail drawings; there's too much risk that a further revision in the detail might not be corrected in the specification (or vice versa).
6. Incorporate a set of general conditions that has been used before on similar projects, with a few revisions or updates as needed to match the requirements of the new project.
7. If this is the first time you are using specifications published by a public agency, review them carefully in case of unexpected requirements or conflicts with processes and methods that you might prefer, or for potential conflicts with the work of other agencies (if they are present or represented on the site being developed).
8. Based on a review of the drawings, details, and specifications, develop a list of all of the construction details that will require shop drawings, and include them in the notes of an appropriate drawing.
9. Be certain that the specifications make clear who is responsible for permits and approvals.
10. Review the specifications and the construction drawings one final time before deciding on any requirements for completion of the proposed work, or certain aspects of it, by a certain date. Sometimes it's appropriate (usually when there is urgency about completing a

project within a set calendar period) to incorporate deadlines, after which liquidated damages will accrue. At other times that's not necessary.
11. Ask a colleague unfamiliar with the project to review the specifications.
12. Be certain that the item numbers of the specifications match the item numbers labeled on the details, called out on the drawings, and listed on the cost estimate entries.
13. Be consistent, but absolute consistency is rarely called for, as every set of specifications usually has outliers. Be diligent enough to find them.

Problem 1
SPECIFICATIONS REVIEW

1. What topics in specifications might be likely to emerge in the future or gain in stature?
2. What topics may become less important?
3. Answer the following questions:
   a. What is the difference between general conditions and technical specifications?
   b. What is a performance specification?
   c. What is a bid bond?
   d. What is a labor and materials bond?
   e. What is a performance bond?
   f. What is a performance specification?
   g. What is a proprietary specification?
   h. What is a reference specification?
   i. What are some guidelines for content and style of specifications?
   j. What is the difference between a change order and an addendum?
   k. What are the three principal types of insurance that all contractors should carry for a landscape architecture job?
4. Review specifications on plant materials in your office or other setting. Answer the following questions:
   a. What are typical ways planting details may not match specifications? How are they similar? How are they different? Which is better? Which would take precedence in a dispute on the site of a project under construction, the specifications or the details?

COST ESTIMATES (201)

     b. Would you change anything about the methods of excavation and setting for the planting of trees, shrubs, and groundcovers, including guying and staking of trees?
     c. What is required for maintenance?
     d. What is the difference between a semifinal inspection and a final inspection for acceptance?
     e. What is the guarantee period? Is it adequate?
     f. Think back to the first problem, on preparing a proposal. What elements might be necessary to consider that are not covered in these specifications?
5. What is your reaction response to the "specification" given to Noah at the *beginning of this section of the book, the quotation from the Bible*? What type of specification was he given? What is the length of a cubit?

Problem 2

OUTLINE AND DETAILED SPECIFICATIONS

For the project you have studied in one of the other exercises, develop a set of outline specifications covering demolition, construction, planting, inspections, and so on. Pick two of the outline specifications and flesh them out—that is, revise and update them as complete specifications. The specifications, in final form, may be descriptive, performance, proprietary, or reference; however, both cannot be the same type. It is all right, even encouraged, to model your specifications after examples you find in your research? Be certain to indicate your sources. (Depending on time constraints, the instructor may require one specification rather than two.)

PART 3: COST ESTIMATES

Documents That Connect to Drawings

Cost estimates must grow seamlessly from the drawings showing the design. When the design drawings are conceptual or schematic, then the cost estimates are preliminary, with considerable contingencies to cover unknown costs; when the drawings are advanced or final, then the cost estimates must be highly detailed, with no contingencies. By the time of the final contract documents, the cost estimates are so fully integrated with the construction drawings and specifications that they can be considered part of them. For public works projects or other work for a government agency, the cost estimates must be based on standard specifications provided by the agency, or occasionally custom-written ones prepared by the consulting landscape architect and other professionals. Often the specification may have a highly specialized technical name, and the entry in the cost estimate should repeat the same name. In the private sector, there is a great deal more flexibility in the listing of items in the cost estimate; again, they should generally repeat the categories and entries of the specifications, but many of these specifications will have been written and refined over time by the landscape architect and include custom materials, processes, or products not usually incorporated in public works. Whatever the nature of the project, public or private, the landscape architect should endeavor to create a cost estimate that evolves from the drawings.

Specialty Items

Many landscape architecture projects include specialty items for which the work is done by a subcontractor or other specialist, such as a swimming pool, tennis court, water feature, playground area and/or structure, a group of sculptures, and so on. Although the construction of a swimming pool, tennis court, or other feature involves some work by many different contractors/trades, it's better to organize the cost estimate so that all costs related to the swimming pool, tennis court, or other specialized feature are recorded in one coordinated list of items. If the swimming pool must be put on hold or is phased, then its lump sum cost is readily apparent and the drawings related to it are also a distinct group. This level of organization greatly eases the strain on the landscape architect if there must be discussion of phasing or reducing costs. The total cost as well as its component costs are line items in the cost estimate, which may be put on hold or not.

Even with work in which the landscape architect expects adequate funding, it can be helpful to break down the costs of different areas of work, such as planting. For a project that includes three distinct gardens, for example, within the cost estimate it is helpful to see the total planting costs for each of these gardens, rather than combining all of the planting costs into one total. Even if each garden includes some of the same species, the cost estimate is more effective and useful for review if there are two separate breakdowns, one by type of garden and one for each species.

Therefore the landscape architect could see at a glance a separate cost for garden A, garden B, and garden C, and at the same time a unit cost for each plant species.

Separate Unknown Costs

Cost estimates typically include items labeled "LS," *lump sum*, which represents an estimate for many different items of work that would be calculated separately should the project progress to the next phase. Usually at this stage of design, there are many unknowns, so typically the LS is a generous figure. For a master plan or preliminary plan, prior to the development of any detailed drawings, a cost estimate might have a single line item for a swimming pool, which is LS. After review and discussion with the client, if it's decided that the landscape architect should proceed with more detailed design, then the LS item will be broken down into many different components in a more detailed set of drawings and a more detailed cost estimate.

Particularly in earlier stages of design, it may not have been possible to price everything even in the crudest of calculations. Therefore, it's important when preparing cost estimates for conceptual drawings, preliminary drawings, master plans, and so on to include in a clear list at the end of the cost estimate a category of "not included." This list should not be so large as to make the client feel that too much is not known about the project to go ahead with it, but at the same time, the list should reflect a realistic determination of remaining unknown costs. Typical items might include, for example, irrigation, pruning of major shade trees, specialty fences and gates, and so on.

Keep Track of Costs for Permits and Approvals

Another category of costs that can add up but may take a while to determine are the cost of permits and approvals. Depending on the nature of the project and the client, it may be appropriate to include in a preliminary cost estimate a list of permits and approvals that will be required should the project proceed to the next stage. At the conclusion of a review meeting, there could be discussion with the client about who will be responsible for each permit or approval. Sometimes that's primarily the responsibility of the contractor; at other times the landscape architect or owner may need to take charge of that task. By listing on the cost estimate all the permits and approvals, with known or unknown costs, the document can serve as a checklist of work remaining to be done, helping to avoid an embarrassing situation in which some previously unknown but major cost surfaces.

Common Sense and Mental Arithmetic

Prior to starting the cost estimate process and several times during its preparation, review the numbers and rely on common sense results. Make mental notes and quick arithmetic estimates of what you expect major costs to approximate. List them carefully, and when the first printed version in Excel or other format is given to you for review, compare the two. If calculated estimates of items are more than 20% off from your quick-and-dirty calculations, review them carefully. It goes without saying that if the numbers for a line item are off by a factor of 10 rather than 2, then something is seriously wrong. Often this is a matter of carelessness—someone entered the number incorrectly in the spreadsheet. The time to catch such errors is at the beginning of the calculations. Yet this mental review underscores an important principle of cost estimates: don't allow the software, whether Excel or some other version, to take the place of careful review and analysis.

The following preliminary cost estimate might be one used by a landscape architect coordinating work by various contractors who have started to price work designed by different architects, landscape architects, and specialty designers. Such a spreadsheet could also be administered by a builder or general contractor. For this project, a residential estate, a landscape architect might be the principal in charge and would assign to various consultants different major elements of the overall design and would also coordinate what CSI specifications each consultant would be responsible for. There can be some variation, as long as everyone is consistent and is fully informed. For example, if there is a swimming pool with a large wood pergola or there is other major wooden construction, such as a gazebo, specifications for these major elements might be in Division 6. However, if there is only a modest amount of wood in the landscape for the site, it could stay within Division 2, Sitework.

For preliminary estimates in which many or most items are based on drawings and specifications that are not final, so a lot of information is not yet available, it is common to add a contingency amount to the estimate, anywhere from 10% to 15%, which is eliminated by the time final contract documents are completed. The degree of precision in each line item of the estimate must reflect the documents on which it is based. Therefore, in this preliminary estimate, most of the figures are in round amounts. By contrast, in the example shown, several items—Paragraph 02100, Site Preparation (relating to tree removals, stump removals, and tree protection); Paragraph 02925, Landscaping (based on a plant schedule); Paragraph 02926, Transplanting (for six specimen trees on-site); and Paragraph 06400, Carpentry (for pergolas)—are based on more detailed estimates, presumably dependent on advanced design drawings and specifications. The more typical CSI Paragraph 02800, Site Improvements, is not incorporated, so this would be one of those projects where it would be important for all consultants and contractors to follow a consistent organization. This estimate can continue to be updated until it reflects final design and specifications. As the project moves forward, should any change orders be developed and issued, it would be important that the costs be calculated based on specifications based on the same paragraph numbers, even if there was a new line item assigned to each in the overall cost estimate.

Although not incorporated herein, a fee for a site supervisor might be indicated, to cover the cost of a foreman on-site to coordinate all the work for the general contractor, who would do the hiring. This critical employee would also be responsible for keeping the principal designers informed and up to date on all major progress of the work. The estimate also includes a calculation for the cost of general conditions, which is estimated to be 7% of the total cost of construction. This cost may change, of course, depending on the location of the project, the prevailing labor rates, and the complexity of the work. Just as one example, if there are a considerable number of permits and approvals itemized in the general conditions, the cost may be higher. The cost estimate does not indicate any.

SUNNYSIDE UP LANDSCAPE ARCHITECTS RESIDENTIAL PROJECT		
	15-Feb-19	
CSI CODE	LIST OF COSTS	
02000	**SITEWORK**	**$479,965.00**
02100	Site Preparation	
02101	Tree Removal	$8,452.00
02102	Stump removal	$5,256.00
02103	Tree protection	$6,357.00
02140	Dewatering	$17,000.00
02110	Demolition	$28,500.00
02200	Earthwork south of main residence	$110,000.00
02200	Earthwork at garage courtyard area	$12,000.00
02200	Formal parterre/garden area to pool house	$29,400.00
02200	Pool excavation and backfill	$15,000.00
02200	Preparation for major lawn area (to be hydroseeded)	$95,000.00
02500	Driveway (rough-in and paving)	$63,000.00
02500	Footing drains & stone under slab	$16,000.00
02600	Site utilities	$25,000.00
02600	Silt fences	$4,000.00
02700	Sitework—drainage budget	$45,000.00
02900	**LANDSCAPING**	**$830,087.30**
02901	New topsoil & fill	$30,000.00
02902	Tree wells	$0.00
02903	Paving & surfacing	$0.00
02904	Curbs	$0.00
02905	Fences & gates (including deer fence)	$0.00
02906	Irrigation system allowance	$50,000.00
02907	Site walls (in Masonry, below)	$0.00
02908	Belgian block	$0.00

SUNNYSIDE UP LANDSCAPE ARCHITECTS RESIDENTIAL PROJECT		
02909	Finish grading	$0.00
02925	Landscape—planting schedule (see plant lists/bids)	$646,853.00
02926	Transplanting 6 specimen trees on-site	$38,549.00
02970	Landscape maintenance for one year post-planting	$64,685.30
03000	**CONCRETE**	**$404,000.00**
03100	Footings—material	$53,000.00
03150	Footings—labor	$124,000.00
03200	Walls—material	$142,000.00
03250	Walls—labor	included in above
03300	Floors—material	not included
03350	Floors—labor	not applicable
03400	Rebar & mesh	$37,000.00
03500	Terrace slabs (includes turn down slabs below frost line)	$48,000.00
03600	Concrete stoops	not included
04000	**MASONRY**	**$679,000.00**
04200	Brick masonry, garage courtyard	$232,000.00
04400	Granite masonry, moon garden	$31,000.00
04405	Limestone masonry, white garden	$72,000.00
04406	Marble masonry, reflecting pool with pergola	$137,000.00
04407	Imported Italian stone, herb garden area	$42,000.00
04500	Restoration of terraces/walls adjacent to existing bldgs.	$165,000.00
05000	**METALS**	**$0.00**
05100	Structural steel framing	$0.00
05300	Steel decking	$0.00
05500	Metal fabrications railings—(in Div 6. as wood)	$0.00

SUNNYSIDE UP LANDSCAPE ARCHITECTS RESIDENTIAL PROJECT		
06000	CARPENTRY	$316,141.00
06100	Rough carpentry	$0.00
06130	Heavy timber construction	$0.00
06300	Millwork	$0.00
06400	Carpentry—wood pergolas in landscape designs	$316,141.00
06450	Carpentry—other (wood railings)	$0.00
07000	THERMAL & MOISTURE PROTECTION	$0.00
07100	Foundation waterproofing	$0.00
07100	Foundation insulation	$0.00
07150	Building insulation and soundproofing	$0.00
07600	Gutters & downspouts	$0.00
07700	Flashing (roof)	$0.00
07810	Skylights	$0.00
08000	DOORS & WINDOWS	$0.00
08100	Metal doors	$0.00
08150	Entry door	$0.00
08240	Window screens	$0.00
08245	Storm windows	$0.00
08300	Garage doors (3)	$0.00
08350	Automatic door equipment	$0.00
08360	Automatic gate equipment	$0.00
08750	Weatherstripping	$0.00
09000	FINISHES	$0.00
10000	SPECIALTIES	$0.00
11000	EQUIPMENT	$0.00
12000	FURNISHINGS	$0.00

SUNNYSIDE UP LANDSCAPE ARCHITECTS RESIDENTIAL PROJECT		
13000	SPECIAL CONSTRUCTION	$217,000.00
13100	Sauna	$0.00
13200	Steam room	$0.00
13800	Reflecting pools	$22,000.00
13801	Formal fountain	$25,000.00
13999	Special construction—poolhouse	$170,000.00
14000	CONVEYING SYSTEMS	$0.00
15100	HVAC	$9,900.00
15080	Humidifiers	$2,400.00
15090	Oil tanks (2) in basement	$3,600.00
15010	Propane tank, pipe, and pad	$2,500.00
15110	Oil tank piping	$1,400.00
15200	WELL	$23,550.00
15210	Excavation for well	$12,000.00
15220	Pump, tank, and pipe line to house	$9,000.00
15230	Water treatment	$1,800.00
15299	Well: annual inspection	$750.00
15300	SEPTIC SYSTEM	$47,000.00
15350	Cast iron pipe to septic	$2,000.00
15399	Septic system allowance	$45,000.00
15400	MECHANICAL AND PLUMBING	$11,500.00
15404	Hot water heater	$3,500.00
15405	Septic system hookup	$2,000.00
15408	Water supply hookup	$1,500.00
15409	Pipe insulation	$2,500.00
15410	Roof drains	$2,000.00

SUNNYSIDE UP LANDSCAPE ARCHITECTS RESIDENTIAL PROJECT		
15500	FIRE PROTECTION	$0.00
155501	Fire alarm (see Electrical)	$0.00
15502	Fire sprinkler	$0.00
15599	Fire protection—other	$0.00
16000	ELECTRICAL	$135,500.00
16200	Generator (for fire sprinklers)	$3,500.00
16600	Lightning protection	$12,000.00
16999	Electrical—landscape lighting allowance	$120,000.00
	SUBTOTAL NOT INCLUDING GENERAL CONDITIONS	$3,153,643.30
01000	GENERAL CONDITIONS: 7% of total construction costs	$220,755.03
	Estimated construction costs to date	$3,374,398.33
	Overhead & profit (15%)	$506,159.75
	Subtotal estimated budget	$3,880,558.08
	Contingency (10%)	$388,055.81
	Total Estimated Budget	$4,268,613.89

DOS AND DON'TS FOR COST ESTIMATES

1. Subdivide the cost estimate into the same divisions of labor reflected in the drawings—for example, demolition, pavements, walls, grading, planting, lighting, and so on. Follow CSI format.
2. Many government entities have their own set of specifications. Prior to starting the cost estimate, review with the government agency's project manager or appropriate representative what the requirements for the cost estimate are to be for any particular phase nearing completion. Ask for a cost estimate template or example.
3. Review all the details in the drawings, and verify that there is an item in the cost estimate that represents each detail.
4. Just as the drawings are organized by trades or scope of work, follow suit with the cost estimate. Where possible, there should be an entry or series of entries in the cost estimate that match the scope of work shown on each drawing.
5. Don't create a degree of precision in the cost estimate that exceeds what is shown on the drawings. Use commonsense rounding. For example, for a final cost estimate prior to bidding, consider rounding unit costs of particular plant materials, both the quantity and cost, to the nearest half or whole unit. Once bids are received, the estimate might be recompiled with specific unit prices provided by the contractor whose bid is being reviewed.
6. Use a set of drawings/prints for marking and measuring all quantities which are incorporated into the cost estimate. Show calculations on the same set of drawings, with a notebook of backup if necessary. Prior to finalizing the cost estimate, the staff compiling it should be able to check and verify every item and quantity incorporated.
7. Print the cost estimate at a legible size. As with drawings, anticipate if any reduced-size sheets will ever be generated, and be certain that at whatever size the cost estimate is generated, it will be fully legible to everyone who must review it.
8. Use item descriptions that match specification numbers or other labels on the drawings.
9. Have someone who is not familiar with the project review the entire cost estimate and drawings, to verify that the whole package is coordinated, legible, and understandable.
10. By the time of issuance of the final cost estimate, prior to bidding, eliminate all contingency items. If permitted by the client/owner, include only those that are justifiable for specific reasons.
11. At the same time that you rely on detailed estimates tied into calculated quantities of a number of different components, each measured carefully, learn to do eyeball estimates as a way of checking and verification. Contractors do them all the time. Landscape architects should learn that skill as well.

Problem 1
DETAILED COST ESTIMATE

For a project you have studied elsewhere in this book, develop a detailed cost estimate for two major areas, such as a new swimming pool and renovation of a rock garden, or a new tennis court and a new perennials garden. (Depending on time constraints, the instructor may request one area rather than two, or on your own, for study purposes make this decision.)

Each cost estimate topic should be broken down by categories, such as demolition and construction. Include a contingency of 10%. Eighty percent of items must be based on unit costs measured and calculated. The remaining 20% may include lump sums.

NOTES

1. Jennifer Gray, "Water and Sustainable Design," Sustainable Build, July 10, 2017, http://www.sustainablebuild.co.uk/SustainableDesignWater.html; Jennifer Gray, "Sustainable Sewage Design," Sustainable Build, March 3, 2016, http://www.sustainablebuild.co.uk/SustainableDesignSewage.html.
2. "ADA Standards for Accessible Design," U.S. Department of Justice, Civil Rights Division, https://www.ada.gov/2010ADAstandards_index.htm; Francis D. K. Ching and Steven R. Winkel, *Building Codes Illustrated: A Guide to Understanding the 2000 International Building Code* (New York: John Wiley & Sons, 2003), 135–216 (these two chapters include means of egress, exit access, exits themselves, accessibility, parking and several other topics).
3. Ching and Winkel, *Building Codes*, 3.
4. Ching and Winkel, *Building Codes*, 4.
5. Ching and Winkel, *Building Codes*, 4.
6. Ching and Winkel, *Building Codes*, 150, 159–160, 190.
7. Ching and Winkel, *Building Codes*, 195, 207.
8. Ching and Winkel, *Building Codes*, 207.
9. Some of this section appeared in the *Handbook of Landscape Architecture Construction*, edited by Maurice Nelischer (Washington, DC: Landscape Architecture Foundation, 1986), and it is reused with their permission and updated. Three key references used by the author originally in 1985–86 in compiling information for this section were Robert Abbett, *Engineering Contracts and Specifications*, 4th ed. (New York: John Wiley & Sons, 1963); Jack R. Lewis, *Construction Specifications* (New Jersey: Prentice Hall, 1975); and Harold J. Rosen and Tom Heineman, *Construction Specifications Writing: Principles and Procedures*, now in 3rd ed. (New York: John Wiley & Sons, 1990). See Chapter 10, "Resources."
10. Construction Specifications Institute (CSI), "Standards," www.csiresources.org/practice/standards.
11. CSI, "Standards"; U.S. Green Building Council, www.usgbc.org.

CHAPTER 7

Areas of Practice

Landscape architecture practice today encompasses many different topics, so broad in scope, that some areas of work hardly seem to be related to each other. As with other professions, such as medicine, many designers choose to specialize. There is still a need for general practitioners, but such people must have a wide array of skills to stay abreast of current knowledge, standards, and techniques, and must also be able to shift gears constantly in responding to the demands of one type of practice or another. Like medicine, each specific area of practice has become so demanding and requires such a degree of knowledge and experience that it's increasingly difficult not to focus on a few areas of practice. The risk of such specialization is that sometimes one doesn't see the forest for the trees. There are so many types, varieties, qualities, and textures of "trees" that one no longer perceives the forest. The general practitioner, in whatever field, coordinates and links one type of practice to another. For landscape architects the risk in specializing is losing touch with general design principles at the same time he/she may be gaining additional specialized skills. What follows is a discussion of six major types of practice. The topics are somewhat arbitrary; there could easily be others added. The goal is to describe for the reader the essential requirements and skill sets for each area of practice.

PART 1: SITE PLANNING

Site planning governs how land is used for sites with multiple uses that must harmonize despite different footprints, requirements, and impacts. As development projects have gained in complexity, so has the art of site

Figure 7.0 Sun Trust Plaza, at the intersections of Peachtree, West Peachtree, and Baker Streets in downtown Atlanta at 6 p.m. in June 2009. Is this a successful and welcoming outdoor space?
Photo by Steven L. Cantor

planning. Site planners often start by working at a fairly small scale in order to study sitewide issues, such as grading, landforms, drainage, and land use, and gradually move to a larger scale as the work progresses to smaller parcels of land within a larger development. For example, a landscape architect can study a large tract of land at a small scale for the purpose of creating a residential subdivision, and then, after all large-scale issues have been resolved, focus on an individual lot within it for a single-family residence. Master plans are the bread and butter of site planners. An artfully created master plan can continue to function as a development and design guide for many years after its initial creation.

Kevin Lynch's *Site Planning,* defines it as "the art of arranging structures on the land and shaping the spaces between," a process both so complex and so important that it requires the participation of architects, engineers, landscape architects, and city planners. It combines technical skills with a moral aim in order to create spaces that enhance everyday life.[1] As our understanding of the world increases as a result of scientific knowledge and more advanced technology, it is incumbent upon us to be as wise, prudent, and practical as possible in making decisions about complex and diverse uses of valuable and often irreplaceable land and resources. How can landscape architects use site planning skills and processes to achieve the best possible results? How can professional practice be both a design method and practical guide to achieving both an esthetic and practical goal?

Design Process

There are many paradigms for the design process; the renowned planner Kevin Lynch, for example, cites eight.[2] But one that is typical incorporates four interactive phases: site analysis, programming, synthesis, and implementation. In our modern world, there has been a proliferation of tools and techniques to allow and encourage more and more detail and precision in measuring and analyzing the landscape. It is incumbent upon design professionals, a moral imperative, to fully utilize the tools and expertise at our disposal. So the challenge is to pick the correct seeds to bear fruit, and not overlook important sources of information, necessary measurements, and scientific inputs as well as esthetic concerns that might contribute to a complete and well-informed result. So how can you, as a project manager or other major player in a design project, contribute to an effective solution?

Site Analysis

The site analysis must respond to the intended use. In a world in which technical data is widely available in digital formats, there is perhaps a risk of too much information for analysis. The scales at which the site will be studied for particular design applications may suggest some of the analyses that need to be carried out. For example, a 200-acre site for a golf course, where much attention must be paid to the contouring of the land and to technical grading standards for tee areas, fairways, and greens, and in which substantial grading may be necessary to achieve challenging pars with deftly located sand traps, might need to be studied more closely than the same site if it were to be used for a soccer or baseball field. The need to balance cut and fill on this site might dictate the degree to which site analysis would examine the types of soil horizons and whether there would be unsuitable soils to haul off and the concomitant need to import specialized soils or other materials for greens and sand traps. Information that used to require hours of digitizing or hand drafting, like calculating and generating slope maps, can now be located and analyzed routinely. The challenge is to focus the analysis on the slopes that have the most relevance to the project functions anticipated for the site.

The design process involves defining the problem, refining and developing a design through various schematic, conceptual, and design development drawings, and then creating a set of final drawings. Multiple rounds of testing various creative design solutions will help ensure the best possible result.

Programming

Programming requires the designer to research the needs of the client for a particular site and convert these requirements into spatial characteristics. Corbusier is credited with saying, "To design requires talent, to program requires genius."[3] The more detailed the architect or landscape architect can be in developing the program as a menu of spatial requirements to fit skillfully, artistically, functionally, and practically onto the site, the more effective the design may be. The more detailed and precise the development of the program to be applied to the site, the more that the design issues can solve themselves. A solid program leads to clear design. Programming may also involve doing enough review and testing to verify that the spatial requirements of the list of activities to be accommodated on the site are too large to fit comfortably. In such cases, the landscape architect must find ways of helping the client set priorities among the wish list, and reduce in size or eliminate some items on the wish list so that there is still an effective design response and a result the client will embrace.

There is a wealth of data available from the sciences: ecology, biology, botany, geography, and others. It must be reviewed, analyzed, edited, and applied as appropriate for the site being studied and its anticipated uses. The skill involved often has to do with learning and knowing what data is important and what should be emphasized, as well as what can be thought of as background and extraneous.

Site planning is also an art, as the designer must always be aware that even as abstract forms are being studied, they will be realized on the site in real dimensions. More than other smaller-scale types of design, site planning requires constant shifting from the large-scale and general to the smaller-scale and specific.

In some ways having a career as a landscape architect implies a moral imperative for the best use of the land. (See also the section on ethics in this book, Part 1 of Chapter 9.) Since site planning projects often encompass large tracts of land and may involve long-term decisions about land use, as well as some types of design that may be implemented in a first phase of construction, the moral imperative gains in significance. If poor decisions, lack of understanding of natural systems, a design that does not fit the site, and poor-quality grading and construction all occur for one single-family residential site of a few acres, at least this scar might be ameliorated by implementing a cohesive design that responds to the site in adjacent lots. When these same problems occur over a much larger site, the impacts are obviously worse.

Design Principles

No matter how complex the site and its challenges, honor time-tested principles of design: unity, variety, diversity, contrast, symmetry, asymmetry, rhythm, harmony. These terms have been part of the vocabulary of design for centuries or decades, yet they have survived because they still have unique applications. Unity is achieved through repetition of the same element. Variety and diversity are achieved by contrasting with unified elements. Symmetry and asymmetry are achieved by different ways of organizing spatial relationships according to a central axis. Rhythm implies the character of movement over time across a site or within a design, and requires close attention to the spacing of its key elements. Finally, harmony suggests that multiple elements are in equilibrium, resulting in a sense of balance.[4]

Since site planning involves working at a larger scale, the landscape architect is often involved with studying systems, such as circulation and utilities, at a scale that is not considered in more detailed, smaller sites. The landscape architect must often draw upon the skills of specialists who may be expert in a particular application, such as horizontal and vertical alignment of roads and paths, or large-scale parking, or a storm drainage network. There is also consideration of the materials to be used in implementing the design as it starts to unfold, and often a hierarchy is developed so that different uses and functions are associated with a particular material. Major roads may be wider and paved with concrete; secondary roads may be asphalt, minor roads may be gravel.

Study Costs

Another aspect of site planning involves a detailed study of potential costs, which become critical as informed decisions are made about what components of a site plan will be constructed. The costs of using particular materials such as types of paving, the sizes of trees, and the extent of grading can be studied over an entire site, and contribute to decision-making about whether to implement an entire system or just one part of it.

DOS AND DON'TS FOR SITE PLANNING

1. Avoid rigid criteria. No matter the depth of an analysis, its data is only as good as the science it is based on. Remember that hundred-year storms can occur in consecutive years. The site contour delineating the maximum elevation of flooding for a hundred-year storm within an urban region in which a lot of development has occurred upstream may be revised as early as the next year.
2. For each site, consider which site analysis and design criteria are the most important; don't treat them all as equal. Consider also if some are interdependent.
3. If small is still beautiful, then restraint is wise. Be wary of pushing the carrying capacity of the land to its limit. If there has been even one miscalculation or misplaced assumption, what will the impact be?
4. Study each system—circulation, utilities, planting, and so on—independently, but also in terms of how it interacts with other systems.

5. Implement a modest first phase that allows for testing how to improve subsequent phases of design with reduced impacts and lower costs. A bold design ineptly or imperfectly implemented over an entire site is potentially disastrous. Another way of thinking of this: to try and fail is to learn, but not to try is to suffer the inestimable loss of what might have been. In New York City, the famous High Line park was implemented in three phases. This was necessary for fundraising but had the advantage that the later two phases benefited because planners were able to adjust some design details of the first phase that were awkward.
6. Give emphasis to sanitary sewer systems. If septic systems are involved, determine where additional septic fields could be located in the event of expansion of the sanitary needs (for example, if another story is added to several buildings on a site, with additional bathrooms) or if, for reasons yet to be determined, a site selected as a leach field is found to be no longer suitable. If the design will tie into existing sanitary sewers within a governmental district, be certain that there is available capacity and consider whether some backup system is needed in the event of sanitary/storm sewer overflows or temporary breakdowns in sewage treatment.
7. Study carefully all the major forms conceived of in the design and verify that they reinforce the program and patterns of the proposed use of the site.
8. The more complex the design, the more critical it is to share information and be certain that all key designers are on the same page.
9. Even if one designer or famous-name personality is in charge, collaborative efforts are essential. So many specializations are required that it's rare for one person to have a comprehensive knowledge of every discipline that needs to be studied in relation to a particular site and program.
10. Take notes and keep minutes. A record is needed of key decisions made, the design criteria being applied, the evolution of the design. No one's memory is perfect; to at least have a summary of what is achieved and agreed upon at each major decision point is critical.
11. Know the contour signatures of major landforms, such as ridge, valley, and convex and concave slopes, and find ways to integrate them into the design. Studying sections of USGS maps, perhaps enlarged or reduced to the scale at which you're working, is a good method to understand how the vocabulary of contour signatures fits together.
12. Create graceful and bold landforms that interconnect. The scale of site planning allows the development and refining of grading over a large enough area to sculpt landforms.

Beijing Vanke, China

Landscape architect Lihong Zhang has designed complex projects throughout China, such as large subdivisions in Shanghai, community development in difficult sites, urban design, and recreational facilities. He has extensive training and experience in the United States as well. The project shown here is typical, in that his scope of work is conceptual design, design development, and extended design development drawings (construction drawings without engineering input) for a high-end condominium community development including clubhouse landscape, commercial street landscape, and all open spaces within the community. (See Figures 7.1a and 7.1b).

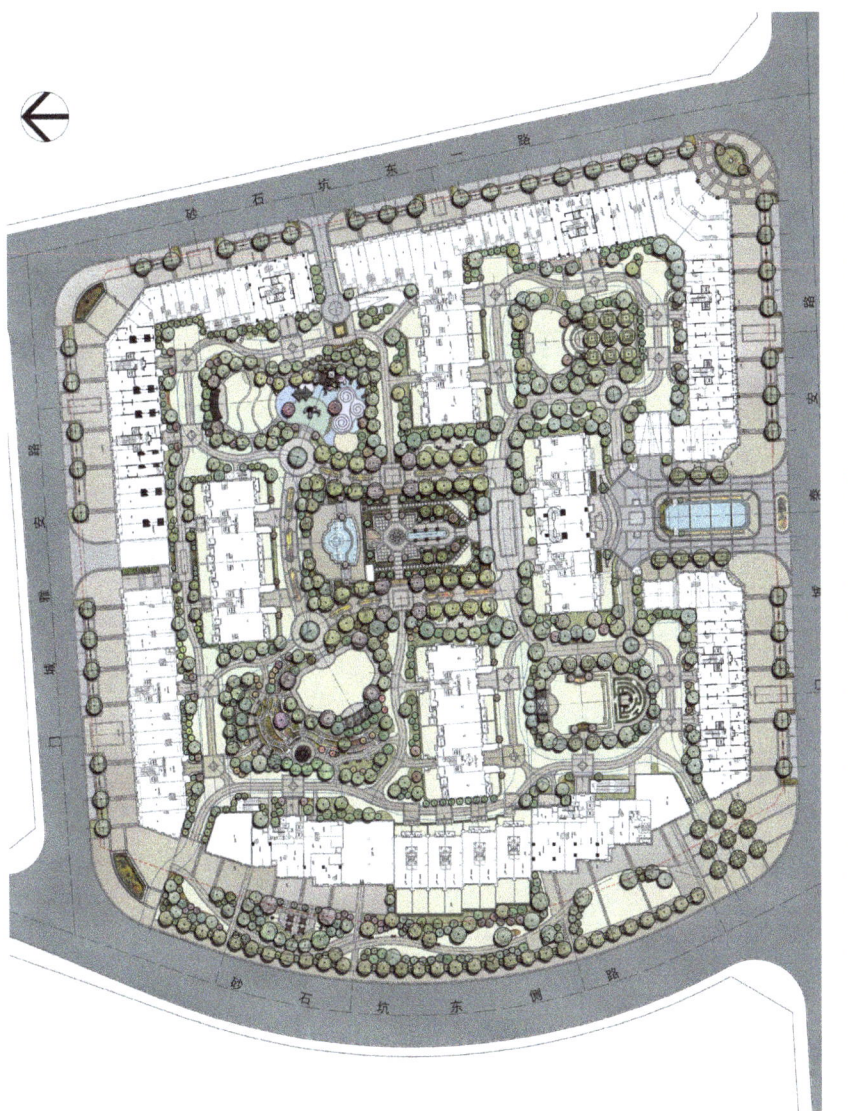

Figure 7.1a Site plan for the Project: Beijing Vanke - The Legend of Chang An in the Yong Ding, Men Tou Gou District, Beijing; the project area is about 12 acres.
Lihong Zhang

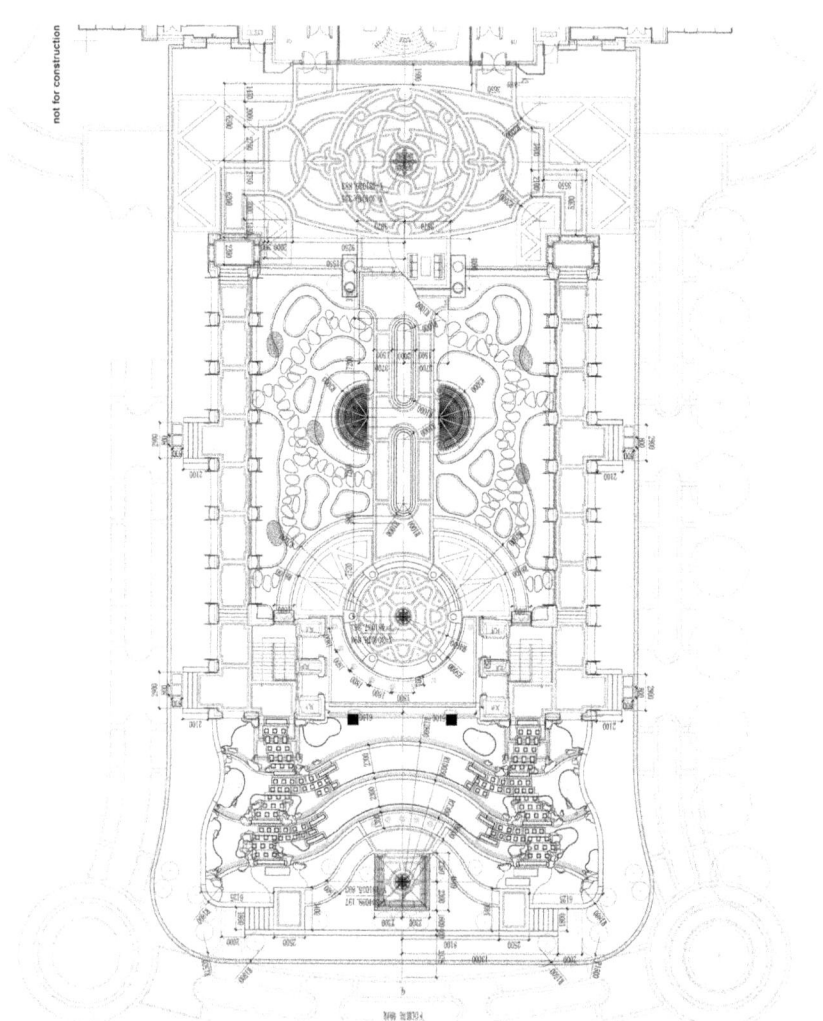

Figure 7.1b Layout plan of a detail area enlargement at metric scale
Lihong Zhang

Problem 1

THE SHIFTING ONE-HUNDRED-YEAR FLOOD PLAIN

(In the following problem, any *terms in italics* may be changed to suit the particular instructor or team leader, or your own personal, work or studio setting. Typical changes might be specific people, such as clients or contractors, plant materials, or the locations of sites, people, institutions or companies.)

Create or find a map of a site up to 4 acres (1.6 ha) in size in which the one-hundred-year storm contour line is present within the topography. This contour could be part of the edge of the *floodplain of a lake*, or it could be some distance away from the line of flow of the current of a *major stream or river*, or it could be a drainage easement that finally meets a *lake or river off-site. The site may be anywhere as long as it complies with these and other conditions as noted.*

Highlight in a bolder line or even with color the hundred-year flood plain contour. The land uphill from the water body should gradually rise another 20 feet (6.096 m) in elevation within the site at a gradient not exceeding 10%.

Site a *single-family residence* or *other building* on the site and show the driveway, a terrace, and a *few amenities, such as gardens and outdoor informal lawn areas* for recreation. The finish floor elevation (FFE) of the residence should be 10 feet (3.048 m) above the contour of the hundred-year flood plain. If the building is not a residence, site *other appropriate amenities or ancillary facilities*.

Assume that upstream of the site there is a lot of dense urban development, resulting in large areas of impervious pavement and increased runoff into the stream, river, or lake. The Army Corps of Engineers once certified the elevation of the hundred-year flood plain contour in relation to any water body over which they have jurisdiction. Now this is the purview of the Federal Emergency Management Administration (FEMA). Assume that every ten years the hundred-year storm elevation is revised upward 1 foot (0.3048 m) vertically.

Shade in on your site plan the increased area of flooding to the hundred-year storm for the next fifty years in ten-year intervals. Then do one last map showing the elevation of the hundred-year storm flood plain contour and the area affected after a hundred years.

Look again at your site plan for the residence. Based on the flooding criteria, what could you do initially as a landscape architect and with the assistance of an architect and/or engineer to mitigate the impacts of the rising hundred-year storm line?

PART 2: SINGLE-FAMILY RESIDENTIAL DESIGN

History

Landscape architecture in America began in the nineteenth century with Andrew Jackson Downing's efforts to standardize an aesthetic for residential estates and to encourage the public, or at least those well-off enough to own property, to enhance their properties. Downing recruited Calvert Vaux (1824–1895), a British architect and landscape designer, to come to America and collaborate. Had Downing (1815–1852) not died tragically in a riverboat disaster in which many leading citizens drowned, his career no doubt would have had greater impact. Even so, he set the stage for a new industry. Prior to his death, Downing introduced Vaux to Frederick Law Olmsted (1822–1903), whose career was still beginning. Much of Olmsted's work to that point had been journalistic, including his long accounts of travel in the southern states, in which he wrote about the impacts of slavery. The Greensward Plan, developed by the experienced Vaux and the novice Olmsted in 1858, was announced as a winner of a design competition for what became Central Park. This ignited both of their careers, and they collaborated on many spectacular projects until Vaux's death. They pioneered a naturalistic and picturesque style of landscape for parks and institutional settings. Vaux provided elegant architectural designs for buildings, bridges, fountains, arches, and other built features of their many collaborative efforts.[5]

Contemporary Landscape Design

Today's modern landscape architect is indebted to both. Olmsted was prolific and worked on projects of almost every type: large city parks, national parks, university campuses, world's fairs, residential estates, and subdivisions. He delved into just about every aspect of the developing profession. Yet in the twenty-first century, increasingly landscape architects tend to specialize. Residential design is the bread and butter of many firms and individuals. Often many of these designers do not delve into other applications. Similarly, many practitioners who concentrate on public works and public projects tend to avoid private clients. One reason seems to be that, as in medicine, every type of work is exceedingly complex and a specialist is needed, so "general practitioners" are unlikely. Another reason is simple economics: to be profitable in a competitive economy, each firm must focus on what it does best. Interdisciplinary firms that employ architects, landscape architects, and engineers seem to be giving way to firms specializing in one of the major disciplines.

The following discussion will focus on residential landscape design and also touch on some of its requirements that make crossover work into the public sector (and vice versa) less likely.[6] Landscape architecture for residences typically includes two areas: new homes and changes to existing homes. Two corollary realms are residential subdivisions and apartment complexes, in which the landscape architect studies all at once at a larger scale the intricacies of laying out a road system with building lots and siting a whole series of single-family residences at once or dealing with the needs of a large group of residents in multistory buildings, but my focus here is on single-family residences. The landscape architect may be a consultant from the beginning, when a client is purchasing property for a residence, and work in tandem with the architect in developing a comprehensive design. The other typical option is that a homeowner wants to expand the home—for example, adding architectural elements such as a recreation room or a den, or an addition for an elderly parent, with concomitant amenities such as gardens. Such changes can be driven also by a desire for additional or different use of outdoor space—for example, the client wants to add a swimming pool and cabana, a tennis court, a specialty garden, and so on, which may or may not include additional architecture. Often these expansions to existing properties occur after the sale of a home to people who envision additional uses for the property compared to those who previously lived there.

Sequence of Services

There are of course many different ways that a new residential property, addition, or renovation may take shape, and it is beyond the purpose of this book to go into great detail about design styles and approaches. However, for anyone considering residential landscape architecture as a career, it's important to emphasize the process. Perhaps, some of the most important aspects are communication and programming. If the landscape architect perseveres in developing and maintaining clear communication with the homeowner, whether an individual, a couple, or another combination of people, so that it is clear to all what the owner desires for what cost, then almost everything else can fall into place with time, patience, and the use of skills in landscape architecture.

A typical sequence of work is the following:

> *Initial meeting.* Discuss the scope of services with the prospective client and develop an agreement (contract) to work together.

Survey and site analysis. Evaluate the site. (Please see "Part 1: Surveys" in Chapter 5.)

Programming. Discuss what the client wants and at what cost.

Design. Develop conceptual, design development, and construction drawings with cost estimate and specifications.

Implementation. Guide the implementation through construction inspections and additional meetings. (Please refer to "Part 4: Site Inspections" in Chapter 8.)

The impetus for an initial meeting may begin with an email, a phone call, or a reference. Typically, the landscape architect meets with the prospective client, shows some examples of the firm's work, develops a preliminary understanding of what the client is seeking and the budget the client has available both for design services and for implementation, and an agreement is reached for a scope of services spelled out in a letter or contract. At this initial meeting, the landscape architect learns which other consultants, such as architects or engineers, may be involved and who will be the lead designer, and starts to establish rapport with the client. No further work is done until there is a contract for services signed by both parties. Often a retainer fee is due upon signing of the contract so that the landscape architect has some leeway and flexibility as design services begin to gather information, think of different design possibilities and coordinate with other members of the design team. Even if there are other consultants involved, it is best that each have a separate contract with the client, so that the landscape architect is caught up only in payment for the firm's services and not the services of others as well.

Survey

The quality and level of detail of the survey should be based, even at an early stage in design, on what the landscape architect anticipates or believes, based on preliminary discussion or phone calls, that the client will want to implement. If the landscape architect expects primarily a landscape development project, with a lot of gardens but minimal impact on the residence, and minimal grading, drainage, or construction, then a great deal of topographic information may not be necessary. If underground utilities are present, they should be located, as it's urgent to avoid excavating and exposing utilities that serve essential functions. The architecture of the home should be located, but to pin down every window and room may be more than is necessary. Similarly, if the existing trees form a forest edge or continuous canopy that is beyond the bounds of where the gardens are to be developed, then to have the surveyor locate each tree and identify its caliber and species may not be necessary, and just a detailing of the location of the forest edge or the first trees in that forest edge will suffice. On the other hand, if a new series of gardens may include some of the specimen trees on the site or the edge of the forest, then their location becomes critical.

Similarly, if from the beginning it is realized that the client wants to add a swimming pool, tennis court, or other major use to the site, then much more information is needed on the survey, including detailed topographic elevations and contours, plus above- and below-ground utilities. Since many pool designs depend on their integration with the residence, so that people inside don't necessarily see the swimming pool but sense once they step outside how to gain access, then detailed information about the finish floor grades, doors, and windows of the residence, particularly on the side that will be the connection to the pool, would be important to pin down, as well as any major trees near the anticipated site of the pool or other major features being proposed. A landscape architect experienced in a particular type of practice will develop over time a menu or checklist of all of those site features and other data that should be included in a topographic survey for a residential design project. At the conclusion of each phase of such a project, someone should conduct a review to determine if any important data is missing and can still be obtained, and to determine whether this data is unique to the particular project or should be added to the checklist for future residential design surveys.

Preliminary meetings with the client are very important, as the landscape architect will want to determine whom he should talk to, discuss the design with, and send invoices to. Sometimes there is a division of labor, where one of the homeowners is the person amenable to design discussions and the spouse or partner pays the bills. It is important to define roles clearly. Then have a discussion about what the client wants; keep it all-inclusive, although it's fine to draw out what the client's priorities are. Generate as precise and specific a wish list as seems reasonably possible, as the more thoroughly one can describe, understand, and explain options, the more effectively the design can respond to the program. The final program should include not only a list of functions but also their spatial characteristics and a clear sense of how they might relate to one another. Functional relationship diagrams are an excellent way to depict graphically the required relationships that ideally should fit on the site.[7]

Based on this meeting, tailor the survey to what you know about the client, and then use a few prints of it to take to the site in order to analyze its principal features, highlights, and difficulties, and even start to sense where different proposed elements might fit. This is also the time to realize that some of the client's preferred activities may not fit on the site at all or without major changes. For example, a tennis court might fit but would require regrading a large hill, with the impacts at a minimum being some massive retaining walls and considerable grading. Or either a swimming pool *or* a tennis court would fit, but not both. There may be room for some parking but not all of the additional spaces the client wants.

Some conceptual plans may be developed to show the client various options. Another meeting occurs to show these options, at which time the client selects one preferred option or a combination of a few, and then the design moves forward with more emphasis, detail, and clarity, and with an approximate budget in mind. Discussion must occur about preferred materials for walks, pavements, parking, various amenities (clay or paved tennis court, the stone or concrete for retaining walls, swimming pool terraces and finishes, etc.) and of course the planting. As the design process ends, the client will be presented with a set of construction drawings, showing demolition, grading, layout, details, and plantings for all of the preferred or first-phase elements in the program. With the client's final approval, the project goes out to bid. Sometimes in lieu of bidding, the client gives approval to negotiating directly with a preferred contractor as long as the final cost for construction is within a range that has been agreed upon. More typically, particularly for projects over a certain minimum budget, there is bidding. The landscape architect reviews the final bids and negotiates on behalf of the client with the preferred bidder (sometimes this is not the low bidder) and a schedule for implementation is set. Usually there is separate negotiation with the client by the landscape architect for site inspection services and visits to verify that the design is being implemented in accordance with the drawings. Other tasks during this phase may be review and approval of shop drawings, tagging of specimen trees at nurseries, and review of mock-ups of sections of retaining walls, pavements, or other elements of the design prior to implementation of these elements for the entire site. The specifications will usually spell out a pre-final inspection and final inspection both for construction elements and for plantings. At the conclusion of the final inspections, a guarantee period starts, and there may be provisions for a company to take over long-term maintenance of plantings or other elements on the site. Occasionally the company that does most of the construction and planting may have a vested interest in continuing to maintain the property, but usually the contractor is more focused on other projects, so the goal should be to have a seamless transition between the contractor of record and maintenance by the company hired at the end of the process to take over indefinitely for maintenance as the original contractor withdraws.

Sustainable Design

Do not overlook modern trends; perhaps the most important is sustainable design. Rather than just reducing heat loss and the need for air-conditioning through the use of the best insulating materials and the most efficient air conditioners, it is now possible to make the entire residence have a net energy usage of zero. Although advances in the design of conventional septic systems allow them to be placed on sloping sites and within sparsely forested areas and to last longer, there are now methods of natural filtration through gravel and sand beds, lagoons, and vegetation, along with the complete recycling of gray water for irrigation and other uses. Composting toilets are common. Landscape architects must continue to prod clients to seek out sustainable design solutions in residential design, particularly where there may be ample space to try some innovative solutions.[8]

So, what can go wrong? What professional practice headaches may occur, and what can be done to minimize them?

DOS AND DON'TS FOR RESIDENTIAL DESIGN (PROJECT MANAGEMENT)

1. Research and understand all code requirements and make it clear to the client from the beginning what is permitted and what is not permitted. Often the setback requirements can remove substantial areas of the lot from use for important functions like parking, swimming pools, and tennis courts.
2. Be willing at an early stage to tell the client that there is or is not room on the site for all that is requested.
3. Do not allow the client to anticipate for a long time that all the facilities desired can somehow be accommodated when it is clearly not possible due to simple rules of geometry and setback requirements.
4. In verifying the survey, pay careful attention to the species, size, and condition of all major trees. If necessary, hire an arborist to evaluate any trees about which there might be questions. The time to realize that a garden planned around a major oak tree is not manageable is before any time has been invested in designing it, not after design drawings have been completed and it's determined that the tree is hollow and on the verge of collapse.
5. Develop and document all communications with the client. This may be harder than it first seems. If, for example, one family member is primarily involved in attending all design meetings while another family member pays the bills, each nevertheless needs to be

informed of decisions and costs. It is prudent to include all responsible parties at some of the meetings, so that any key individual does not find him or herself confused or uneasy about what has previously been agreed upon. Concise sets of minutes documenting the topics of discussion and decisions reached are critical to ensure the continued smooth progress of the work.
6. Keep track of all proposed costs of construction. At a preliminary level of design, it is desirable to show or at least discuss options with a range of costs—for example, a garden with primarily larger-size trees and a few options with smaller sizes of plant materials—in order to pin down the budget constraints for each phase of the work.
7. Try to gauge as the work progresses the degree to which the client can understand and appreciate the content of various drawings. Sometimes perspective drawings are very helpful and are more readily understandable to a client than even well-rendered sections and elevations.
8. Consider building a study model of critical areas, especially if the client is having difficulty deciding among options or does not seem to be able to visualize what is proposed.
9. Develop lists of construction materials, finishes, styles, and plant materials that the client embraces and those that the client abhors. If, say, the client has a preference for blue, purple, and white blooming plants and objects to yellows, follow those parameters. If the client prefers bluestone paving to granite, follow that preference.
10. Don't promise the moon (even if you're designing a moon-viewing garden). Tone down statements or expectations of fabulous results. Even the greatest designs can take some months or years to be realized, as it requires time for construction materials to weather, for plant materials to grow, and for other elements, like lighting, to be adjusted in the field.
11. Keep track of costs, both for design services and for what is being implemented. It is possible that the costs may exceed what is initially discussed. Yet with the support of documents, such as minutes of previous meetings, it's usually possible to document the sequence of events that has led to the potential for a large cost overrun, and reach a decision about whether to proceed or pull back.
12. Understand and explain to the client the sequence of operations that must occur on the site, and why. The client must approve this prior to the start of the work on-site. At an existing residence, homeowners may have all sorts of plans for activities in a certain week or month that will conflict with the work of contractors. The contractors must schedule the work based on a clear knowledge of these impediments.
13. For major items of construction, such as large terraces, long walkways, swimming pools, retaining walls, and so on, incorporate language in the construction drawings and specifications that requires shop

drawings and mock-ups of sample sections for review by the designer and client. The shop drawing gives the design team one last chance to review the intricacies of particular design materials, patterns, and how things fit together. The mock-up gives the client one last chance to see what something will look like. The time to find out that the client does not at all like the pattern or range of hues of a bluestone pavement pattern is when the sample stone has been delivered, or at the latest after a small sample has been built, not at the point where the stonemasons are finishing the installation of the last few pavers.

14. Require the client to try out with their vehicles staked-out mock-ups of parking and driveway areas. These can easily be adjusted while they are being laid out but become a challenging difficulty if the pavement has already been built.
15. Have regular meetings on the site with all contractors for coordination. Encourage the client or a representative to attend all of these meetings.
16. Use some color-coordinated system to stake out all existing and proposed utilities. Prior to actual construction, there should be an evaluation to verify any potential conflicts and decide the safest way to resolve them.
17. Don't joke about any design or construction topic that might be misinterpreted, no matter how obvious you think the humor is. You may end up with something larger than a foot in your mouth.
18. Protect large existing trees anywhere near construction areas with solid wood or other fences that will deter anyone and their vehicles or construction equipment from approaching and will ensure that drivers feel resistance if they encounter the fence. Ideally the barrier should coincide with the dripline of the tree, but if this is not possible, at least protect a minimum radius underneath the tree. Include the services of a certified arborist if there is any dispute about the area to be protected.
19. Sometimes drainage problems do not show up until demolition or grading operations have started. Try to anticipate where problems might occur. Incorporate language in the specifications to provide for this situation.
20. Some jurisdictions will define wetlands, easements, or other zones in which any disturbance, grading, parking, or storage use is forbidden. The boundaries of such areas within the site should be clearly marked and staked.
21. For pavements adjacent to one another, forming continuous areas, such as asphalt pavement, granite curb, bluestone pavement, and concrete sidewalk, excavate to the same depth so that there is a uniform impact of frost heaving and settling. If the subbase for one material is excavated a few inches more or less than for an adjacent one, undifferentiated settlement may occur, resulting in uneven surfaces.

22. The development of a set of as-built drawings is critical. Often the locations of critical utilities must be adjusted in the field. Sometimes specimen trees, walks, or other elements are placed in locations other than those shown on the drawings in order to avoid obstacles or utilities. If some emergency arises years later and the contractor who installed the design does not have total recall, if he even happens to be available, precious time can be lost and dangers created by having to excavate or search in the wrong location. A set of accurate marked-up drawings should be kept in the field by the general contractor or whoever is in charge, and at the end of the project these should be finalized and stored as archival records, so that they can be referred to if a problem arises.
23. Provide or require in the specifications some way to require ongoing maintenance after the guarantee periods in the construction contracts have expired. The specifications may include such requirements, so that when the project goes out for bids the client can receive an estimate of the monthly maintenance cost. The specifications should also require the creation of one bound document, perhaps in a three-ring binder, that includes all the manuals and operation requirements for all equipment to be used on the property for ongoing maintenance operations.
24. With each new project undertaken and completed, keep a checklist of problems and their solutions, from things as simple as a plant material found to be susceptible to a disease or insect to changes in specifications that offer better protection to the client or greater ease of installation. Keep updating specifications to incorporate such changes.

DOS AND DON'TS FOR RESIDENTIAL DESIGN (DESIGN CONSIDERATIONS)

1. Many single-family residences have a blocky garage facing the street with a wide, straight driveway providing the vehicular access. If at all possible, and in consultation with the architect, try to avoid a site placement where vehicles entering the property drive in a straight line to the garage. If there is room on the property, a parking court and entry to the side is far preferable. Whenever possible, and when the dimensions of the site permit it, locate the entrance to the garage on the side so that the driveway curves toward the entrance and leaves an attractive architectural facade that can be enhanced with planting. The plantings can help frame a view toward the main entrance or other principal features of the architecture.

2. In siting swimming pools, remember that except in some southern or tropical climates, it is rare that pools will be in use all year round. Don't make the swimming pool the centerpiece of the landscape design. Who wants to have a Thanksgiving party looking out on a pool cover?
3. Carefully analyze parking needs. Some large families may want to accommodate one car per person plus a few more for guests, so parking spaces add up quickly. Consider where some parking overflow areas might occur, perhaps not fully paved or treated with some sort of checkerboard grid that will allow for the weight of the vehicles while also supporting grass or groundcover.
4. Remember that 90-degree parking is the most efficient, but varying degrees of one-way angular parking are useful when there are long, narrow spaces or other spaces that allow for spreading out the parking function.
5. Give some emphasis to asymmetrical lines. People don't typically live in symmetry. One side or end of a swimming pool likely needs more use than another. Each side of a residence will have different types and intensities of use.
6. Limit the areas and expanses of lawn, and encourage the use of mulched areas, wildflower mixes, and meadows, all of which have a much lower need for irrigation.
7. Pay attention to transitions, where *one type of material* meets another. Avoid a uniform width of pavement, sidewalk, or stepping-stone path where it meets another usage. Contrasts in adjacent pavements are important.
8. Be certain that there is at least a 1% minimum gradient away from the residence in all directions. If berms or mounding occurs for screening or aesthetic reasons, be certain that on the side of the landforms nearest the residence that there is adequate space for swales and drains.
9. Even for a small family, but of course more urgently for a large family, design clear separation from conflicting uses. Don't put the kids' playground or basketball net next to the kitchen garden, where stray bouncing balls risk turning ripening tomatoes into tomato sauce.
10. Even if there will be frequent movement from the residence to certain outdoor spaces such as a terrace, consider a few steps down to it from the residence, so that the house is not constantly victimized by wind gusts blowing in leaves, debris or rain.
11. Build steps of contrasting materials from the walks leading to them. Yes, one can use a system of both concrete steps and sidewalks, but varying the material makes each show up more clearly in the landscape, all the more so if it's illuminated.
12. Coordinate with the architect and engineers so that control switches for outdoor lighting, irrigation, stereo speakers, and gas grills are located in one conveniently accessed secure space inside the house,

preferably with clear views so that the person activating any of these functions can see them start to operate without having to go back and forth indoors and out.
13. Place pool equipment such as pumps in a location where the noise they generate is absorbed by mounding, vegetation, or acoustically absorbent walls. At the same time, they must be located so that they are readily accessible for maintenance.
14. Whether there is a series of individual air-conditioning units or a larger cooling tower, be careful where it is sited, so that both its appearance and the noise it produces are screened. Again, access for maintenance must be considered.
15. Turnaround space for delivery vehicles must be accommodated. If space permits, a loop driveway/access is preferred. If a loop is not possible, then design some manageable way for a delivery truck to arrive and drop off packages, furniture, or equipment without having to back out of the property.
16. Swimming pool sites with full sun are preferred. Tennis courts or other court games may benefit from some shaded areas.
17. Inquire about and write down the client's preferences about plant materials. If there is a strong preference for white, yellow, and blue flowers, don't create a focus of purple.
18. Know what allergies to plants or insects any member of the family or permanent staff has. Aside from obvious plants to avoid, like poison ivy, if anyone has allergic reactions to bee stings, then great care must be exercised to avoid plants which are strong attractors, at least in some areas of the garden or perhaps the entire site, depending on the age of the person and the seriousness of the risk. It is common to have special play areas for children, so it's all the more important for these areas not to include poisonous plants, no matter how decorative or beautiful, the blooms or fruit of which children might try to eat, and to avoid plants that could attract bees or other dangerous insects.
19. Lyme disease and deer depredations have changed lifestyle habits and patterns of use in the American landscape, particularly the Northeast. Consider carefully whether and where to include deer fencing and how to minimize casual exposure to deer ticks.
20. Family pets are still major figures in a household. Anticipate how and where to place fenced enclosures for animals, and how to prevent their exposure to ticks, fleas, or (in some locations) coyotes or other dangerous animals.
21. The use of exotic materials may be intoxicating but is often problematic. Verify that any unusual material can be tested according to clear standards and the results verified prior to incorporating into a project. Be aware that what is classified as "invasive" varies from jurisdiction to jurisdiction. Imported stone from remote regions, no

matter how beautiful, comes with risks: what happens if it arrives in shattered pieces or, once installed, does not have adequate structural strength?

22. Some clients are very attuned to sustainable design practices. In such cases, from the beginning consider a range of options, dependent on your own priorities as a landscape architect as well as the client's interest, from a LEED certification process to the incorporation of particular elements such as green roofs, bioswales, rainwater storage, drip rather than spray irrigation, and the use of photovoltaics. It is now possible to make the entire residence have a net energy usage of zero. See http://www.passivehouse.com.

23. Many plants are drought-tolerant once established, but very few are when first planted. A week-long period of high temperatures and no rainfall, which is not uncommon in most temperate climates, can entirely desiccate a new garden. If an irrigation system is not provided, at least install ample spigots for manual watering with hoses and quick-coupling valves for connecting the hoses.

24. Barring the highly unusual project in which a client wants a rigid rectilinear geometry, develop a composition mixing curved and straight alignments for the driveway, the edges of the lawn, and planting beds. The ideal way to integrate a formal architectural element into a natural or created landscape is through the use of curves and organic shapes.

25. Pay close attention to transitions, where one *design element* meets or embraces the next. For example, an expanse of lawn that aligns with or is parallel to a face of the house or architecture on the side closest to it, but then curves or evolves organically farther away from the residence, is quite effective.

26. One challenge of residential landscape architecture is to create smooth, compelling, and enchanting sequences of movement between indoor and outdoor spaces.

27. Don't overlook time-tested, traditional methods of achieving balance. The concept of the golden mean (a ratio of 10:16), dating to the ancient Greeks, still achieves graceful and attractive forms, whether for a terrace, an architectural element, a swimming pool, or other major features.

28. Develop and keep testing through practice a vocabulary of minimum and maximum gradients for your residential landscapes, including lawns, pavements, cut-and-fill slopes, and so on. Some typical examples are the following:
    Lawns. Maximum 1% to 3% for formal lawns for games and entertaining, 5% to 8% for transitional areas.
    Pavements. Maximum 1% to 3% for formal spaces and sidewalks; steeper, up to 8%, for access routes and connecting systems.

Minimum gradients for formal lawns and pavements tend to be about 1%, occasionally less down to about 0.5% if exacting controls can be maintained to minimize puddling.

29. Become well versed in the requirements and standards of the Americans with Disabilities Act. The population is aging, so demand will increase for housing and activity areas for elderly residents. Just as architecture may accommodate them with single-story buildings with minimal elevation changes, and smooth transitions between inside and outside, landscape architects must anticipate carefully designs of carports, landings, terraces, walks, and other elements. (See the ADA criteria spelled out in Chapter 6.)
30. Protect and save all major existing trees wherever possible, and try to make them part of the final landscape.

Barlow-Hill Residence

The Barlow-Hill home was originally built in 1903 and was featured in a book on Hamptons houses published in the 1980s. Like many of its vintage, it was added to over the course of many years, and after almost a century of use, it needed a firmer foundation to accommodate upper stories. Architects Eric Woodward and Rich Anderson of Southampton and builder and engineer Ken Wright of Bridgehampton reinforced it from below with a new steel frame, while meticulously maintaining its historic exterior. The entire house was shifted approximately 12 feet to accommodate a carport. A rear courtyard and light well with concentric planting beds were developed to bring light to the new basement levels. While the entire home was renovated, most of the mature plantings were heeled into a temporary nursery at the edge of the site and then relocated under the direction of landscape architect Steven L. Cantor, with the expert assistance of landscape architect Christina Lynch. Mr. Cantor replaced the single-lane, back-out driveway with an expanded loop drive with overflow spaces, and new plantings were added by the designers to harmonize with the transplanted materials. The single largest transplanted material was a saucer magnolia, *Magnolia soulangeana*, although the exact locations of the transplants shifted somewhat to accommodate the new location of the residence. (See Figures 7.2 to 7.11).

Figure 7.2 The 100 year old house was propped up on a steel frame prior to its being moved and to allow for the construction of a new basement.
Photos by Steven L. Cantor

Figure 7.3a A specimen *Magnolia soulangena* with almost 40-foot canopy was transplanted to on-site nursery.
Photos by Steven L. Cantor

Figure 7.3b Newly transplanted three years later.
Photos by Steven L. Cantor

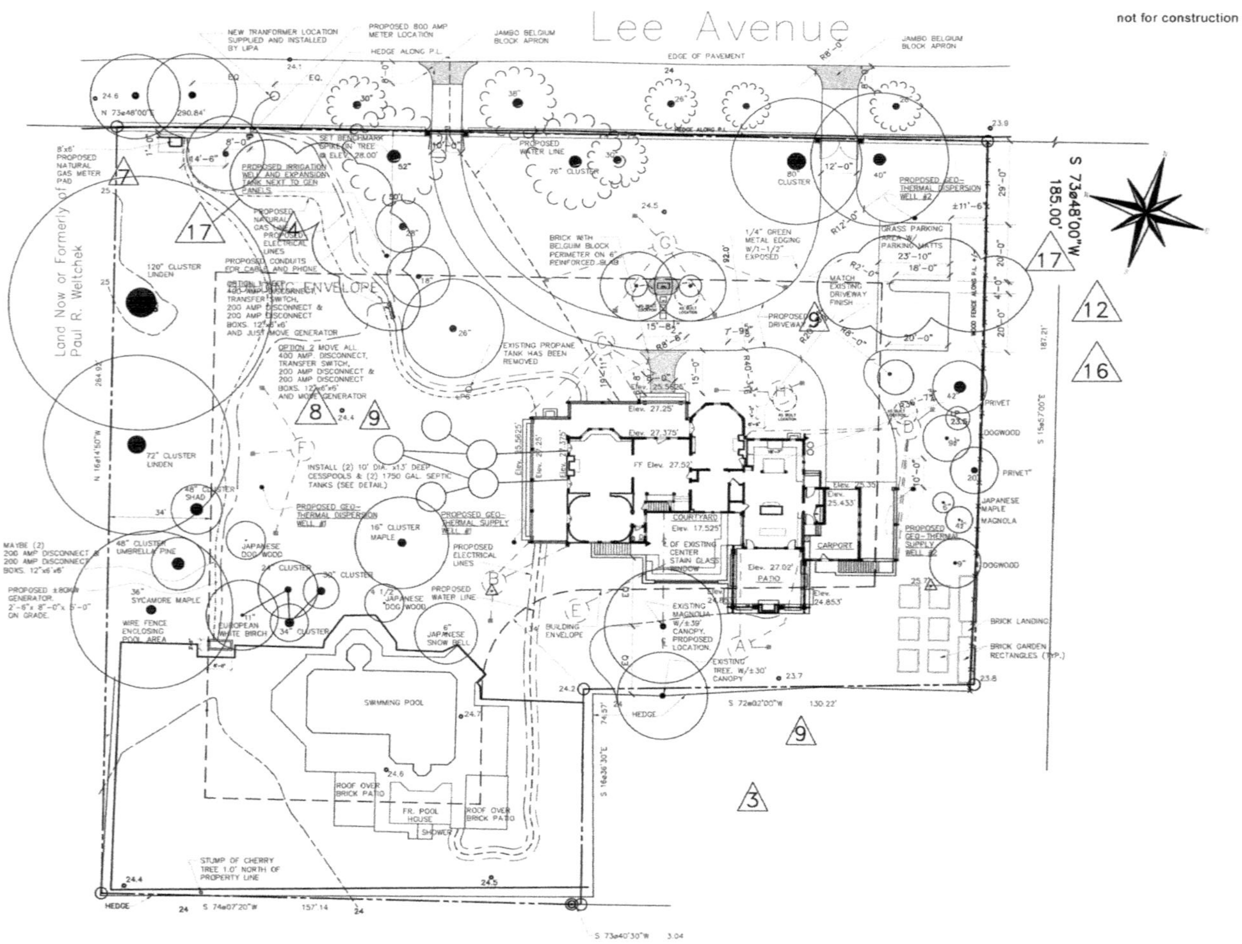

Figure 7.4 Illustrative site plan shows arrangement of driveway and major trees and pavements.

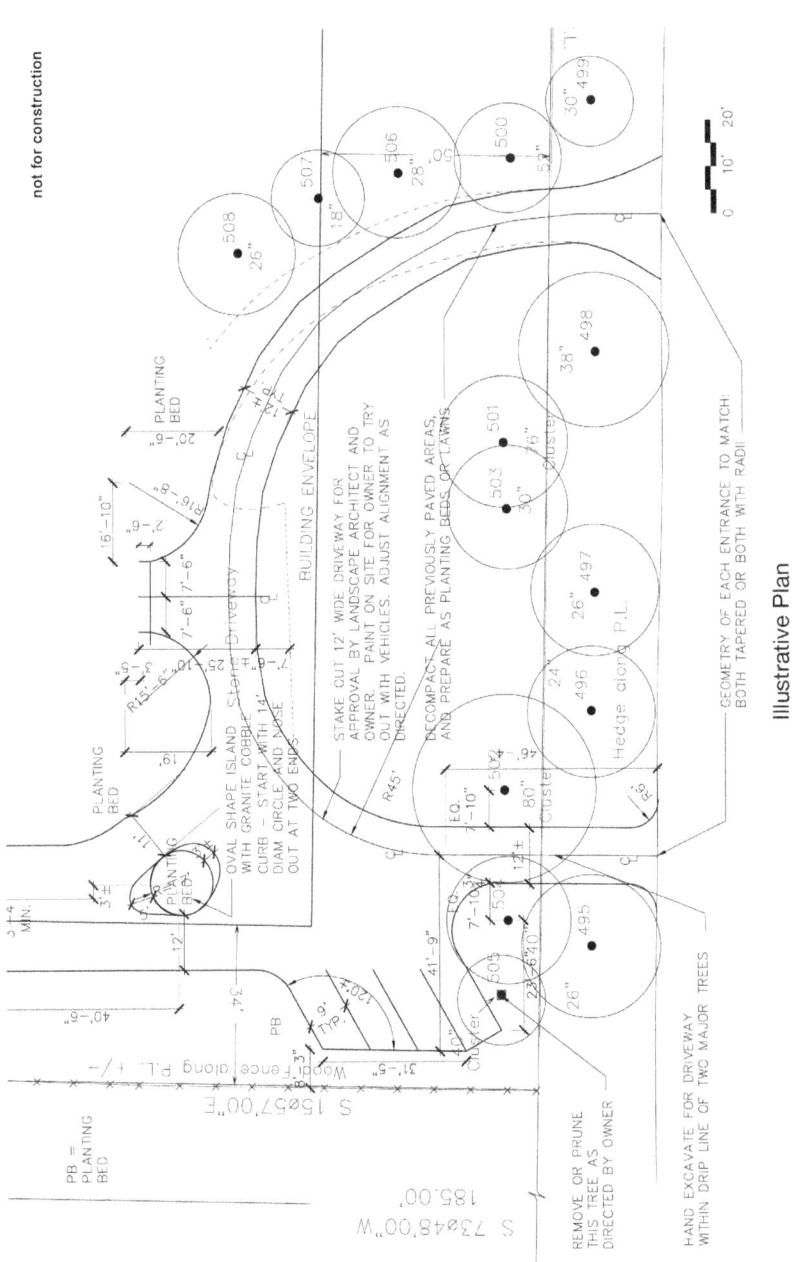

Figure 7.5 Entrance site plan shows detail layout of new entrance loop drive.
Drawings by Steven L. Cantor; Architects Eric Woodward and Rich Anderson of Southampton and builder and engineer Ken Wright of Bridgehampton

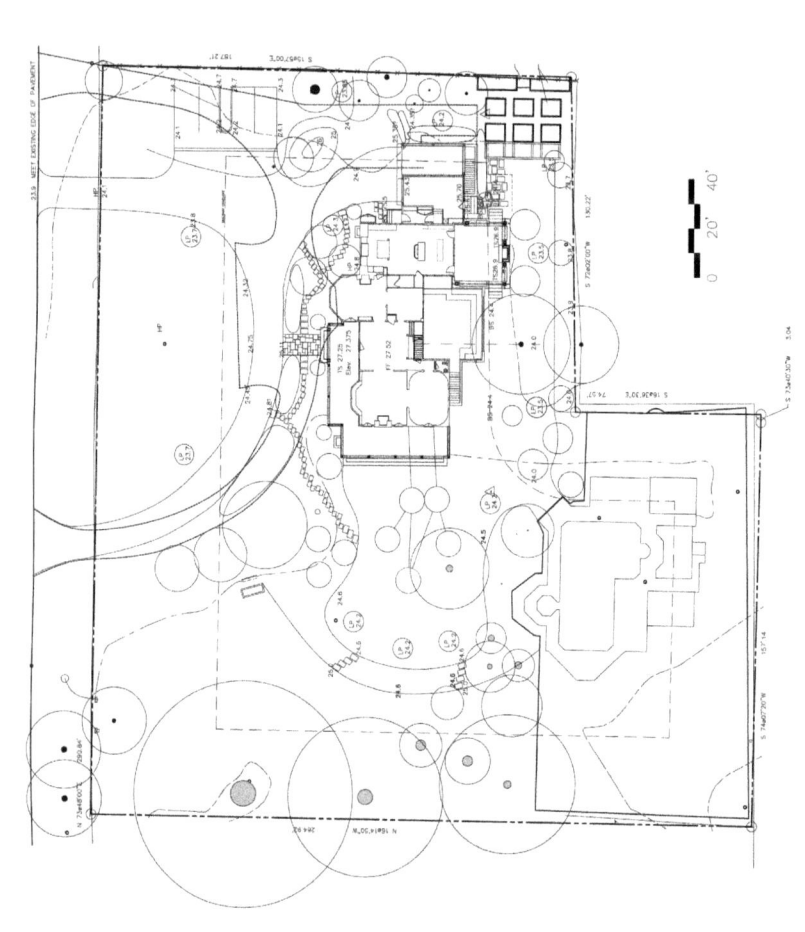

Figure 7.6 Grading and drainage plan for generally flat site.
Drawings by Steven L. Cantor; Architects Eric Woodward and Rich Anderson of Southampton and builder and engineer Ken Wright of Bridgehampton

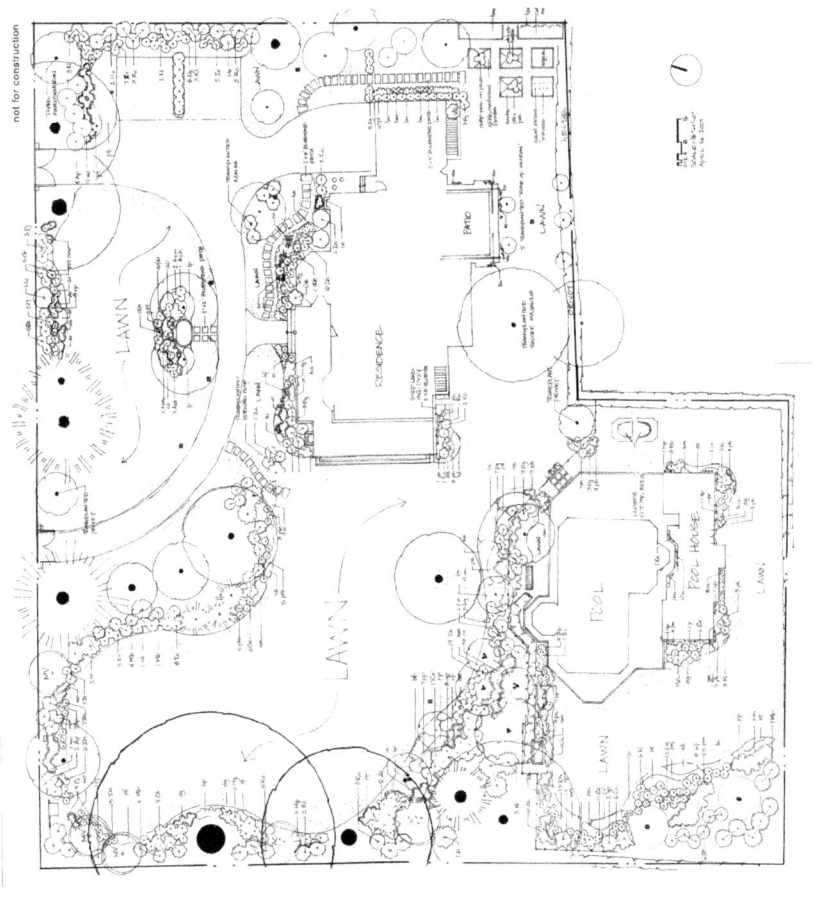

Figure 7.7 Planting plan of entire site by landscape architects Christina Lynch and Steven L. Cantor. Drawing by Christina Lynch and Steven L. Cantor.

Figure 7.8 Side view of the entrance loop drive with plantings.
Photos by Steven L. Cantor

Figure 7.9 Transplanted peegee hydrangeas and seating area in the front yard.
Photos by Steven L. Cantor

Figure 7.10 The front entry is gracefully off center.
Photos by Steven L. Cantor

Figure 7.11 Sunken courtyard in the rear of the residence (pavement design by architect).
Photos by Steven L. Cantor

Problem 1

RESIDENTIAL SITE AND FEE

(In the following problem, any *terms in italics* may be changed to suit the particular instructor or team leader, or your own personal, work or studio setting. Typical changes might be specific people, such as clients or contractors, plant materials, or the locations of sites, people, institutions or companies.)

A prospective client, *a couple* with a two-bedroom, two-story house of 5,000 square feet (465 square meters) with a 2,500-square-foot footprint (232.5 m²) and with an attached garage of about 450 square feet (42 m²) on a 1-acre (0.4 ha) site in *Caramel County, XY* has called you and is asking for a proposal for design services. They *are both professionals in their late thirties with busy lives and no children*, and they like to entertain.

1. The house is in good condition both structurally and physically. *Victorian-style detailing dates from the 1920s.* It's perched atop a hill yet visually appears to be falling off.
2. There are overgrown old gardens over about half of the site, some on level ground, some on rolling topography, and some within a rock garden area; some of the plants are worth salvaging. There is limited pavement, which is uneven and poorly drained.
3. There is a total elevation change of *20 feet (6 m)* within the site.
4. Views from the house are very nice in two directions, but in the other directions, there is a view of a neighbor's house and the street. Although the street is quiet, it is not particularly attractive.
5. There are eleven mature trees on the property, including several *oaks* and *maples*, as well as two evergreens and a few non-native species. Two trees have large limbs growing over the house. There has been no maintenance on the trees in some time.
6. Each of the owners has one car, and they want to be able to accommodate three additional on-site parking spaces for guests near the garage, which faces the street. There is a broken-up driveway that's a straight shot from the street.
7. They want some lawn.
8. They want to minimize maintenance even though they like to garden.
9. They want a garden for vegetables and a garden for cut flowers and perennials.
10. They want to consider a swimming pool and possibly a tennis court.
11. They want to have outdoor space for entertainment or parties and some private outdoor paved space.
12. There is no irrigation, but *there is a city water line that enters the house at the garage with a standard water meter and backflow preventer, as per code.*

13. The house has central air-conditioning, and compressors are located on a large concrete pad in the rear, which needs screening.
14. There is a city storm sewer in the street, and also one 10-inch-diameter (25 cm) line into the property with a catch basin in the lawn and a manhole in the street where the sewer and line connect.
15. They want to incorporate sustainable design principles.
16. They are uncertain what to spend or what to do.

TO-DO LIST IN STUDIO

1. What is the existing site like? What do you need to know about the site?
2. Draw a quick site analysis.
3. What is the program? Is it reasonable?
4. Who are these people? What sort of information about each person is important to know?
5. Discuss what should be in a complete proposal.

DEVELOP THE FOLLOWING

1. Submit a final site analysis drawn to scale with your assumptions as designer. Include on it relevant details about the client.
2. Prepare a proposal for providing landscape design services (two pages maximum, double-spaced).

Problem 2
RESIDENTIAL UPGRADE

(In the following problem, any *terms in italics* may be changed to suit the particular instructor or team leader, or your own personal, work or studio setting. Typical changes might be specific people, such as clients or contractors, plant materials, or the locations of sites, people, institutions or companies.)

You are called to meet *a couple (further define them)* who wish to upgrade the landscape of their residential property, about ½ acre (0.2 ha). They have two young children and will probably purchase a dog from a breeder in the next several months, as the children are reaching an age where they would both enjoy a pet and have enough sense of responsibility to take care of it. The site is a flat (or hilly) rectilinear lot, about 200 feet (61 m) long by 110 feet (33.5 m) wide, with frontage along a residential street along the shorter segment. The front setback is 40 feet (12.1 m), the rear setback is 25 feet (7.6 m), and the side setbacks are 15 feet (4.6 m). The two-story residence, with a footprint of 3,000 square feet (279 m²), is well built and well maintained, and has a two-vehicle carport with overhanging roof attached to the residence. Primarily in the rear yard, there's a series of gardens, overgrown and in disarray: (1) formal terrace with plantings, (2) rock garden, (3) native plants, (4) fenced vegetable garden, and (5) shade garden under a grove of four mature deciduous trees. The couple also wants to incorporate a tennis court and a swimming pool. Both parents work, although one spends more time at home with the children whenever possible. The couple indicate a strong interest in gardening, but it seems that they rarely have enough time for it.

Consider the following:

1. As a landscape architect or designer, how do you develop a fee for working with this couple that meets their needs and at the same time ensures for yourself a profitable relationship with them? Develop a proposal, three pages maximum, to outline your scope of services and your fees.
2. Will all the features the prospective client wants fit comfortably in the site?
3. What types of drawings or other documents would be useful to you as a landscape architect in studying the site and anticipating their needs?
4. What types of drawings or other documents would be the most useful in showing the prospective client ways to comply with their program requirements and what is possible with their property?

Problem 3

ROLE-PLAY FOR RESIDENTIAL DESIGN FEES AND SERVICES

(In the following problem, any *terms in italics* may be changed to suit the particular instructor or team leader, or your own personal, work or studio setting. Typical changes might be specific people, such as clients or contractors, plant materials, or the locations of sites, people, institutions or companies.)

Form teams of three classmates or colleagues. (If the total class or studio size is not divisible by three, create one or two teams of four, and have two people be the client, rather than one individual.)

One of you is the landscape architect, one is a family member or close relative, and one is the landscape contractor. You have agreed to provide landscape design services for X (your sibling, aunt or uncle, cousin, ex-partner), who owns a home on a 2-acre (0.8 ha) lot in *suburban Atlanta, GA* about 10 miles from your own residence. X wishes to add a tennis court and swimming pool and several gardens to the property. You have some concerns about whether both the tennis court and swimming pool will fit on the site. In order to expedite the work, you have selected a landscape contractor, Y, with whom you've collaborated on a number of other residential projects, to do the work, and s/he will negotiate a fee for the design you develop. The contractor has agreed to a markup of 8% instead of the standard 15% for plant materials and labor costs. S/he will also act as the general contractor and will find a contractor to install the tennis court and/or swimming pool.

Discuss how the work will proceed. What will your responsibilities be as the landscape architect? How will you set your fee for design services and site inspections? How will the contractor set the cost for implementation? Present your results to the class.

PART 3: PUBLIC WORKS

Definition

A public work is any work for a public agency: a local, city, or state parks department; a city or state transportation department; a public or quasi-public economic development agency. In general, what distinguishes projects in this area is that they are conceived of and implemented for the benefit of the general public and funding is provided by government, whether local, regional, or national. Public works projects can be as varied as the diverse agencies that seek them: streetscapes, roads, and parkways for transportation agencies; urban and rural parks for all sorts of parks departments; cemeteries for veterans agencies; urban design projects for economic development agencies; infrastructure studies for environmental protection organizations; and so on. Compared to projects in which the client/owner is a single individual or family or a privately owned company with a single or few directors, public works projects require the landscape architect to work seamlessly with bureaucracies. It's not unusual, for example, for an urban design project to be studied, reviewed, and approved by various members of many different agencies, often with competing interests, such as parks departments, transportation agencies, and police and fire departments. Access to the site in an emergency for the police, who need clear and direct access, may be quite different from what the local fire department requires, which includes twenty-four-hour access to fire hydrants and a means of raising ladders against the building to facilitate access for rescue. The type of street curb installed may govern whether or not a fire truck can easily jump the curb and get close enough to a building to hoist the fire truck's ladders.

With most public works projects, a great deal of time is spent in meetings coordinating among agencies and finding common ground so that a design gradually emerges that is approved by and satisfying to everyone involved. Special skill must be applied in writing and disseminating minutes of meetings in a timely manner as well as preparing and distributing revised drawings incorporating all of the changes discussed at each meeting. Budgets for design services are often strained as a result of the bureaucratic requirements, yet the rewards of successfully executing such a project are unusually satisfying. The landscape architect has designed a beautiful setting for not just one individual or family but for an entire community.

Role of the Landscape Architect

Because the client is a bureaucracy, the landscape architect typically spends a great deal of time in preparing and submitting invoices and having them reviewed and approved in a timely manner. Therefore, no matter how glamorous the project or how appealing it is to design for the public, it is essential to have at least one experienced accountant on the payroll who is highly organized and detail-oriented, can submit invoices in a timely manner, and can respond to all the client agency's requests for clarifications.

Sometimes landscape architecture firms may be the prime consultant for a public works project. In other settings, they are a consultant to a larger firm, such as an engineer or architect. The prime consultant receives the invoices prepared by the subconsultants and submits them all in one package to the client agency. The challenge that a subconsultant faces is that if the principal firm does not submit its complete invoices, including the subconsultants', in a timely manner, then it may take a long time for the prime consultant to be paid, and this delays payment to the subconsultant.

Such delays are one reason some firms prefer not to do exclusively public works as subconsultants. They may also occasionally seek some modest projects as the principal, and have a few other types of work, such as single-family residential design, where billing the client and receiving payment is more routine, without long delays.

Public works also include some specialized areas (sometimes involving private enterprises) in which there is specific expertise required of landscape architects—for example, gardens for health care facilities. As the population ages and people are living longer, there is a great need for high-quality residential and health care facilities to provide housing and temporary or permanent care for aged or infirm people. In addition, hospitals specializing in surgery for children or specific types of medicine, such as spinal medicine, are also common. Most of these facilities benefit from gardens and other spaces that welcome visitors and provide respite for patients and family members needing a break from the often regimented institutional environment. Landscape architects focused on this type of practice must be expert in applications of the Americans with Disabilities Acts (see Chapter 6), the special needs of particular classes of patients (such as those living with dementia, people with spinal injuries, children having surgery, and so on), the needs of the professional staff (from orderlies and law enforcement to nurses and surgeons), and the needs of the public (including visitors, such as parents and siblings, who often come to these facilities in a stressed state of mind).

Firms doing public works must recognize that, particularly in the beginning of a professional association with a client agency, a lot of time will be spent learning the particular agency's requirements. For example, it's common for the agency to review the resume of each major staff member proposed for a particular function on the job, and the number of years of experience shown on the resume may be used to justify paying each employee according to various graded categories of professional staff permitted by the client. Often even if payment for a certain phase of work—say, design development—is a percentage of a negotiated total fee, it may still be necessary to submit the time sheets of the employees to verify that the number of hours is documented on the relevant employees' time sheets.

Unique Specifications

Each bureaucracy is likely to have its own design standards, and the landscape architect must become thoroughly versed in these standards, as well as aware if and when some of them may conflict with another agency's criteria. For example, a recently completed, quite successful program called PlaNYC Schoolyards to Playgrounds, developed by the New York City Department of Parks and Recreation (DPR), the Department of Education and the non-profit the Trust for Public Land was the renovation of many public school playgrounds. In return for funding by the Parks Department, the schools agreed to open their renovated playgrounds as public parks after school hours and on weekends.[9] The existing chain-link fence and other details originally built in these playgrounds complied with School Construction Authority (SCA) specifications (now the Department of Education); while new fencing complied with the Parks Department standards. This typically meant that it was hard to repair an existing fence section; new sections of fencing (DPR) could be placed adjacent to existing fence sections (SCA) in good condition, but it was a challenge to upgrade mesh, posts, or top rails in existing fences because these components did not comply with the applicable DPR specifications. In general since DPR was funding the program, it paid for most of the improvements. Through careful coordination, problems like these were solved.

Most agencies have standard specifications that govern the selection and installation of all sorts of details: pavements, fences, benches, and so on. Years of experience with these particular items often result in well-tested elements that are durable, safe, and attractive. If the landscape architect chooses, based on the requirements of the design, to suggest the development of a unique new design element, it typically must be reviewed (as a detail drawing and specification) and approved as part of a long process in which new specifications, based on a combination of existing components already defined in the standard specifications and some new or unique materials or arrangement of materials, are incorporated. One typical example might be play equipment. Many parks departments or school systems will have in place preapprovals for play structures designed and built by certain manufacturers who specialize in this area and whose play systems have been subjected to long reviews, verifying safety for children, the durability of the components, and the wide use and enjoyment of them by a particular age group. So if a landscape architect proposes a new design, no matter how creative, innovative, and attractive, it is likely to be subject to a long review process, during which many obstacles may interfere or delay approval. So the landscape architect must carefully determine, based on the budget permitted for design services for the project, what unique design elements to propose and include in the final design.

Special Challenges

Among the challenges of public works projects is that the available information for a particular site may be limited. Usually when the project begins, the landscape architect or consultant is given a survey, which may or may not be accurate. It's critical for skilled staff to verify the data on the survey—for example, to find the utilities shown and determine their condition and capacity, and to note all major trees. Often as a result of such an inspection, it's necessary to request additional survey information.

Furthermore, many existing utility systems may be partially obsolescent. In most American cities there are districts in which during heavy rains there are periods of combined sanitary and storm sewage overflows. Sustainable design techniques often are required, like rain gardens or bioswales, to capture some of the stormwater runoff and mitigate this problem. Yet the extent of the problem can be so great that mitigation could potentially utilize more of the budget than is anticipated. After any additional survey information is incorporated, the landscape architect should revisit the scope of work to verify that the work proposed is still within the budget.

It is common in public works projects for more graphics, models, 3-D renderings, and perspectives to be required by the client agency in order to demonstrate compliance with requirements and sell the design to the public than would be expected for a private client. Contract negotiations, if possible, should set an upper limit on the number to be included in design services. Sometimes special conditions or additions to the standard agency's contract may involve sustainability, historic preservation, bikeways, or other elements, and the landscape architect should get a sense that the design for such services is possible within the fee awarded.

As with any design, the landscape architect must carefully review the proposed contract prior to beginning work on the design to verify the list of all deliverables—drawings, cost estimate, renderings, and so on—so that it's clear what is required. Often these are listed within the contract and not subject to much revision. Unlike private contracts, in which the landscape architect can write and negotiate language tailored to the specific requirements of the project, with public works projects there is much less room for negotiation.

Long Time Frame

Typically public works projects require a longer time frame for review because many different agencies must participate. Their work may be complicated by the opinions expressed by politicians or other officials and by members of the public. Public hearings are often required to present the project to whatever cross section of the affected community shows up to participate, and many critical comments may be received at such forums. This process may recur indefinitely until there is a consensus of an acceptable design for an acceptable budget. So the challenge for the landscape architect is not only to keep this process moving forward but to anticipate to the extent possible the number of hours required to receive approval. Since the review and approval process is usually much longer than would be expected for a single-family residential estate, a roof garden, or many other types of less complex projects on smaller sites, the landscape architect must verify that the fees being paid for design services allow for multiple reviews, meetings, and revisions.

Additional Work

Perhaps one of the most critical aspects of the contract for design services for a public works project is the section that governs "additional work" or work "outside the scope of services." For many public works and urban design projects, when the contract begins it's not very clear exactly what will be designed; often this takes a period of preliminary investigation, site analysis, and design to determine. The landscape architect must be certain that the contract contains enough flexibility to be able to negotiate additional fees when major changes or additions occur, whether in an initial phase of the project or in later phases. The most obvious scope changes are adding to the size of the site or adding specific design components that were not part of the original scope of work. The more specifically the scope of work is defined, the more helpful it is to the landscape architect in such negotiations. Another typical problem may be that the contract does not define the deliverable documents in a quantitative way, so that rather than stop at two illustrative renderings of the preliminary design, the agency representative feels justified in requiring the landscape architect to produce alternative versions indefinitely, until approval is received.

Public works projects often require a range of permits and approvals. Some are required during the design phase and others during the construction phase. For example, the local Department of Environmental Protection (or equivalent agency) may require proof that the storm drainage resulting from the proposed design is comfortably handled by the proposed or existing storm drainage system. There may need to be review and approval by the local police and fire departments. If the project is adjacent to a stream, river, or other body of water, requirements relating to the Army Corps of Engineers regulations on navigable waters may kick in. There may be a local public design or art commission that must sign off on any work in which public funds are being used. Often natural wetlands are protected by law; if there is no alternative to filling in a certain area of wetlands, the landscape architect must demonstrate this, and show an area where a wetland could be created in order to mitigate the impact. Prior to starting work, the landscape architect should make an estimate of the number of permits and approvals that must be obtained during the design process, and verify that the fees allocated for these services are adequate.

At the same time, the final construction documents should spell out for the bidding contractors a list of all permits and approvals that must be obtained prior to the beginning of construction. Often these are routine but still take time. For example, if there is asbestos abatement, there is a clear-cut, well-defined process to follow. However, different agencies (say, city, state, and federal) may have totally different requirements for what is needed to obtain a permit for demolition, construction, or major earthwork operations. The landscape architect must know what is required and be able to list all of these requirements in the specifications or other parts of the construction documents.

Cost Estimate

The cost estimate is a critical part of any public works project. Usually the client agency has a required format—for example, an order in which the different elements must be shown. Most often each entry of the cost estimate is keyed to a specific specification number, a detail for which is shown in the construction drawings or referenced to the agency's standard specifications. New item numbers are assigned to the unique details that have been incorporated. Usually the agency has a clear record of the costs per unit of most standard items according to their specification. The landscape architect must spend the most time reviewing and determining the costs for unique items added to the project. Usually in early phases of design, a contingency of as much as 15–20% is added to the cost estimate to cover the cost of unknown items or details for which a specific cost has not yet been determined. As the project advances through different phases and finally reaches the bid phase, the contingency is reduced to zero.

The cost estimate is critical to show that the design can be achieved for the total price indicated. By the time of bidding, it's usually a legal requirement that the bid must be within 10–15% of the final cost estimate. In such cases, if the bid is no more than 10–15% higher or 10–15% lower than the cost estimate, a quick review can be made to determine the reasons for the overage or underage, and adjustments made to the design, if necessary, in a straightforward manner to keep it within budget to the satisfaction of the client agency.

However, if the low bid or average of bids ends up exceeding by more than 10% to 15% of the total of the final cost estimate, it is often the case that the landscape architect on his or her own time must review the bids and the cost estimate to determine what changes must be incorporated in order to meet the budget requirements. This is often a stress-provoking and tense process, so it's all the more reason that every effort should be made to prevent this from happening. It implies that one or more mistakes were made in the design phases, which reflects poorly on the skills of the landscape architect. Still, if this type of review is required, once the adjustments in the design documents have been agreed upon and/or the bidding contractor agrees to the cost changes, then the project may move forward.

Bid Process and Specifications

Both the preselection of bidders and the bidding process for public works projects for public agencies are arduous. Since the budgets for construction are often substantial, there can be a tremendous level of competition to win a bid. Some agencies permit a preapproval process, whereby a company wishing to bid on a particular project must submit proof of expertise (as represented by relevant projects) and experience (as represented by a certain minimum number of years' experience with the required type of construction and key people with specific skills and qualifications) to be considered for a project. When preapproval is permitted, it can be helpful in weeding out those firms that do not have the minimum qualifications and experience necessary for the work proposed.

Specifications usually include many standard sections provided by the client agency (such as local or state or national transportation departments, departments of environmental protection, parks departments, or school systems) in which standard procedures are provided for many operations, based on the agency's long experience with that type of construction. The landscape architect provides new specifications for unique elements incorporated into the design, and quantities or clarifications to standard specifications for other less unusual items. For example, the specific plant list may be entered into a chart or table that's part of the agency's standard specifications for plant materials and planting. The general conditions are often quite extensive, so the landscape architect should review them carefully to understand the impacts these specifications may have on the scope of work for which s/he is responsible.

The bid process is highly regulated. All bids must be received by a certain time at a certain place. Often proof of delivery, such as a postmarked receipt or signed receipt from the client agency, is required. Usually there is a public opening of the bids, often at the client agency's offices or some other public setting, in which the bid is read out loud and recorded. In most public works projects for government agencies, the agency is required to accept the lowest qualified offer. If there has been a prequalification process, it's even harder to reject the lowest bid, since each firm bidding has already been certified as being capable of doing the work. However, under certain, rare circumstances the design team, usually consisting of the landscape architect and representatives of each agency involved in the work, may challenge the low bid. These circumstances may include a situation in which the low bid is so much lower than the other bids that it seems impractical or even impossible that the work could be done for such a low price. Another condition that might allow for rejecting the low bidder would be if evidence is submitted of any irregularity or legal difficulty the bidder may have encountered on similar jobs. Since the low bidder is so often awarded the contract, it is all the more important that the landscape architect's set of construction documents, including specifications, is prepared with completeness and precision. A common technique by some unscrupulous contractors is to bid low and then, upon being awarded the bid, search the drawings, details, and specifications for discrepancies or errors so that requests for change orders can be submitted to justify extra fees.

Fees

Many public works contracts have predetermined fees in which the total payment is based on the total estimated cost of construction and determined by a formula. There will usually be a manual or other document for the consultant in which the fee is broken down and the deliverables for each phase of work are stipulated. Often the percentages for each phase of work are predetermined. For example, a typical breakdown might be as follows:

Total estimated cost of construction = $1,000,000
Total fee is 12% of projected cost = $120,000

Schematic phase, 20% = $24,000
Design development phase, 30% = $36,000
Final bid documents phase, 40% = $48,000
Construction administration, 10% = $12,000

In some cases (though not typically), the client agency is willing to negotiate a fee for construction administration after the project is bid, which is much fairer to the landscape architect than having this fee locked in from the beginning of the job. This fee is the most difficult to predict, so postponing its calculation is helpful. There are many factors that help to determine the amount of time that will be necessary for construction administration. For example, how complex will the work be? If the design incorporates complex layout and grading and uses some nonstandard construction details, this could require more time on the site to review and approve the layout and grading, and more time with shop drawings and approval of special details. It's not usual for the landscape architect to know in advance which contractor will be implementing the design; it depends on the lowest bid. Therefore, there could be a learning curve in working both with the contractor and with the client agency to make the construction administration process as smooth as possible. It is also usually not possible to know in advance all the permits and approvals that either the landscape architect or the contractor will have to obtain. Therefore, if the fee for construction administration is negotiated after the contract documents are complete and ready for bidding, it's much easier to make an estimate of these costs and incorporate them into the fee. Client agencies vary as to whether the bidding process is included in the final contract documents phase or the construction administration phase of the project. If it's the former and there are challenges to the low bidder or the bids come in more than 10–15% higher than permitted, a considerable amount of time and effort must be spent to reconcile the differences and finalize the selection of a contractor at a point when the landscape architect's fee may be dwindling. If it's the latter, then the landscape architect may be able to negotiate a more realistic fee based on the detailed knowledge of the contract documents and the credentials and submittals of the contractor. Even though a construction administration fee of 10% to 15% is often standard, it is very difficult for the landscape architect to manage capably all that may occur during this phase for this particular fee. One option is to try to stipulate that the fee will cover a period of construction with a set calendar, so that if the work continues after the end date, there is the potential to negotiate an additional fee. Another helpful method is for the landscape architect to itemize all of the permits and approvals, shop drawings, and site inspections that will be reviewed and approved during this phase, and if items not on this list spring up, she may suggest to the client agency that they represent additional services. Another factor is the distance of the construction site from the landscape architect's office and how much travel time is involved in each round trip from the office to the site; some contracts allow charging for travel reimbursement, while others do not. It is better to know this ahead of time.

Another factor that may have an impact on the flexibility of the fee structure is the relative importance of the project to the client agency. If the proposed project is greatly valued by local politicians, key personnel at the client agency, or other major personnel in collaborating agencies, then it's much more likely that the landscape architect will be able to negotiate higher fees, so long as it's possible to guarantee high-quality results. Unfortunately, some contracts may lock in certain fees or requirements that are onerous, costly, and impractical. It is up to the landscape architect, where possible to negotiate alternatives to mitigate such elements to assure a successful contract.

DOS AND DON'TS FOR PUBLIC WORKS PROJECTS

1. Have sustained expertise in the area of work—housing, infrastructure, roadwork, and so on—that the client agency is seeking. Have the ability to wait at least several months for payment after invoices are submitted.
2. Be highly organized, with junior staff who can readily do tasks such as site analysis, data gathering, preliminary design, minutes of meetings, and early drafts of construction drawings with relatively limited supervision from senior staff, who can save most of their time for important meetings, design reviews, and oversight.
3. Investigate whether other projects, perhaps privately funded ones within your firm, can absorb some of the cost of taking on a major public works responsibility, as if there is a pro bono aspect to this work.
4. Learn quickly the design vocabulary of the client agency's detail library and specifications and how to apply them to different site situations in order to solve problems.
5. Review costs provided by the client agency for comparable projects to get a clear sense early in the project of what the budget will pay for.

6. Review very carefully the scope of work and the limits of work of the site. Be as familiar with what is *not* included as with what *is* included. As the design starts to evolve, keep in mind whether certain design elements or areas of design should be considered as additions to the scope of work.
7. Familiarize yourself thoroughly with the client agency's specifications regarding standardized items, and determine whether any consideration should be given to proposing and designing unique elements not included in the specifications.
8. Evaluate whether any phasing of the proposed work should be considered or required.
9. Inquire about the degree to which the Americans with Disabilities Act applies to the work.
10. Detail and implement the design in order to improve the quality of life and safety of those using the site.
11. Evaluate which amenities are the most important to provide to those using the site.
12. Study how the project links to any other major projects under construction or proposed by the same public agency or some other public works agency, and their schedule for implementation.
13. Inquire what reviews and approvals will be required for each phase of the design to be developed and how they will be paid for. Verify whether your firm and/or the client agency will be responsible for the presentations, documents, or other aspects of the permits and approvals. Distinguish between *design* reviews and approvals (your responsibility) and *construction* reviews and approvals (the contractor's or client agency's responsibility).
14. Ask if there are any restrictions on the use of construction materials or plant materials as a result of any local, state, or federal laws or other regulations.
15. Assign someone as project manager who has strong communication skills and can prepare minutes skillfully and efficiently.
16. Anticipate a design process with many collaborators and other public agencies involved.

Browns Race

A street in the High Falls Historic District follows the old water raceway that powered industry in this neighborhood of Rochester beginning in 1815. A 250-foot long (76.2 m) re-creation of the race had been installed in the 1990s but was later abandoned. Working with a local nonprofit organization, Greentopia, LDGN developed a design for the Race that references the historic nature of the site and its location next to the Genesee River and the High Falls. The new watercourse references the local geography and has a variety of water effects to engage the user along the entire length. Sustainable design includes a rain garden, rainwater harvesting, low-wattage lighting, and use of native plants. The Race is the first phase of a visionary capital project that will transform the Middle Gorge of the Genesee into a world-class showcase of sustainable and environmentally sensitive development in harmony with the natural beauty of the Gorge and High Falls. (See Figures 7.12 - 7.15).

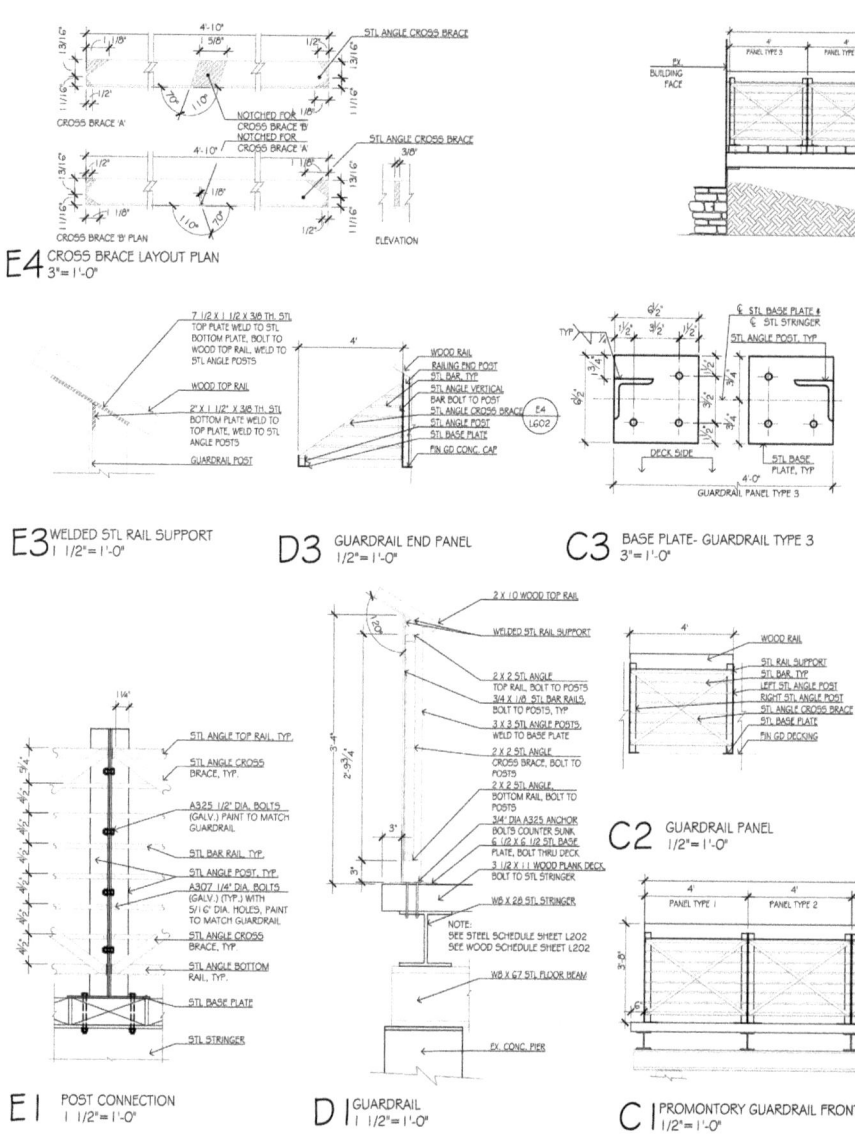

Figure 7.12 Browns Race, Rochester, NY, guard rail details.
Courtesy of Jeff Dragan, LDGN Landscape Architects, DPC

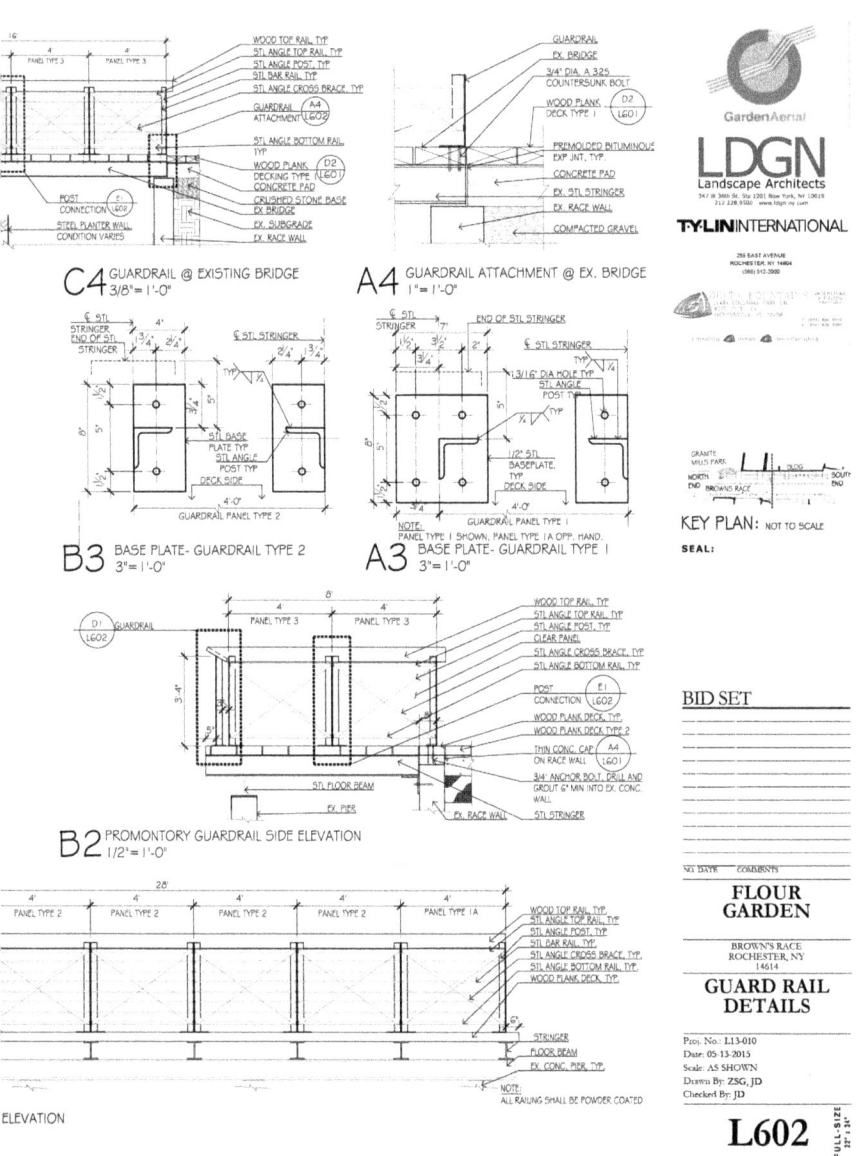

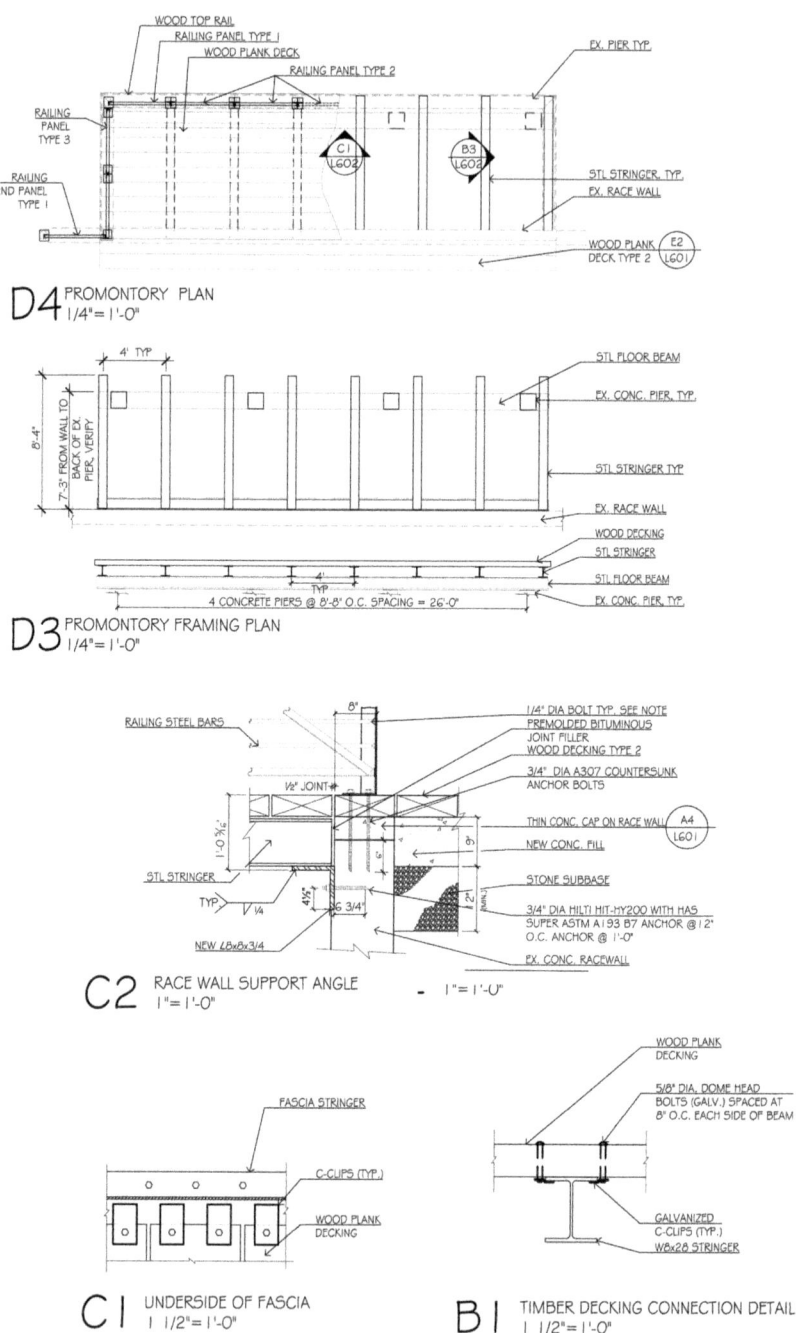

Figure 7.13 Browns Race, Rochester, NY, promontory details.
Courtesy of Jeff Dragan, LDGN Landscape Architects, DPC

not for construction

266 EAST AVENUE
ROCHESTER, NY 14604
(585) 612-2000

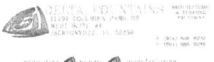

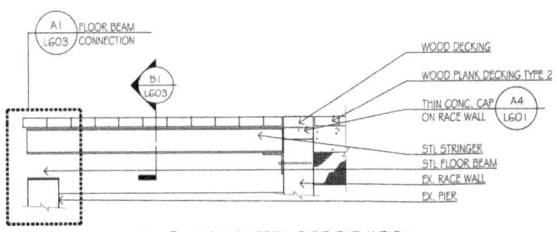

B3 PROMONTORY SIDE ELEVATION
1/2"=1'-0"

KEY PLAN: NOT TO SCALE

SEAL:

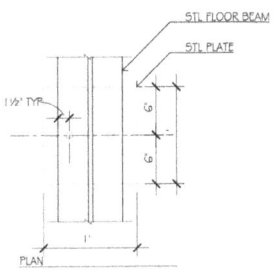

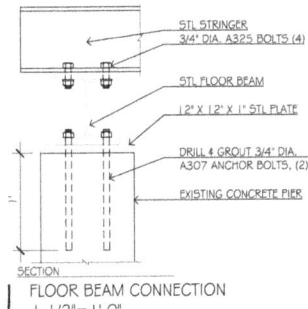

A1 FLOOR BEAM CONNECTION
1 1/2"=1'-0"

BID SET

NO.	DATE	COMMENTS

FLOUR GARDEN

BROWN'S RACE
ROCHESTER, NY
14614

PROMONTORY DETAILS

Proj. No.: L13-010
Date: 05-13-2015
Scale: AS SHOWN
Drawn By: ZSG, JD
Checked By: JD

L603

FULL-SIZE
22" x 34"

Figure 7.14 Browns Race, Rochester, NY, illustrative sketch.
Courtesy of Jeff Dragan, LDGN Landscape Architects, DPC

Figure 7.15 Browns Race, Rochester, NY, as-built installation.
Courtesy of Jeff Dragan, LDGN Landscape Architects, DPC

Problem 1

PUBLIC HOUSING APARTMENT LANDSCAPE AND INFRASTRUCTURE PROPOSAL

(In the following problem, any *terms in italics* may be changed to suit the particular instructor or team leader, or your own personal, work or studio setting. Typical changes might be specific people, such as clients or contractors, plant materials, or the locations of sites, people, institutions or companies.)

As a landscape architect, you are called to interview with two representatives of the *New York City Developing Authority (NYCDA) in the Bronx* because they wish to study some landscape and infrastructure issues at an existing apartment building. It's a five-story brick building constructed in *1935 in an Art Deco style*, and it has a full basement with laundry and storage facilities. In the spacious lobby there are three elevators reaching all floors and the basement. From the top floor there is a wide staircase to the roof, although the doors at the top are locked, as no access is permitted. As one enters from the street, there's a fairly standard layout, with three steps up to an entrance plaza with full-length planters flanking it. These planters, which are three feet wide and extend over the basement, have well-established shrubs and perennials, but the plantings appear crowded. Along the sidewalk at the street, there are five standard *New York City Parks Department tree pits* with some straggly shade trees in fair condition with no tree guards, nor any substantial perennials underneath them, so the surfaces are weedy or bare without mulch. The center of the lobby has a wall of windows with views of a rear courtyard, but access to it is from either end of the hallway that runs symmetrically from the central lobby. The courtyard is surrounded by 8-foot-high (2.4 m) retaining walls but nevertheless receives several hours of sun in the summer. The concrete pavement is in poor condition and drains poorly.

The higher-ranking *NYCDA officer, Ms. Hernandez*, with some commentary from her subordinate, *Mr. Smith*, describes issues at the building. She is very welcoming and thorough. The other three people present, residents of the building, comment only occasionally, if at all; a few check messages on their smartphones during the meeting, and one is actively texting. Ms. Hernandez indicates that they wish to have a unified landscape at the front of the building, including the street trees; to consider some beautification of the rear courtyard; to correct drainage problems; and to take a look at the roof to see if it might be possible to consider a roof garden or possibly even a green roof on all or part of the roof. There is a continuous parapet that a contractor told them meets city standards.

NYCDA prefers to work with a contractor who has done a lot of work at the building and is still working there. Although this may be an unusual arrangement, the contractor they prefer has an excellent work record, and so they expect approval. In the interest of a very efficient and expedited operation, they wish to negotiate a contract with this preapproved contractor that contains a liquidated damages clause requiring completion of all work within one year. A representative of the contractor, *Ms. Jones*, is present and introduces herself; she says that she can provide most subcontractors or other specialty trades that you might need, gives you her card, and leaves. It's further explained by Ms. Hernandez and Mr. Smith that everyone is very busy and therefore it is difficult to have meetings with all key people present, all the more so since the building is some distance from *Manhattan (where the NYCDA and the contractor are located)*, so they would prefer to have most communications by email, although they might be able to set up some occasional conferences by video or Skype if necessary. The two *NYCDA* representatives show you a few documents: an old plat map of the site, a faded plan showing the tree pits at the street, and an old blueprint of the roof that is almost illegible, so it's difficult to see what the structural design is. Mr. Smith explains that since the building is old, there is no available plan of the entire building or site. He opens a loose folder of random documents, skims through it, and gives you his and Ms. Hernandez's email addresses and phone numbers and the name and phone number of the head of maintenance at the building, who was not present at the meeting.

It's early in the evening of a summer day in mid-June, so there is still plenty of light. Ms. Hernandez and one of the residents leave, but Mr. Smith and the other two residents guide you around the site. You have a digital camera with you, and you are given permission to take some images. There is one hose bib near the street and one in the rear courtyard. You notice three or four surface drains of varying types. In the rear courtyard is one area with a large puddle centered around an area that does not drain at all, and another, larger puddle around a drain that appears to be clogged. There are five scattered plastic chairs, a wobbly plastic table, and a trash can. There is some surface debris—blown leaves and some cigarette butts—but things are relatively clean. You are told that the building superintendent (who works with the head of maintenance) lives in the basement apartment and is very capable but doesn't speak English well.

Occasionally, his two children, a young man and a young woman in their twenties who still live at home, help out, and they speak good English. In the planting beds near the entrance, there is considerable mortar debris, and it is explained that re-pointing of the masonry has just been concluded; the contractor is the firm whose representative you met earlier.

Mr. Smith asks you for a proposal for services. *NYCDA* is unsure what sort of budget they have, but they want you to recommend what you feel is a reasonable program of activity. You are welcome to consult with Ms. Jones beforehand. *NYCDA* meets monthly, so it would be helpful to them to have a proposal from you no later than three weeks from now, since they'd like to have a chance to review it prior to the next meeting and let you know whether they'd want to meet with you again. You thank them, spend a little more time at the front of the building looking at the street tree pits and the entry planters, and head home. You would very much like to provide good services for this building and think that if you are successful, this could lead to many other similar opportunities throughout the borough.

The next morning, a Thursday, you think of four questions and call Mr. Smith, but you reach his voicemail, so you leave a message.

1. How deep are the planters and what sort of drainage layers are in place?
2. When were they planted?
3. Is there any as-built structural plan of the roof?
4. How much use does the rear courtyard get?

The following Tuesday he responds by email, and answers three of your four questions. The planters are 3 feet deep (0.91 m), but he doesn't know if there are drainage layers. They were planted four years ago. The courtyard gets intermittent use at present, but there is no objection to more use. He doesn't answer the third question, so you call him back, and again reach his voicemail, so you email him back. The following Thursday, he responds via email to the third question: to his knowledge, there is no as-built plan of the roof. You remember at that point that you forgot to ask if there's any document that shows the location of drainage pipes in the rear courtyard, but you expect the answer is likely to be no.

What are some of the design questions that concern you? What are some of the professional practice concerns you have? What are your minimum requirements in order to be able to work for the *NYCDA* on this project? How do you proceed? Are there conditions or terms that you might require to be changed before you would agree to work on this project?

What if this were a privately owned cooperative apartment building instead of a public project and you were dealing with a committee of shareholders (that is, a president, vice president, and three or four other officers)? What if you were dealing only with one owner or a representative of a corporation? What changes, if anything?

PART 4: ROOF GARDENS AND GREEN ROOFS

Landscape architects focus on roof gardens as part of their practice for many reasons. As urban areas are built more and more densely, which means that available open space for outdoor recreation is very limited, roof gardens increasingly become desirable, protected oases for residents of apartment and institutional buildings. Just as specialization in residential design or public works requires detailed experience and qualifications developed over many years of diverse projects, the design of roof gardens requires landscape architects to concentrate on a range of special conditions and applications. Firms that master all of the complex factors and requirements may find a lucrative niche in the design market. But such projects are not for everyone. Because they are often intricate designs and require contractors with specialized skills, roof gardens must be developed, built, and maintained by firms that have sufficient resources to review the applicable building codes in detail; develop detailed drawings for pavement, planters, seating, fencing, lighting features, water features, and other special elements; inspect the installation on a regular basis; and be available when the client may want design changes or another phase of work.

Live and Dead Loads

Load requirements are divided between dead loads and live loads, but even with clear definitions such as the following ones, there can be room for different interpretations by building inspectors or reviewing personnel.[10]

1. Various combinations of dead loads, live loads, seismic loads, and wind loads are to be applied in the design of structural systems.
2. Dead loads are determined by the actual weights of materials and construction. Fixed service equipment is to be considered as dead load. These fixed elements include plumbing, electrical feeders, HVAC systems, and fire sprinklers. The weight of plant materials is to be considered as dead load, and this weight is to be calculated based on saturation of the soil.
3. Live loads are produced by the use or occupancy of the building.
4. Landscaped roofs are to have a uniform designed live load of 20 pounds per square foot ($0.958kN/m^2$).

To summarize, the dead load of the roof refers to the actual weight of the materials used to construct it and those built elements that can be considered as permanently fixed in place. The live load, in contrast, is the weight added to the dead load as a result of the use of the space, such as people moving on and off the roof.[11] However, sometimes there are disagreements about such items as portable planters or other movable objects. Are they dead loads because they are fixed in place at least for a certain time, or are they live loads because they can be moved from one area of the roof to another, or removed altogether? Similarly, the planting itself is usually defined in building codes as part of the dead load, but certain jurisdictions consider it live load because it could be removed and is therefore not permanent. More recent ASTM standards specify live and dead loads and in time may result in more standardized approaches, requirements, and definitions. Building codes will usually specify a load requirement that is a combination of the dead load and live load. Sometimes other loads come into play, such as the impacts of winds, snow, and even seismic factors, so designers of a green roof or roof garden must allow for a safety factor. The total weight of all built components as a result of all loads should be less than the maximum load permitted. In the course of planning the design, it is often necessary to calculate concentrated loads, that is, the maximum load permitted at stress points where, for example, structural beams cross one another or meet columns. At such locations, it is often possible to support larger weights, so trees or other heavier elements are often located over such strong points.

Structural Reinforcement

In cases in which the roof must be reinforced in order to support additional load, the construction changes typically start at the support for the roof deck itself. It is typical of most construction for the most weight to be supported at the parapet walls and at points directly over the intersections of structural beams. Structural engineers are able to determine, based on careful measurements and calculations, how much reinforcement will be needed in order to support a certain amount of additional weight at different locations on either an existing roof or one being planned. The more efficient the design team can be in terms of determining the purpose of the traditional roof garden or green roof being designed, the type of roof garden or green roof installation, and its various components, the easier it will be for the structural engineer to find a solution to the problems of increasing the load-bearing capabilities of different areas of the roof deck as required.[12]

There has been much variability in requirements about the heights and other dimensions and details of railings on steps and roof decks. The IBC has standardized some of the criteria, but the designer must still review applicable local codes. Different local codes may have different height criteria for handrails over steps. Guardrails for multiple occupancy buildings must be 42 inches high (1,067 mm) but lower heights, 34 to 38 inches high, (864 to 965 mm) are permitted for residences. The triangular space between the tread, riser, and rail may allow a sphere no more than 6 inches (152 mm) in diameter to pass. Between 34 and to 42 inches the pattern may be more open, allowing a sphere no more than 8 inches (203 mm) in diameter to pass.[13]

Challenges

Roof gardens typically feature myriad challenges. Wind and exposure to extreme conditions of heat and cold are often existing conditions. There is limited space, often complicated by restrictions on what may be visible on the roof to people on the street. Existing parapets must meet standards of durability and height to avoid accidents. Unlike gardens at grade, there are always structural concerns about the weight of major elements such as stone, soil, and water, which may not exceed certain thresholds, or the roof could collapse. Fire departments often require at least two major means of access, so that if one route is blocked by a fire, people have an alternative means of escaping an inferno on the roof. Increasingly, public buildings and many private ones must (or choose to) be accessible to people with disabilities, so ramps may be required for access with a limited number of steps, a quite challenging requirement when space may be so tight. Special types of insulation and waterproofing may be needed to prevent any problems occurring on the roof garden from causing difficulties to residents or users of the upper floors of a building, Another critical factor is drainage. For existing buildings, it's often impossible and too risky to add drains that connect to pipe systems in the building, as leaks may occur. The concentration of the weight of stored water may also create problems. Once leaks occur, they can be very difficult to repair, because it can be a challenge to find the source. For proposed buildings, the same factors apply, but the landscape architect, architect, and engineer can anticipate drainage, structural, access, and other problems and provide sound solutions from the beginning: a roof that drains well and can support a considerable amount of additional weight in terms of pavements, water, plant materials, people, planters, and pergolas. Parapets for new buildings should be several inches higher than the minimum height requirement to allow for the installation of pavers on the roof, as these elements raise the finish grade of the roof garden as much as six inches higher than the existing grade.

Many clients who want a roof garden also demand privacy, but this is not always achievable if the roof site abuts several adjacent buildings with heights overlooking it. A "fishbowl" almost always remains a fishbowl, so one of the most important factors in developing a roof garden for privacy is that it be located in the highest skyscraper or building within a group of buildings. I once worked on a roof garden project in Manhattan and went to visit the site and take measurements. An old doorman who had been employed in the building for many decades took me to the roof and explained that the penthouse used to be a private trysting place for President John F. Kennedy, but that the Secret Service insisted that he stop using it after some taller buildings were erected in the neighborhood, thereby giving anyone on their upper floors a great view of activities in the penthouse. Curtains were not deemed effective enough to prevent spying. The existing elegant outdoor terrace would be easily viewed from the upper floors of some of these taller stories as well.

Design Process

The design process for roof gardens is similar to other projects, but the site analysis must emphasize specific components. It must study wind and sun exposures, which can be extreme on roof gardens, particular if they are on the top floor of a tall building. Drainage must be clearly understood, and the existing system must be fully functional and accessible for maintenance. The landscape architect must also verify that the site is in full compliance with code requirements for egress; for example, often this means two access points so that there is always an alternative exit if one is blocked. There must be a structural review to verify that the roof is in compliance with the minimum code requirements. At the same time, as there starts to be discussion with the client about program and costs, this is the point when any decision about additional structural reinforcement should be considered.

The rest of the design process—the program, the preparation of increasingly complex design drawings and finally construction drawings, and the implementation process—is quite similar to other design projects.

Construction Materials and Responses

A number of specialized construction materials and design elements are often seen in roof gardens. Often paving on pedestal systems is used to achieve a flat floor plane over a sloping base, as the height of each pedestal can be adjusted. Grid squares of 12, 18, and 24 inches (300, 450 and 600 mm) of strong stone material are the most common, with some variations for rectilinear borders. Common materials are concrete, marble, and slate. The finish must be rough enough not to be slippery. Limestone is often used but usually requires a sealant since the stone is porous. Unusual pavers such as porcelain or ceramic tiles are sometimes used. The larger the grid, the thicker the paver must be to have the structural capability to bridge the wider gap. Organic forms may still be achieved but require more study and careful detailing.

Steps may often be open metal risers and treads to minimize weight and allow free drainage of water. Although it's possible to build a set of more standard stone or brick steps with mortar, the additional weight would likely require more structural support.

Even well-drained standard loamy topsoils can be too heavy when saturated with water, so many commercial lightweight soil mixes, weighing a third to a half as much as regular topsoil, are used. Still, despite the use of many weight-saving methods, it is sometimes necessary, particularly with existing buildings, to reinforce a limited area of the roof with steel in order to support additional weight.

Materials for Planters

Planters may be constructed of metal, wood, and/or plastic. The woods used must tolerate exposure to the elements, as well as intense sunlight, bitterly cold temperatures, and high winds as well as moisture. Teak and redwood are very popular, although sustainable design practices make finding an appropriate wood product more challenging. Even the sturdiest planters are often designed and built with liners, such as polyurethane, metal, or fiberglass, that fit snugly into the wood and protect it from moisture damage. This method has the added advantage that the liners may be lifted out with the plants still installed in them if it's necessary to do repair or maintenance work on the planters themselves. (See Figures 8-14 and 8-15).

Drains must be protected at all times. If a pavers-on-pedestal system floats over the roof, there is less risk of clogging of the drains. They should be equipped with covers or grates with adequate spacing of openings to facilitate drainage. Irrigation systems are often necessary on roof gardens because the intense and variable solar exposures during drought periods will desiccate many plantings, particularly newly installed ones. Better to have the capability of occasional waterings, manual or automatic, during such periods rather than risk losing all the plants. The pavings-on-pedestals systems give the advantage of hiding the installation of irrigation systems and other utilities.

Pergolas and other decorative structures, as well as awnings and umbrellas, can make a roof garden more appealing, enjoyable and colorful. Yet such elements must be unusually durable due to the extreme conditions they may encounter. All such elements must be tied down or anchored, and be adjustable or removable during intense windy periods or storms.

Special water features may be highly desirable accent elements, but they must be designed and coordinated carefully with the engineer and architect to verify that the risk of leaks or water overflow is minimal, and that the entire water element and its utility systems comply with building codes.

Particularly for public buildings, but often for private ones, two means of access are required. Gates should include locks, so that people may be prevented from entering the roof garden when it is closed or during periods of inclement or impending dangerous weather. The height of the parapet is usually specified as about 36 inches (900 mm) above finish grade. The height of security railings or guardrails must be carefully evaluated to prevent any risk of accidental falls and to verify that it dovetails with the parapet details. Special features such as stereo speakers and lighting are effective elements to incorporate, but their location must be carefully evaluated in order to minimize the impact on neighbors.

DOS AND DON'TS FOR ROOF GARDENS

1. Carefully evaluate all existing conditions, particularly the parapets, drainage, and utilities such as water supply and electricity, prior to starting any design work.
2. Verify the building code requirements for structural design limits—for example, 100, 125 or 150 pound per square foot (488, 610 or 732 kg/ m^2)—and allow at least 10% room for error in the design.
3. Don't assume that the same structural limits apply to the entire roof. Work with the engineer to know the variation in limits. Often near structural walls, such as the exterior walls culminating in parapets, a greater weight can be accommodated.
4. Address and repair major drainage problems, access limitations, deficiencies in parapet height, and durability problems—or at least make sure their budgets are clearly determined, planned for, and approved—prior to any design studies.
5. It may help to consider a roof garden to be like a stage set—subject to intense periods of use followed by long periods of inactivity, with many elements that may need to be disassembled and put in storage only to be re-erected repeatedly.

6. Be especially careful with any water features, as they require verification of weight restrictions and applications, leak-free installation and use, and permits and approvals by many parties, from the owners of the building to local or regional governmental entities.
7. Don't experiment with plant materials that are not at least one (and preferably two) planting zones hardier than the zone typical for the location being designed for.
8. Anchor everything to avoid possible dire results such as severe injuries from falling, breaking, splintering, or wind-blown materials. Since conditions on roofs can feel so delightful so often, it is easy to underestimate the severe damage a sudden strong wind can cause.
9. Require shop drawings detailed to exacting standards and scales for anything that must fit snugly on the roof: planters, furniture, awnings, pergolas, etc.
10. Work only with contractors who have several years of verified, demonstrated experience on similar projects. Roof gardens are not a setting where the landscape architect should try out someone new.
11. Cleanup of the site on a daily basis is essential. A severe windstorm or other weather event may occur at any time.
12. Incorporate into the specifications and final contracts long-term maintenance arrangements. A few weeks of intense sun with no watering can fry even the hardiest of plantings, all the more so when they have just been planted and are acclimating.
13. Rely primarily on hardy perennial plants as annuals on a roof garden are unusually risky. Limit annuals to kitchen herbs and a few color beds where space allows.
14. Custom designs (for planters, pavers, and other elements) may achieve spectacular results where budgets permit them, but modular systems are much more cost-effective.
15. Although a seamless connection to the outdoor roof garden from the living space of a residence or office space of an institutional setting is desired, consider a few steps down from indoors to outdoors, to keep out blowing leaves and windswept rain or snow, except in the most sheltered locations.
16. Use lighting to dramatize design effects such as plantings and also to emphasize safety at entrances and exits.
17. Use simple and bold hedges and fences or lattice elements for screening.
18. Provide at least one or two vandal-proof and freeze-proof hose bibs for outdoor use, even if there is an irrigation system. Manual watering during sunny winter days when the irrigation system is deactivated is often required, and they can be used to hose off pavement surfaces.
19. Set at least some of the outdoor lighting to be activated from inside.
20. Use sealants on pavers, particularly the first year after installation.

KLM Equities Sun Roof

This dramatic Manhattan roof garden is a collaboration between Philip Evangelista, who did the planning and execution of the design, and Mark Davies of Higher Ground Horticulture, who did the planting design and installation. Situated in the northwest corner of the roof, the elevated section of the deck is cumaru (*Dipteryx odorata*, also called Brazilian or golden teak), a lightweight, strong, and durable tropical wood, which is set on a reinforced structural steel base so that the deck can support the weight of a range of furniture and planters, as well as many people and outdoor activities. Located on Broadway near East 21st Street, the building tenants enjoy the dramatic views of such city landmarks as Hudson Yards, the Flatiron Building, and the 35XV Building. The roof garden is on two levels. The lower level is a terrace of porcelain pavers less than 1 inch thick, chosen for their high strength and light weight. Visitors climb eight steps to the upper deck and arrive at an elevation which meets the minimum required code height in relationship to the existing parapet of the building. The planting concept is to use evergreen materials on the lower level and grasses and flowering materials on the upper deck. A challenge was to incorporate the stanchions for the davit system used to inspect and maintain the facade on a regular basis; they fit neatly underneath the wood deck. The total area of the elevated deck is about 800 square feet in a trapezoidal shape.

Due to high winds and exposure, it's necessary to carefully select plant materials for hardiness. Specimens of river birch are planted in sunken planters flush with the deck so that as visitors walk at grade the experience is of being on ground level. *Fargesia* sp. (bamboo), *Panicum* 'Shenandoah' (switchgrass), and *Nessella tennissima* (Mexican feathergrass) are among the unusual and effective choices of plant materials. A Greenscreen modular trellis system is used along the north wall of the lower terrace to give a decorative covering of *Hedera helix* (English ivy). On the upper terrace smaller Greenscreen panels are planted with *Hedera helix* and three species of clematis: *C. virginica, C. tangutica*, and *C.* 'Betty Corning.' This rich combination completely cloaks the existing building wall. (See Figures 7.16 to 7.21).

Figure 7.16 KLM Roof Garden, NYC under construction.

Figure 7.17 KLM Roof Garden, NYC under construction.

Figure 7.18 KLM Roof Garden, NYC completed as-built.

Figure 7.19 KLM Roof Garden, NYC completed as-built.

Figure 7.20 KLM Roof Garden, NYC completed as-built.

Figure 7.21 KLM Roof Garden, NYC completed as-built.
photos by Steven L. Cantor

Green Roofs

A green roof may be considered a special type of roof garden, one designed primarily for its environmental benefits, although some green roofs may be used by people for outdoor recreation and related activities.[14] Green roofs first evolved over a century ago in European cities with dense populations and therefore a dearth of open space: the roofs of many buildings became colonized for stormwater storage, using specialized plant materials that could tolerate extreme conditions.

Some of the earlier green roofs, many of which still function, were each unique with different components, both layered materials and plants, from a wide array of sources. By contrast, the modern green roof is typically a sandwich of layers, all of which are usually provided by the same manufacturer, to provide insulation, waterproofing, water storage, root barrier, and finally a growing medium, which is a mix of locally derived inorganic substrate combined with a unique recipe of organic material. The plantings will take hold and gradually provide a continuous carpet of growth. Each manufacturer may have a proprietary stake in the substrate and plantings, and these will vary according to the site, its conditions and the depth of the substrate layer.

Green roofs are typically defined as *extensive*, where the layer is under 6 inches (15 cm), or *intensive*, where the layer can be considerably thicker, up to several feet deep, and thereby able to sustain some large shrubs and even trees. Intensive green roofs often require reinforcement of the roof to support additional weight in existing buildings or expensive reinforced concrete and steel columnar supports for new buildings. Some intensive green roofs mimic what we would call a modern roof garden and can accommodate many people, a range of recreational activities, water features, and many other amenities. Extensive green roofs, which are far more common due to their practicality and lower costs, have relatively thin layers, and often do not have the structural capacity to support more than a small number of horticulturalists or other staff for maintenance. Such green roofs are usually inaccessible and are designed to be seen from above like decorative elements in the landscape, a series of floating carpets alive with an array of plant materials of varying colors and textures. Pure sedum roofs have been the standard for decades, but many green roofs now incorporate additional species to complement the sedums, provide biological diversity, and perhaps harbor insects and birds, even beginning to mimic a natural habitat. A third category of green roofs, increasingly common in North America, is *semi-intensive*, a relatively new type, in which 25% of the green roof area is above or below 6 inches in depth. Such installations may have characteristics of both extensive and intensive designs because of the plant materials used or the thickness of the systems.

Stormwater Overflow Storage

When a green roof is first conceptualized as a design goal, an urgent question is to ask about its primary purpose. Perhaps the most important one may be to store a certain volume of stormwater. During heavy or even moderate periods of rainfall, many cities are burdened by combined sewage-stormwater overflows in which the storm sewers are filled beyond capacity and overflow into the sanitary sewers. The unfortunate result is a discharge of large amounts of untreated sewage into rivers or other bodies of water. A cascade of problems ensues: the water quality deteriorates, available oxygen is depleted, fish and other riverine wildlife die, and water purification costs rise. Green roofs can be designed with the capacity to store stormwater long enough so that by the time it is released into the storm drainage system, there is no longer the danger of combined overflows.

When such methods are applied over a large urban area, there can be dramatic effects. The city of Portland, Oregon, instituted a green roof system, which they call ecoroofs, in response to a mandate from the federal government to protect the Willamette River, which runs through the heart of the city yet is a salmon-spawning stream. The capacity of a new stormwater system could be reduced as a result of the savings brought about by the widespread implementation of an ecoroof system. Other cities across the United States and Canada, such as New York City, Chicago, Toronto, and Vancouver, to name only several, have implemented, regulated, or required green roofs of different kinds for different areas of the city. The New York City Department of Environmental Protection (DEP) set up a highly successful grant program that contributes to paying for green roofs in neighborhoods with serious combined overflow problems if the owners of affected buildings create designs that can store the water from a 1-inch storm (that is, it must retain all of the first inch of rainwater during a storm).

The Grange, Brooklyn, New York

One of the most successful has been the Grange, in the Brooklyn Navy Yard, home to several hundred businesses, including some of the city's most innovative manufacturers, artists, and nonprofit organizations. In recent years the Brooklyn Navy Yard Development Corporation (BNYDC) set the goal of becoming the greenest model of industrial facility redevelopment in the country. Starting in 2011, landscape architect Elizabeth Kennedy, of EKLA Studio, helped BNYDC secure a grant to retrofit the top of an old industrial warehouse building to support a 60,000-square-foot

(5,574 square meters) green roof. Her design now serves as a DEP Office of Green Infrastructure standard for retentive agricultural overburdens. The Grange annually harvests 20,000 pounds (9,071.84 kg) of salad greens, kale, eggplants, cucumbers, basil, and cherries, and it hosts an apiary that yields 1,500 pounds (680.39 kg) of honey per year for restaurants and farmers markets in the region. (See Figures 7.22 - 7.27).

Some cities now require that government-funded buildings include some green roof capacity. The Leadership in Energy and Environmental Design process is often used for documentation, and some government agencies require LEED ratings for new buildings.

Additional Benefits

Many other benefits are associated with green roofs. They moderate temperatures on a roof, often keeping them within a range ideally suited for the most efficient photovoltaic installations. "Photovoltaic panels are often installed on roofs where the solar radiation is the most intense, thereby assuring a high degree of efficiency in converting solar energy to electricity. These panels work best, however, within a certain range of temperatures, without large fluctuations; green roofs help modulate the temperatures within the range that best suits the photovoltaic installations. Green roofs tend to stay much cooler during the day than a standard roof—an advantage for photovoltaic installations."[15] On May 9, 2018, the five-member California Energy Commission unanimously voted to require that all new homes must have solar power.[16] This decision, to take effect in two years in a state whose economy is the fifth-largest in the world, will have a dramatic effect on residential architecture. Landscape architects might have an equally dramatic impact on ecological design in this region of the country by proposing ways to integrate green roofs into the solar power requirement. (See Figures 7.28a and 7.28b).

The layers of a green roof provide natural insulation, so the building is less cold in winter and less hot in summer. Green roofs implemented on a regional scale mitigate the urban heat island effect and have potential for easing the dangers of global warming. Green roofs are a natural fit for schools, because so many school buildings have flat roofs, where installation is the easiest, and they present teachers, students, parents, and local community leaders with a great opportunity for education and science projects. The plant materials on green roofs help filter the air, reducing air pollution. An emerging trend in green roof design is to provide alternative habitat for endangered species or plant communities. In Great Britain the green roof movement there started as a way to provide habitat for the black

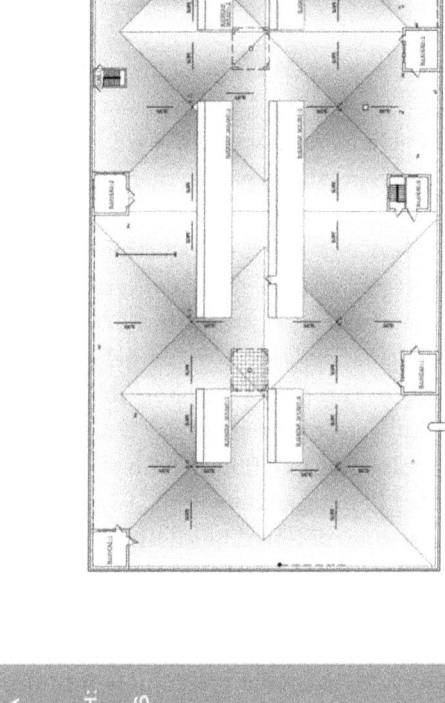

Figure 7.22 The Brooklyn Navy Yard, the Grange: Roof Plan Contributing Area.
Elizabeth J. Kennedy, Landscape Architect, EKLA Studio, MBE, WBE

Roof Plan - Retention Area

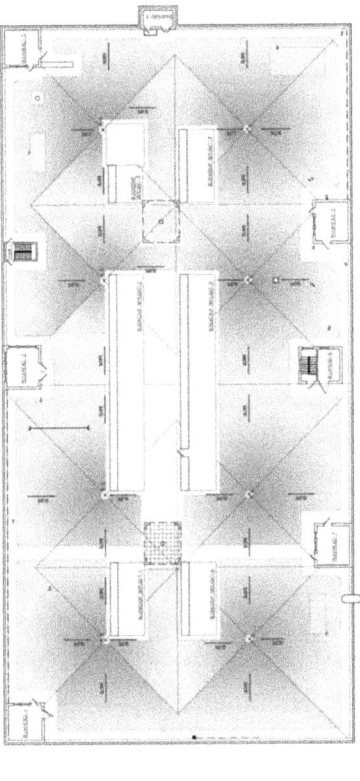

RETENTION VOLUMES @ 1 INCH:

OVERBURDEN ±42,700 SF AREA

3560 CF/26,630 GAL

0.08 CF/SF

REMAINING CONTRIBUTING AREA ±25,700 SF

2140 CF/16,010 GAL

0.05 CF/SF

Figure 7.23 The Brooklyn Navy Yard, the Grange: Roof Plan Retention Area.
Elizabeth J. Kennedy, Landscape Architect, EKLA Studio, MBE, WBE

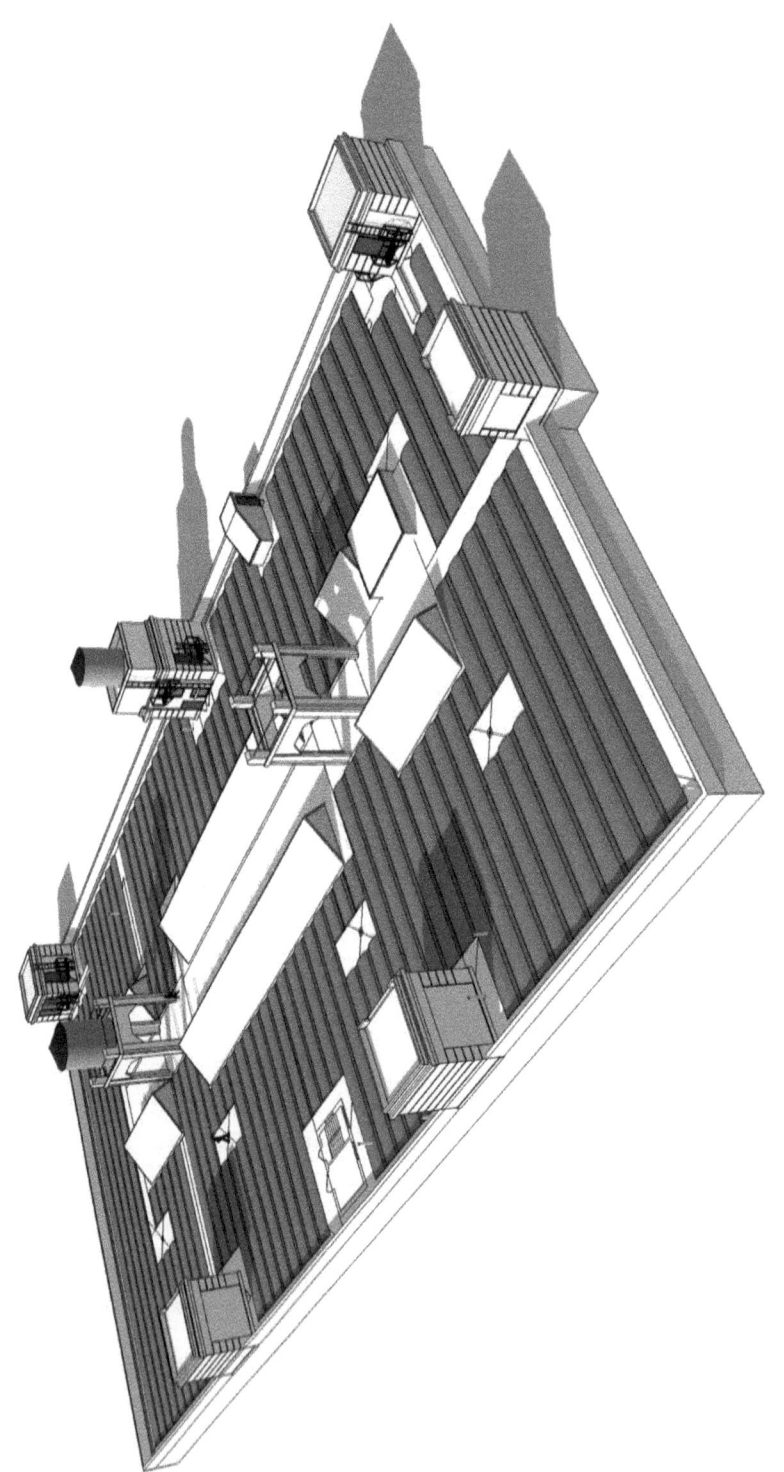

Figure 7.24 The Brooklyn Navy Yard, the Grange: Axonometric view.
Elizabeth J. Kennedy, Landscape Architect, EKLA Studio, MBE, WBE

Figure 7.25 The Brooklyn Navy Yard, the Grange: as-built view.
Elizabeth J. Kennedy, Landscape Architect, EKLA Studio, MBE, WBE

Figure 7.26 The Brooklyn Navy Yard, the Grange: as-built view.
Elizabeth J. Kennedy, Landscape Architect, EKLA Studio, MBE, WBE

Figure 7.27 The Brooklyn Navy Yard, the Grange: as-built view.
Elizabeth J. Kennedy, Landscape Architect, EKLA Studio, MBE, WBE

Figures 7.28a and 7.28b Green roofs in Europe, two installations, Neubauer and Leipzig Biocube.
Zinco, Amsterdam, Europe©

redstart, a species of bird, much of whose habitat has been threatened or destroyed by encroaching development. Living roofs, as they are called in England, as initiated by Dusty Gedge and others, provide a safe habitat for colonies of birds.[17] Increasingly, green roofs provide an ideal setting for experiments, since monitoring operations can be set up rather easily and different variables (depth of substrate, particular plant species, amount of irrigation) can be tested and controlled with limited interference from people or unwanted guests. In Switzerland there are successful efforts by such scientists as Stefan Brenneisen to create habitats for invertebrates like spiders and beetles, which in turn attract other animals, such as birds looking for a meal.[18]

Modular Systems

Although by far the greatest number of green roofs are layered systems, in which many of the components are unrolled like large industrial carpeting, there are also modular systems. Modules may be square or rectilinear planters that fit together into a large interlocking network. They have become more popular in the United States, perhaps because people and potential users want to see one area implemented successfully before giving the approval for a larger system. It's easy to add to such a system. Another advantage of a modular system is that the plantings can be pregrown under ideal nursery or outdoor conditions and brought to the roof as construction concludes, resulting in an instant finished product.

The costs of green roofs remain higher than the costs of conventional roofs. However, as the public and people insist on better environmental quality and the demand grows, prices will tend to moderate. A number of manufacturers based in Europe now have subsidiaries in the United States and Canada (and a few in Mexico) who focus on certain applications in particular regions of the country. Competition and demand will gradually result in more effective green roofs and competitive prices.

Germany is home to Forschungsgesellschaft Landschaftsentwicklung Landschaftsbau (FLL), an organization that rigorously tests materials and components of green roofs (filter fabrics, root barriers, insulation) as well as a wide range of plant material. Every several years, it produces a new edition of its guide (translated into English as *Guidelines for the Planning, Execution and Upkeep of Green-Roof Sites*), and it's an essential resource. Even though there can be climatic variations between Germany and the United States, Canada or elsewhere, the results of the German plant tests are a strong indicator of what to expect in other settings. It has also set industry standards for the last fifteen years, such as definitions of key terms.

For example, it is appropriate to refer to the growing medium as a *substrate* rather than topsoil, as it often consists of primarily inorganic materials (frequently drawn from nearby regions) that will not decompose over time and therefore will not require replenishment; a small percentage of organic materials, nutrients, and other materials may be incorporated into the substrate, according to a precise recipe, to create the final growing medium. Sometimes confusion arises when some manufacturers or suppliers use *substrate* to refer to their particular growing medium recipe. Read their specifications carefully.

USPS Morgan Processing and Delivery Center, New York City

An earlier project designed by Elizabeth Kennedy is the green roof on the United States Post Office's Morgan Processing and Delivery Center in Manhattan. Built in 1933 and designated a historic landmark, the 2.2-million-square-foot (204,380 square meters), seven-story building is located in Midtown. Its green roof—the first completed for any U.S. federal building, and at almost 2.5 acres (1.5 hectares), the largest in New York City—was designed to provide facility employees a pleasant space for a breath of fresh air while taking in panoramic views of the skyline and the northern New Jersey shore.

The roof is extensively planted as a large field of succulents, with bands of lush native grasses, perennials, and flowering shade trees. Site furnishings constructed of Forest Stewardship Council–certified wood and contemporary lighting complete the modern, minimalist design. The overburden will protect and extend the life of the waterproofing by up to fifty years, and it reduces stormwater runoff contributions to NYC's sewer system by as much as 75% in the summer and up to 35% in the winter. (See Figures 7.29 to 7.31).

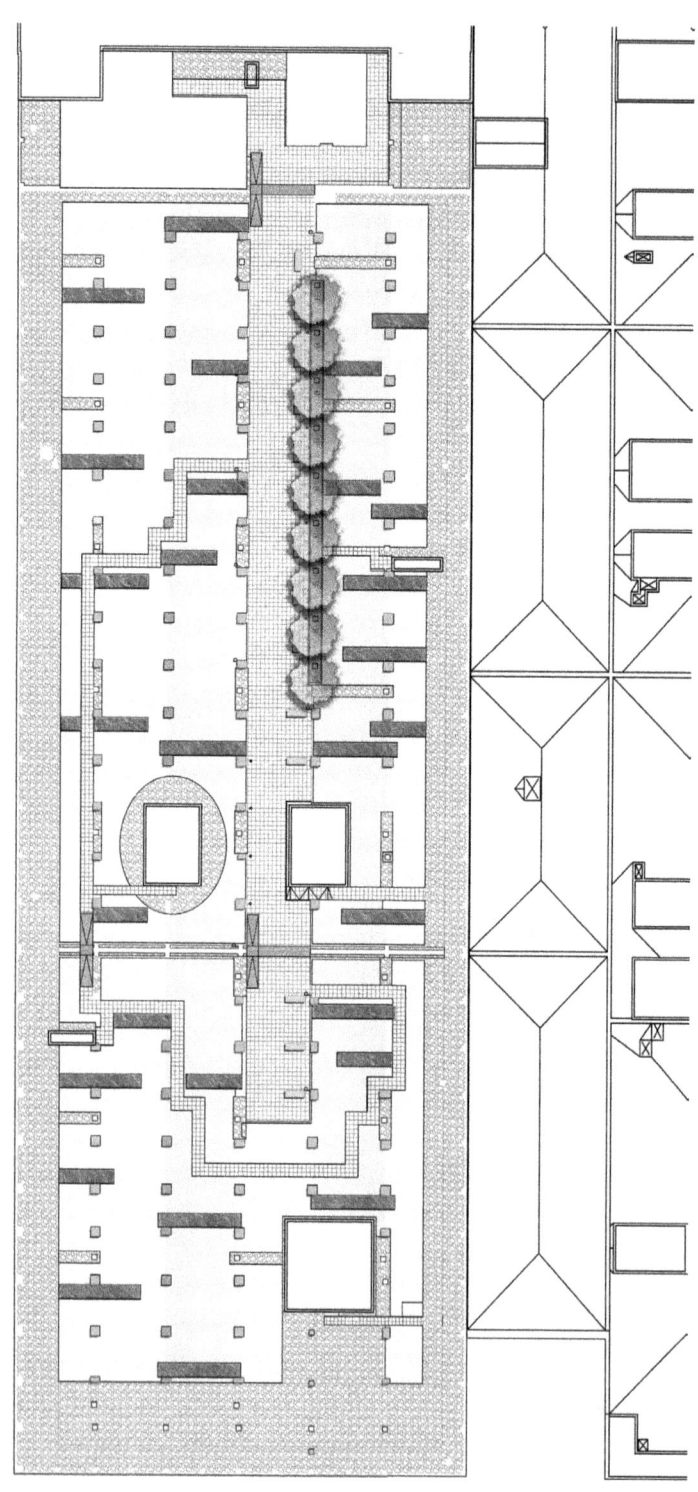

Figure 7.29 United States Post Office Morgan Processing & Delivery Center, Manhattan New York City: Illustrative Site Plan.
Elizabeth J. Kennedy, Landscape Architect, EKLA Studio, MBE, WBE

Figure 7.30 United States Post Office Morgan Processing & Delivery Center, Manhattan Elevation.
Elizabeth J. Kennedy, Landscape Architect, EKLA Studio, MBE, WBE

Figure 7.31 United States Post Office Morgan Processing & Delivery Center, Manhattan, New York City: Aerial view as-built. Notice Hudson River in the background.
Elizabeth J. Kennedy, Landscape Architect, EKLA Studio, MBE, WBE

DOS AND DON'TS FOR GREEN ROOFS

1. Determine the purpose(s) of the proposed green roofs at an early stage. A green roof need not provide every conceivable benefit to be successful, cost-effective, and beautiful.
2. Include an engineer as part of the design team and verify the structural load requirements at an early stage.
3. Interview representatives from several green roof manufacturers before deciding on the system(s) that seems most suitable for your setting.
4. Include a diversity of plant materials, but verify that they are suitable for the setting and environment.
5. To encourage a diversity of plant and animal communities (including insects and invertebrates) on a green roof, vary the height of the substrate layer and model the surface with undulations.
6. Provide irrigation for the first year and have the capacity for manual watering during periods of extreme drought.
7. Develop and train a skilled maintenance staff (at least one person) that knows which plants are weeds and must be removed; and which plants are desirable and should be encouraged.
8. If a favored species dies, try something else. For example, there are thousands of species of sedums, many of which resemble one another yet have vastly different horticultural requirements. There are many other species that resemble sedums, so some combination will succeed.
9. Think of the green roof as the guide to its own horticulture, that is, allow whatever plants grow naturally to become well-established and weed out the ones that die.
10. If suitable (see item 1), provide plants and substrates that encourage colonization by insects or other animals like birds.
11. Don't apply herbicides and insecticides.
12. Fertilize modestly. Intensive green roofs are more likely to need some fertilization, but the goal must be to allow the plants to adapt naturally to the size and depth of their growing space.
13. Include leak detection systems where feasible and affordable.
14. Set up an educational program (lectures, displays, graphics) that explains to the public the systems of the green roof and its advantages.
15. Consider an alliance with a group of independent or university scientists to undertake measurements and monitoring of the green roof's key functions. How much water does it store? How much filtering of dirt and soot from the air occurs? How much are temperatures moderated?

16. Design at least some of the green roof so that once it is planted, the goal is to let it be, other than standard maintenance and weeding.
17. Even for extensive green roofs, try to upgrade the structural capacity enough in a small area to allow for a seating and viewing area.
18. Connect to and integrate with the architecture of the building.
19. Incorporate green roofs in contrasting microclimates and exposures. Not every green roof can be on a south-facing site with wide exposures and a flat roof.
20. Consider sloping green roofs.
21. Incorporate photovoltaics. These may be a considerable coverage or just a few panels that might provide enough electricity to operate lighting or irrigation.
22. Now that the green roof industry has advanced well beyond infancy in the Western Hemisphere, consider performance-based specifications for some designs rather than detailing every square millimeter of a roof space.
23. Don't reinvent the wheel. There is reliable data about the feasibility of a great many species and varieties of plant materials in the German FLL standards, now available translated into English.

Problem 1

GREEN ROOF AS HABITAT FOR ENDANGERED SPECIES

One emerging trend in green roof design in the last decade is to develop a habitat—that is, to use the space on the roof of a building as a laboratory in which to attempt to recreate an environment that may be threatened by encroaching development at ground level, so that students of the natural world may study how best to protect endangered species and habitats and how additional sites might be created. A corollary concern involves the risks of such design and research efforts: are they too complex to consider?

The following readings describe different projects and settings in Europe, North America, and other locations as an introduction to this topic. The most recent article is from the July 7, 2017, issue of Audubon Magazine, in which Brooke Borel describes ongoing research efforts by teams of scientists to combat predation by invasive mice on endangered ashy storm-petrels on the Farallons, rocky islands off the coast of San Francisco. She also documents predation by other species of mice, the descendants of stowaways on nineteenth-century seal hunting ships, who have thrived on the island of Gough, 1,700 miles west of South Africa, by dining on the chicks of Atlantic petrels and Tristan albatrosses, reducing the survival rate of the chicks to about 10% and placing the long-term survival of these two species of birds at risk.

Brooke Borel, "How Genetically Modified Birds Could One Day Save Island Birds," *Audubon*, Summer 2017, http://www.audubon.org/magazine/summer-2017/how-genetically-modified-mice-could-one-day-save.

In Steven L. Cantor, *Green Roofs in Sustainable Landscape Design* (New York: W. W. Norton, 2008), read the following sections: chap. 2, "The Green Roof Design Process," 38–61; excerpts from "Green Roofs in Europe" on University Hospital, Basel, Switzerland (spiders, insects, endangered bird species), 93–98; the black redstart in London, 113–119; Oak Hammock Marsh Conservation Centre, Manitoba, Canada, 175–182; River Rouge Ford Plant, Dearborn, MI, 193–196; Neuhoff Meat Packing Plant, Nashville, TN, 197–200. You may wish to browse or skim other sections of interest.

O. E. Wilson, *Anthill* (New York: W. W. Norton, 2010); this is a novel.

Propose a design atop a building of your choice for a green roof, ecoroof, or living roof to be a habitat for a particular species or group of species of insects, spiders, beetles, or small animals. Provide a plan to scale, and section or perspective. Include a plant list.

Assume everything is seeded or plugs, unless indicated otherwise, or you wish to propose other methods for clear reasons which you provide.

PART 5: ENVIRONMENTAL ASSESSMENT

The golden age of environmental assessment began with the passage in 1969 of the National Environmental Policy Act (NEPA) by Congress, signed into law by President Nixon.[19] This law established the Council on Environmental Quality, which became the arbiter of the impact of federal actions on the environment. Eventually, several states, such as California and New York, established their own environmental quality organizations and regulations. NEPA required that any project over a certain size that was deemed by the federal government to have a potential impact on the environment must be carefully studied in order to evaluate the impacts and how they might be mitigated. An environmental assessment is the process of following this series of complex tasks. An environmental impact statement is a published document in which all the existing conditions, proposed actions, and their potential impacts are described, analyzed, and predicted. Public review is required; public hearings are held and submission of written comments and questions is solicited, so that when the final environmental impact statement is published, it is a complete record of the project's scope, budget, and impacts, with recommendations of how to proceed.

Complex environmental assessment projects require the services of many consultants, such as civil, structural, and sanitary engineers; scientists; economists; traffic planners; and support staff. The client may be a public agency charged with studying a specific proposed action, such as the creation of a housing development for a large property, or it could be the company/owner wanting to move forward with the development with all approvals in place. In either case, the landscape architect must develop skills in communicating well with all sorts of individuals (including citizens and specialists) and special interest groups, and must have the patience to explain the same information over and over again for public hearings or other community meetings.

Some useful terms:[20]

> Adverse environmental impacts that cannot be avoided if the proposed action is implemented
> Categorical exclusion (CATEX)
> Development plan preferred alternative
> Draft environmental impact statement (DEIS)
> Environmental assessment (EA)
> Environmental impact assessment (EIA)
> Final environmental impact statement (FEIS)
> Finding of no significant impact (FONSI)
> Lead agency
> Mitigation measures
> New York State Department of Environmental Conservation (NYSDEC)
> Non-point-source contaminants
> State Environmental Quality Review Act (SEQRA)
> Waste assimilation capacity study (WACS)

Typically, the requirement for an environmental impact assessment is triggered by law when a proposed action of over a certain minimum size is studied or initiated. Depending on the state authority or federal authority involved, different thresholds require the developer to carry out an environmental assessment. Just as an example, in New York State, an EIA is triggered by the New York SEQRA; the federal law has different requirements. California's law, the California Environmental Quality Act, is similar. The lead agency is the government or other organization empowered by local, state, or federal law to approve the project. An environmental assessment is technically a preliminary study to determine if the proposed action has enough impact (crosses a threshold) to require a full-scale environmental impact statement.

All correspondence or inquiries concerning the EIS should be addressed to the lead agency, which is then responsible for forwarding all legitimate concerns and questions to the consultant preparing the EIS. The consultant, such as a landscape architecture firm, is usually hired by a developer who owns the land or intends to move forward with the project once it's approved. The principal consultant, as assisted by the developer and other consultants and with input from the lead agency, prepares a DEIS. Public hearings and other review processes occur, and based on public comments and written statements submitted to the lead agency, a FEIS is prepared, which incorporates changes to the plans, additional studies as called for in the questions previously submitted. At each stage of the EIS, the lead agency usually sets a deadline by which time all comments and inquiries must be received by the lead agency. The FEIS incorporates responses from the consultant and its team to all the questions and comments to the satisfaction of the lead agency. Depending on the number of questions (from citizens and other interested parties attending public hearings or reading the draft document) and the extent and types of additional reviews required and generated as a result of the DEIS, it can take months or several years to complete the FEIS. Then the lead agency votes to authorize the project to move forward, or in the event of further questions or concerns, withholds approval pending further study, tables the project, or some combination of the above. The whole point of the EIA process is to study the potential impacts of a proposed major development in order to be certain that, prior to its approval, all major impacts have been studied and means of mitigating them incorporated into the final approved plans.

The table of contents of an EIA suffices to show its comprehensive nature. The following is an example of a particularly thorough one involving Rushmore, a major residential project proposed for the town of Woodbury, New York.[21] Rushmore went through a long EIA process under the direction of Raymond J. Heimbuch, PE, of Clarke + Rapuano, Inc. As a landscape architect/project manager there who worked on this project under his direction, I still remember him years later as perhaps the most knowledgeable, accomplished, and versatile engineer with whom I ever worked. Other EIAs would not need to be so complex; some, however, might even be more complex. Their content and organization will vary depending on the nature and scope of the project, its location and setting, and the requirements of the assessment.

SAMPLE TABLE OF CONTENTS OF ENVIRONMENTAL ASSESSMENT REPORT[22]

1. SUMMARY
2. PROPOSED ACTION
Introduction
Description of the Project
3. ENVIRONMENTAL SETTING
The Region
County
Town
Travel Patterns
Mass Transit
Municipal Fiscal Conditions
Town, School, and County Taxes
Sewer and Water Districts
Fire District
Municipal and Community Services
Sewage
Water Supply
Electric, Telephone, and Natural Gas
Fire Protection
Police Protection
Solid Waste
Hospitals, Health Care, and Ambulance Service
Education
Recreation
Socioeconomics
Population
Employment and Income
Retail Commercial Services
Housing
Market Feasibility of Project

Land Use Planning and Development
Present Land Uses and Zoning
Typical Development Trends
Town Master Plan
County Development Policies
Transportation
Existing Streets and Highways
Existing Traffic Volumes
Railroad Travel
Bus Service
Accident Data
Air Quality
Air Quality Standards
Existing Background Levels
Noise
Noise Level Standards
Ambient Noise Survey
Topography, Geology, Soils, and Groundwater
Topography
Geology
Soils
Groundwater
Water Resources
Hydrology, Watersheds, and Flooding
Wells
Water Quality
Town Aquifer and Recharge Area
Major Streams
Wetlands
Aquatic and Terrestrial Biota
Aquatic Biota
Terrestrial Biota
Aesthetic, Cultural, and Historic Resources
Aesthetic Resources
Historical and Cultural Resources

4. POTENTIAL ENVIRONMENTAL IMPACTS
The Region
County
Town
Travel Patterns and Mass Transit
Municipal Fiscal Conditions
Town, School, and County
Sewer and Water Districts
Fire District
Municipal and Community Services
Sewage
Water Supply
Electric and Telephone
Fire Protection
Police Protection
Solid Waste
Hospitals, Health Care, and Ambulance Service
Education (School District)
Recreation
Socioeconomics
Short-Term Employment
Long-Term Employment
Population and Housing
Retail Commercial Services
Land Use Planning and Development
Displacement and Acquisitions
Relationship to Adjacent Land Uses
Zoning Patterns
Compatibility with Town Master Plan
County Development Policies
Transportation
Methodology
No-Build Conditions
Build Conditions
Traffic Impact Analysis

Mass Transit
Accidents
Adequacy of Proposed Streets
Air Quality
Impacts During Construction
Impacts Due to Traffic
Noise Environment
Potential Receptors
Vehicular Traffic
Noise Predictions
Criteria for Abatement
Long-Term Impact Without the Project
Long-Term Impact with Full Development
Short-Term Impacts During Construction
Topography, Geology, Soils, and Groundwater
Topography
Geology
Soils
Groundwater
Water Resources
Hydrological Impacts
Water Supply
Water Quality of Town Streams
Impact on Town Aquifer and Recharge Area
Impact on Wetlands
Aquatic and Terrestrial Biota
Biota—Vegetation
Biota—Wildlife
Esthetic, Cultural, and Historic Resources
Esthetic Resources
Cultural and Historical Resources
5. ADVERSE ENVIRONMENTAL EFFECTS THAT CANNOT BE AVOIDED IF THE PROPOSED ACTION IS IMPLEMENTED
6. IRREVERSIBLE AND IRRETRIEVABLE COMMITMENTS OF RESOURCES

7. MITIGATION MEASURES
8. GROWTH-INDUCING ASPECTS OF THE PROPOSED ACTION
9. EFFECTS OF THE PROPOSED ACTION ON THE USE AND CONSERVATION OF ENERGY
10. ALTERNATIVES TO THE PROPOSED ACTION
Do Nothing
Existing Zoning Without Public Sewer and Water
Comparison of Alternatives
11. REFERENCES AND APPENDICES
12. ILLUSTRATIONS

Due to the complex and unpredictable nature of EIAs, it's very challenging for a landscape architecture firm to propose fees for carrying out an environmental impact statement. It's rare that every phase of the work will go smoothly. Sometimes public opposition or requirements imposed by the lead agency will greatly complicate the process. It's not uncommon to uncover in the preliminary stages of study certain issues or topics that were previously unknown yet demand additional study not part of the original scope of work. A site may turn out to harbor an endangered species of plant or animal. The proposed development may place such a demand on existing infrastructure that new sewage plants or other utilities must be incorporated. Archaeological artifacts may be discovered that could trigger requirements for a complete cultural survey. Stormwater needs may tax the capacity of downstream stormwater management systems. Often each of these issues may require expanding the scope of work, including hiring specialized consultants. All of these types of issues arose with Rushmore. Even if a draft environmental impact statement of high quality is prepared, there may be so many questions and comments that it could take the firm months to respond to all of them in the process of developing the final environmental impact statement. The design itself may evolve dramatically during the course of the EIA process so that what is finally approved is dramatically different from the initial concepts. Although NEPA was enacted almost half a century ago, it is hard to imagine that the legislators responsible had the foresight to realize how dramatically the law would affect and improve the quality of the designs it was responsible for mitigating. (See Figures 7.32, 7.33 and 7.34a to 7.34e.)

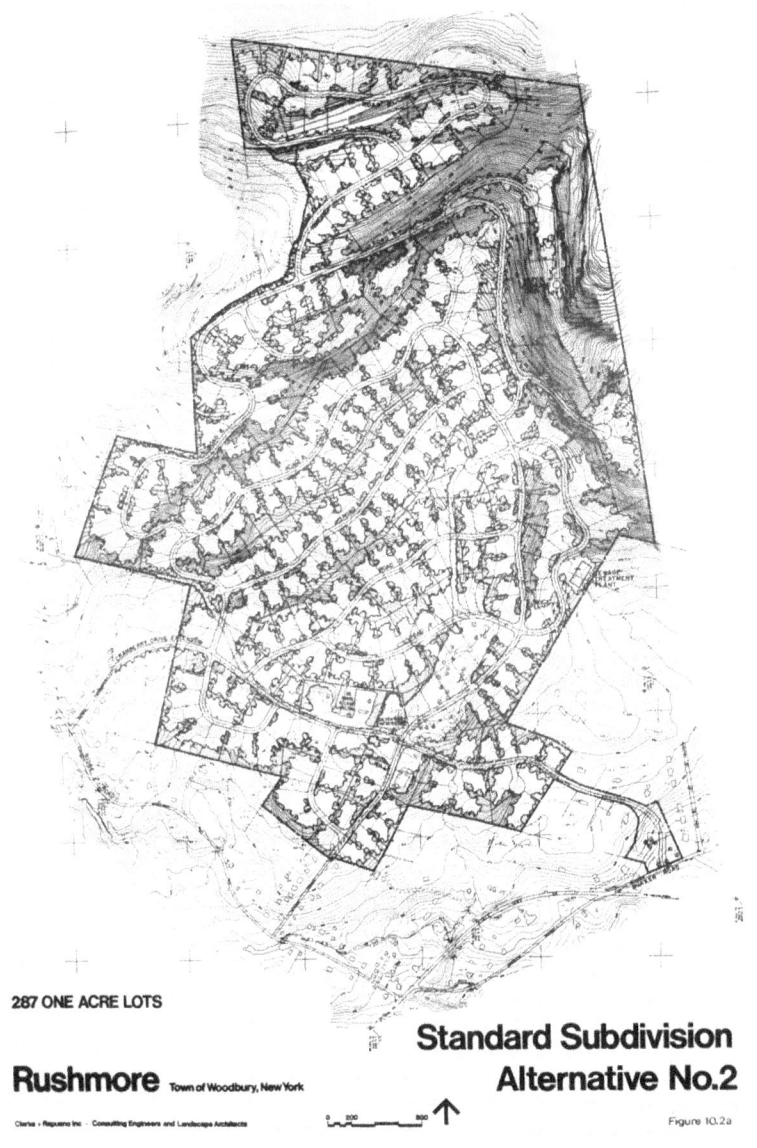

Figure 7.32 Rushmore standard development showing single family lots over all developable land, one of several such studies undertaken.

Figure 7.33 Rushmore, Development Plan Preferred Alternative, showing cluster development on flatter, more developable land with steeper and higher elevations left undisturbed.

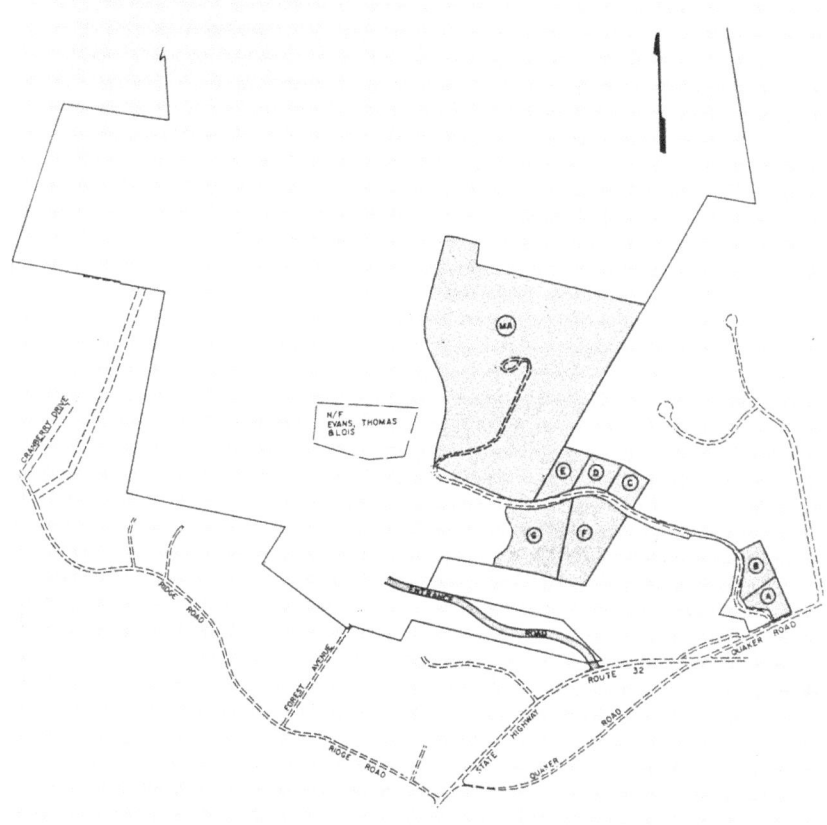

PHASE 1

Rushmore Town of Woodbury, New York

Figures 7.34a–e Rushmore Phase 1 to Phase 5.
Clarke + Rapuano, Inc

PHASE 2

Rushmore Town of Woodbury, New York

Figures 7.34a–e Continued.

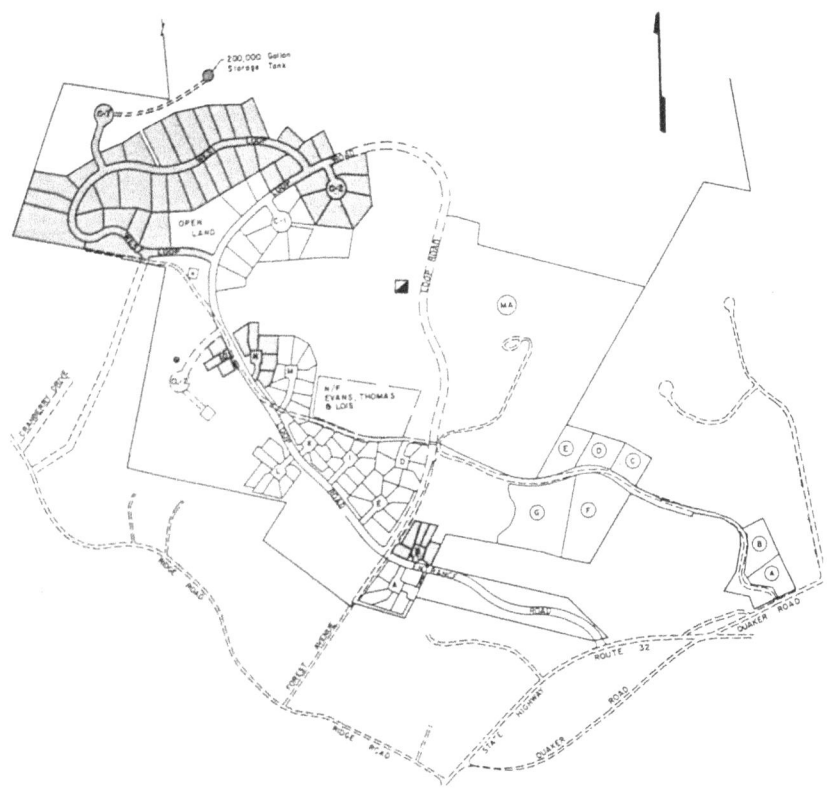

PHASE 3

Rushmore Town of Woodbury, New York

Figures 7.34a–e Continued.

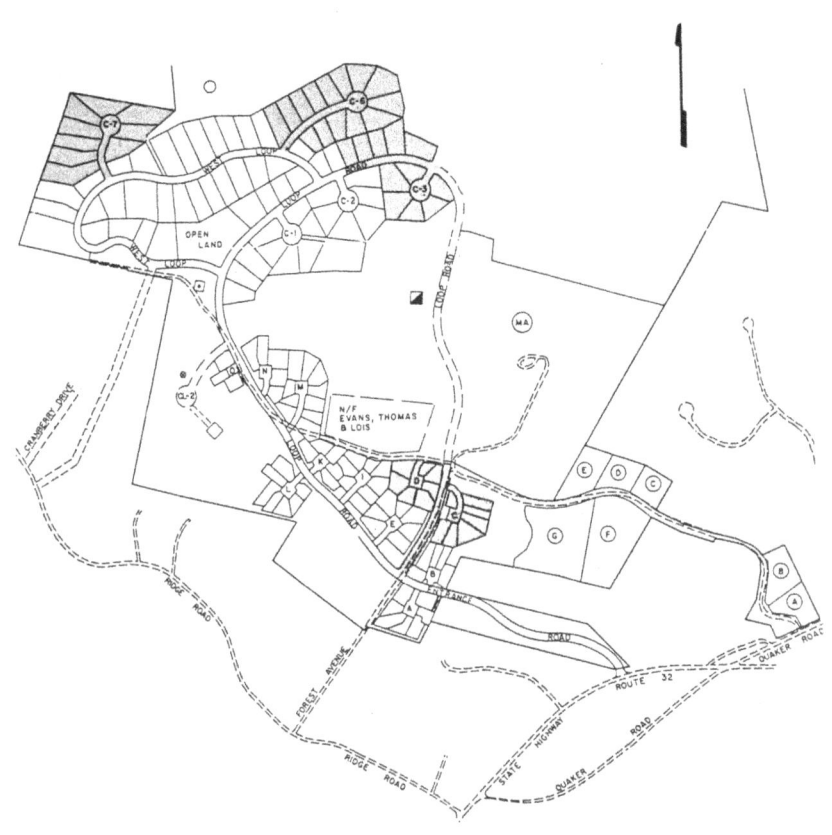

PHASE 4

Rushmore Town of Woodbury, New York

Figures 7.34a–e Continued.

PHASE 5

Rushmore Town of Woodbury, New York

Figures 7.34a–e Continued.

DOS AND DON'TS OF ENVIRONMENTAL ASSESSMENT

1. Know the requirements of the environmental assessment that apply to your project.
2. Don't assume that the new project will be just like the one recently completed.
3. Try to submit and gain approval for fees one step at a time. For example, don't begin by submitting a fee for the entire process. Instead, suggest a fee for the draft EIA, leaving room for extra items of work, and with the understanding that a fee for the final EIA will be proposed upon completion of the draft EIA.
4. Develop a list of consultants with specific areas of expertise with whom you are comfortable collaborating. As you complete an EIA, evaluate these consultants and update your list so it's in place for the next project.
5. Have a preliminary meeting with the representative of the lead agency in order to gauge your firm's and your own comfort level in working with this person. The time to realize that this person is unmanageable as a representative is *not* after you've already committed to work with him/her.
6. Learn to say no to additional studies or additional delays in periods for comment if no additional funding is provided. By the same token, say yes as long as additional compensation is provided and the compensation is not for political purposes—that is, simply to delay completion of the project or confuse issues.
7. Attempt to be the lead consultant, if possible, rather than a consultant to another firm, but if you're a subconsultant, be absolutely clear about your scope of work. Under some circumstances, if the project is unusually complex and your scope of work is quite specific, this is an advantage.
8. If possible, propose separate fees for public hearings, presentations to town boards or officials, or any other public meeting, as there can be so many of them over the course of the EIA process.
9. Do not confuse the EIA process with the actual preparation of contract documents. Sometimes there is a smooth transition from the approval of the final environmental impact statement to the contract documents, but they are still quite different documents, even if the former gives you a head start on the latter.
10. Do not assume that because a FEIS has been finally approved with great fanfare, that the project will move forward to implementation, and your firm will be hired to do the complete contract documents. Sometimes, upon the success of the FEIS, the developer may sell the project to another firm. On other occasions, so much time has

elapsed, that economic or other conditions may have changed to such a degree, that it is no longer viable to implement the project.
11. Anticipate a great deal of writing time for a long, detailed document for public review and comment, with many plans, exhibits, and diagrams.
12. Have a staff employee skilled in presentations available for public hearings or any important meetings where it's critical to present the team's work in the best possible light and at least one other person available to attend to take minutes and/or be able to gauge the reaction of the public or other important people attending. Weak presentations or argumentative individuals can sabotage months of solid work.
13. Ask for feedback from the consultants with whom you are collaborating. Even if you have hired them for a specific scope of work within the EIA, such professionals often have expertise in other fields, excellent observational skills and experience in similar projects which they would likely be quite willing to share over lunch or when they discuss their preliminary findings on their assigned task on an individual basis with you.
14. Strategize with the entire team of consultants as a group on a regular basis. At a minimum, share all new information via digital maps and files and ask for their feedback on a regular basis.

Freshkills Park

A fitting penultimate project for this chapter is Freshkills Park because it encompasses three major aspects of landscape architecture previously discussed: site planning, public works, and environmental assessment. It is the largest landfill-to-park transformation in the world. At 2,200 acres (890.3 hectares), it's almost three times larger than Manhattan's iconic Central Park, and it features state-of-the-art engineering and design. No park of this magnitude had been built in New York City in over a century.

A little less than half of the site of the park is landfill; the balance is creeks, wetlands, and dry lowlands. The landfill operation was officially closed by New York City in 2001. Four major mounds of the landfill were gradually capped in sequence from 1996 to 2018. The environmental impacts of the recovery of methane and leachate processing were integrated into the design so that the land could be reclaimed for recreational purposes. The current design already includes many recreational features open to the public: ball fields, soccer fields, miles of pedestrian and bicycle trails, wetland restorations, roadways, and other facilities. Wildlife abounds, particularly birds and aquatic life, since freshwater, brackish water, and saltwater are all present or nearby. East Park, currently in design, will support renewable energy, kayaking, wetland boardwalks, and recreational trails for walking, running, and biking. There are also ongoing research opportunities aimed at urban ecology and restoration. The combination of its innovative physical design and its potential for renewing links between previously fractured elements of the site promise to transform and reconnect the site to its natural history, local ecosystems, and neighboring communities.[23] (See Figures 7.35 to 7.41).

Figure 7.35 Illustrative site plan Freshkills Park, 2,200 acres.
Courtesy of Freshkills Park and the City of New York

Figure 7.36 View of active landfill, which finally closed in 2001.
Courtesy of Freshkills Park and the City of New York

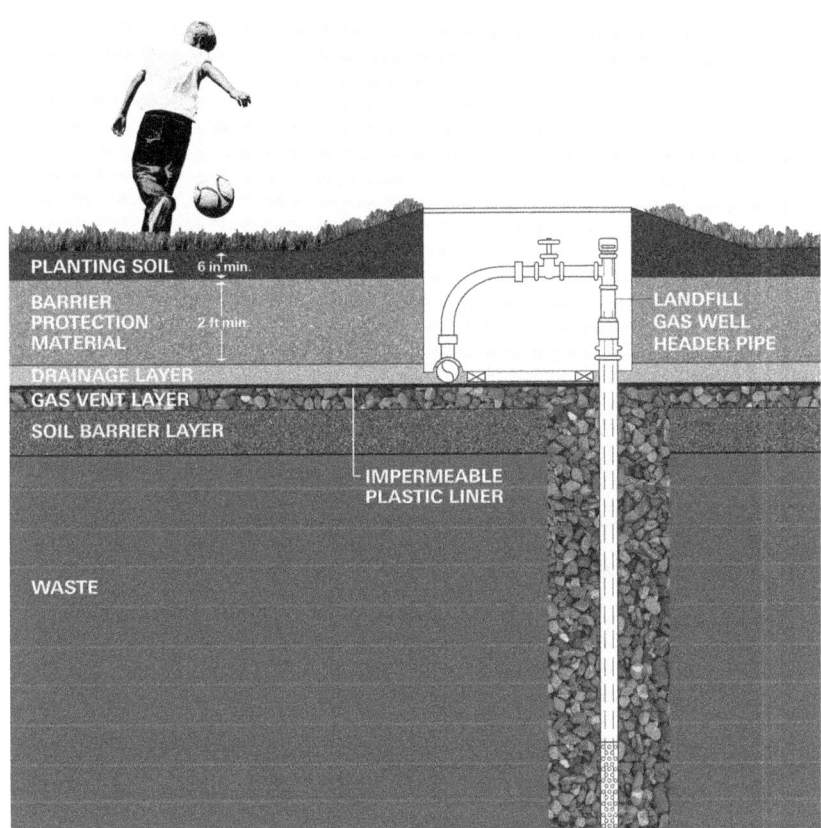

Figure 7.37 Cross section of landfill closing.
Courtesy of Freshkills Park and the City of New York

Figure 7.38 Mature vegetation has become established.
Courtesy of Freshkills Park and the City of New York

Figure 7.39 Flowering meadows abound.
Courtesy of Freshkills Park and the City of New York

Figure 7.40 Four aerial views of the vast landscapes of the park.
Courtesy of Freshkills Park and the City of New York

Figure 7.41 There is a large network of bicycling routes.
Courtesy of Freshkills Park and the City of New York

Problem 1

QUICK ENVIRONMENTAL ASSESSMENT

(In the following problem, any *terms in italics* may be changed to suit the particular instructor or team leader, or your own personal, work or studio setting. Typical changes might be specific people, such as clients or contractors, plant materials, or the locations of sites, people, institutions or companies.)

You are hired by the *City of X* to do a quick environmental assessment of a park project being considered for a 10-acre site (4 hectares), 6 acres (2.4 hectares) of which is over an old landfill that was closed twenty years ago. At that time this site was in the midst of an undeveloped area that was not part of the incorporated city limits. However, since that time the city has grown up around it, and the site is part of a substantial additional area that was added to the city limits a few years ago. Two sides of the trapezoidal site are bordered by roads, and two sides back onto residential developments. The city now proposes to develop the site primarily for two baseball fields or one soccer field, two tennis courts, two basketball courts, and a large informal recreation area. The preliminary program calls for about sixty parking spaces, a set of restrooms, and a small cafe or food kiosk.

You are given copies of the engineer's landfill closing plan, which shows where areas of trash were compacted and capped, and the location of a methane release pipe. You are also given a survey of the site at 1" = 20'. Befitting a former landfill site, the topography is rolling, with four mounded areas. There is some established vegetation, including some mature trees on the parts of the site outside the landfill area. The former landfill itself has a well-established pasture-like appearance with nice wildflowers, but it does not have major shrubs or trees.

Prepare a two-page outline of an environmental assessment that you propose to conduct. Also, prepare a letter to the *City of X* in which you suggest a fee for your services. You may choose to divide it into phases.

PART 6: INTERNATIONAL WORK

In our increasingly complex world where everyone is connected, international practice is becoming commonplace. Design firms often have offices in overseas locations so that staff can respond quickly to the demands of residents and clients in faraway locations. Coupled with the proliferation of the internet and social media, almost anyone in one location in the world may communicate with someone far away. Design videoconferencing and other modes of work unknown just a few decades ago are now common practice. A principal or project manager can communicate from a remote location to colleagues thousands of miles away, and simultaneously share information and the latest design concepts or details.

Basic Requirements

Despite the increasing ease of such work, there are basic requirements. Ideally, anyone determined to work in international practice must be fluent in the language and culture of the country in which he/she is doing work, have a sense of adventure, enjoy exotic settings, be able to travel (often flying long distances), live and work there for considerable periods, have exceptional communication skills to harness and direct the efforts of diverse designers in many different locations, and have a high level of self-direction and informed independence. There was perhaps a time when countries such as China imported Western-educated landscape architects to come to their country as lead designers. Often they were not fluent in Chinese (or other local language), but their mere presence added to the legitimacy of the design team in foreign eyes. As China gained in independence and stature, many of the landscape architects and other design professionals working there are homegrown, and obviously as skilled as anyone, perhaps more so. Given that they are immersed in the local environment and intimate with its complexities, there are perhaps more reasons to embrace them than try to exclude any of them from an international team. Foreigners may be welcome when they have the full range of credentials, including language skills, that are most needed. Such work is not for the fainthearted. Despite our age of instant communications, the pressure of the work in foreign settings and the pace at which major decisions must occur often require landscape architects to make decisions without the input of everyone in the home office. Confident, self-directed people are great assets. Similarly, many Americans may be constrained by expectations of thirty-five- or forty-hour workweeks, which may seem like unattainable luxury goods to striving designers from other cultures who are used to working whenever it's necessary to get the job done. Finally, one must not overlook that many foreign-born professionals have immigrated to the United States or other Western countries, studied and mastered landscape architecture or other professions, returned to their home countries, and excelled as founders of design firms, where they have the advantage of being at least bilingual and totally conversant in Western and Asian cultures. It is remarkable what some of these people have achieved.

Broadening of One's Perspectives

International work broadens one's perspectives. What we consider beautiful and practical may be based on local or acquired tastes originating in our home country that in no way apply to a particular foreign culture, such as China's. Two examples there are favored colors and numbers.[24] Red, yellow, and green are favored, whereas white and black are deemphasized. Blue and gold have more neutral symbolism. Certain colors are traditionally used for buildings of particular functions: banks and restaurants are often painted in red and green, government cars and police uniforms are often in black, associated with authority, power, and neutrality.

Even numbers are favored over odd numbers, with some exceptions. If a color or number in Chinese represents a word in the vocabulary of the language with a favorable or auspicious meaning, this is transferred to the symbolism and use of that particular color or number. For example, the number 8 is similar in pronunciation to *fa*, which means "prosperity" and "happiness." The pronunciation of the number 9 sounds similar to the word for "permanence." Similarly, the number 2 represents harmony, and the number 6 success.

In Chinese culture, some traditions originated thousands of years ago, such as *feng shui*, the art and philosophy of harmonizing elements in the landscape, including buildings. Some people feel that such traditions are grounded in superstition; others find the basic principles and guidelines helpful, like an environmental or ecological shorthand, in considering how the different elements in a landscape may relate to one another or their settings and surroundings. The origins of the practice date back to astronomical and meteorological knowledge that helped site buildings and landscapes to receive favorable light and winds and be sheltered from intense sun or gusts.

Americans risk creating another chapter in *The Ugly American* if they persist in imposing or even suggesting that Western aesthetic principles should preside in various Asian, African, or Latin American settings or landscapes, in which the native population has a long tradition of its own design culture and preferences. China, Japan, India, and Korea all have overlapping yet distinct cultures, which one must learn and respect in order to establish a business there. Similarly, acquiring at least a basic knowledge of the culture of various African, South American, Caribbean, or even European cultures must be the starting point for any landscape architecture or other design services being developed there by guests in those countries.

Potential Conflicts Between Cultures

One difficult aspect of working in a foreign country occurs when the cultural mores and requirements abroad do not match or are in conflict with American or Western values. Think of the roles of women, minorities, the LGBT community, and people practicing a particular religion. In the United States the trend, particularly over the last few decades, is one of inclusiveness—that is, more of all of these types of people are able to work in the design industry, sometimes in leadership roles and certainly as participating members of the design team. Yet if an American-based firm is planning to do design work in Saudi Arabia or other conservative Muslim countries—where existing laws require the separation of men from women, where people practicing certain religions are unwelcome, where women are just starting to earn the right to vote or to drive—it is incumbent on the design firm from the beginning, prior to starting work, to reach an agreement with a representative of the country's government, and/or the company with whom the firm is there to collaborate, on how to handle such interactions. For example, can housing in Saudi Arabia for the design firm's employees include some private apartments that are integrated? How will meals be handled? Can some meetings with the client in Saudi Arabia include members of both sexes, or could there at least be videoconferencing to allow the participation of all of the key people involved in the work? Will all employees be permitted to drive to the job sites or to offices where meetings will be held? Each country is different in terms of such cultural mores and restrictions, and each country should be studied carefully.

How will health insurance be handled, both in terms of continuing overseas the insurance coverage in place for all employees but also the designation of hospitals and doctors' offices where the employees may go if health care is needed in a timely manner? Even compared to the unsettled nature of health insurance in the United States, the conditions, policies, and circumstances abroad are vastly different. It is incumbent on any individual or firm planning to do work abroad to know exactly how health insurance will be paid for and how health services will be provided. There are not always easy answers, yet to commit even willing staff to working for long periods overseas requires a careful resolution of such issues. I once turned down a job offer from a prominent overseas university to teach landscape architecture because its health insurance policy did not cover pre-existing conditions for *three years*, and I was already at an age where most of my anticipated health issues, barring emergencies, were in that category. Although health insurance options and benefits in the United States cannot be said to be absolutely more reliable, I was fortunate at that point in my professional career never to have encountered such a restriction.

It is not possible to predict the stability of foreign governments nor necessarily expect their relationships with your home government to remain stable. Some foreign governments might have the same expectation of yours. It's important to be able to anticipate the impact on your employees, contracts, business, and personal relationships overseas should things change gradually or suddenly. An extreme example might be the arbitrary detention of an employee for political purposes. Plan and hope for smooth sailing but anticipate the unexpected, and be prepared for the emergency that may never happen. Develop relationships with some key local people whom you can contact if necessary. Direct the office manager or director of your foreign office to visit the local consulate's office and learn what to do and with whom to communicate in case some urgent need arises.

Expectation of Speed and Total Immersion

In many developing countries the expectation of speed is a critical aspect of the work. Since labor costs are low, long working hours and extended workweeks are the norm, and a forty-hour workweek is a rarity. Interruptions of family or non-work-related activities may be a fairly standard occurrence. Yet these must be balanced with one's immersion and experience in an often rewarding foreign culture. Yet, particularly for Westerners used to the routines of a more regimented lifestyle, such exposures could be quite challenging. Perhaps such adventures are best suited for younger employees without spouses and children, or older employees (with families) who specifically seek out such experiences.

The policies set by the Peace Corps, one of President John F. Kennedy's successful programs, which has thrived for decades in many foreign countries, provide some guidelines.[25] For example, there are clear policies on fraternization with other employees, careful review of what immunizations might be required prior to traveling to a specific foreign country, regulations about driving Peace Corps vehicles, and certain pre-existing health conditions that might preclude service, since the health care available even within Peace Corps–sponsored facilities in a foreign country may not be adequate to maintain an employee's health. Working as a landscape architect in a foreign country is, of course, quite different, yet the more an individual or firm can at least discuss or put in place flexible, thorough, and practical employment and personnel policies, the more successful the results are likely to be.

Fluency in Language

One helpful policy is to find and designate a bilingual expert as a liaison between the two firms or countries, someone who has the trust and confidence of both constituencies. Such a professional could perhaps be based in the foreign country but have ready access to principals in both the host country and in North America, and be able to preempt the occurrence of major problems by identifying minor ones in their earliest stage, which can be discussed or used as learning tools by all staff. There might be an office website or even a Facebook page devoted to the task of international relations. One other useful procedure is to schedule occasional social events, from dinner or lunch gatherings to joint excursions, that, while meeting the standards of the host country, still allow people from both countries to interact and get to know one another.

An innovation that is facilitating international work is artificial-intelligence-based software. In a twenty-four-hour work cycle, when people need ways to communicate at any time of day to colleagues elsewhere in the world, such software helps people stay ahead. Methods are being developed of translating hand drawings into digital drawings. Just as it is increasingly common in universities for professors to develop AI systems to answer routine questions, there is a trend to develop AI software to answer typical client questions about practice so as to save the design professionals' time for the most important client-designer interface.

ASLA (American Society of Landscape Architects) and AIA (American Institute of Architects) have many professional networking opportunities for members and associates that includes international links that can benefit small or large firms or organizations. They may help in identifying overseas markets and contacts for travelers and professionals seeking work opportunities overseas. These organizations also identify and fund research objectives and programs that advance the professions.

Cookie-Cutter Designs

Because design fees overseas, particularly in developing countries, can be limited, some firms are selling "cookie-cutter" designs. Such firms may find it economical to reuse an old design on a new site, especially if the fee cannot sustain a full design service. Such practice is not necessarily insensitive and does not necessarily produce poor design if a thorough application is done on a carefully surveyed foreign site. The major problem tends to be local context. For example, projects in China (or elsewhere) may look too similar to projects originating in the United States. The client in China (or elsewhere) may not know the best choice for their project, but the designer should be educating their clients to make the right choices to preserve the context of the site, its originality, and its history. Particularly as the International Building Code has come into use, general principles of health, safety, and public welfare are almost universal and may be applied consistently to such projects, regardless of location. If the budget allows it, it is still wise and prudent to use local consultants, using due diligence about who can interpret local law, customs, and regulations. Absent local consultants, there is increased risk of a result that lacks authenticity and joins the world of kitsch, no matter how well built or laid out by American standards.

Legal Protections

A final consideration of international work is legal protections. It is perhaps taken for granted in the United States, Canada, and many European countries that if an individual or firm produces an original and authentic design, it will receive full credit for its effort and hard work, it will be protected by copyright laws and other means from having its work illegally copied or pirated, and there will be substantial penalties should someone try to take credit for it. In other parts of the world, such protections are not as securely in place, so the legal representatives and project management team doing work or collaborating in a foreign country should investigate carefully how the work it will be performing will be protected as its own. With modern technology, 3-D printing, AI advancements, and social networking, there are many opportunities to go from copycat leather handbags to copycat site plans.

Walk DVRC, Hong Kong

The section on site planning earlier in this chapter offers a few examples of the work of landscape architect Lihong Zhang, based in Beijing, China. The architect Vicky Chan, based in Hong Kong, whose work appears throughout this book, designed the project Walk DVRC. It's a scheme to turn two city street blocks, a total of 328 feet (100 m) long, into a temporary park for pedestrians. The park will be filled with local plants, herbs, and cultural references to the food culture of Hong Kong. The aim of this three-month exercise is to understand how people and cars behave in a walkable street that was previously dominated by vehicular traffic in a city as densely populated as Hong Kong. There will be a lot of activities at night, including for children. (See Figures 7.42 and 7.43).

Figure 7.42 Walk DVRC Hong Kong, existing Conditions.
Vicky Chan

Figure 7.43 Walk DVRC, Hong Kong, proposed design.
Vicky Chan

DOS AND DON'TS OF INTERNATIONAL WORK

1. Know the basics of the foreign culture in which you will immerse yourself. You cannot afford to insult by accident an important official or client simply by assuming that what you do in your home office is what you will do in a foreign setting.
2. Don't promise the moon. Achieve in incremental steps to gain confidence in your own capabilities and in the client/firm with which you'll be allied.
3. Incorporate into your design team where possible bilingual members of the host country. Be certain they have enough clear direction to point out errors or call attention to major concerns whenever they become aware of them.
4. Learn what the review and approval process will be, and base your fees on this approach.
5. Respect the local laws and customs of the host country.
6. Be certain to study and find out definitively what limitations the host country may place on your work. For example, in some countries only a firm founded and based in that country may do design and construction there, no matter how qualified a foreign firm is in a particular type of work.
7. Have all licenses in place, permits acquired, and fees paid prior to beginning major work.
8. Develop a library of local resources.
9. Coordinate with the home office to determine how to communicate the exchange of critical information and responses to key questions. The time difference between the home country and the foreign setting may require some people to adjust their daily work schedules to be available at critical times.
10. Include as many people in the host country as possible in your design process.
11. Go out of your way to generate goodwill. This is good business practice and will pay off in future projects.
12. Learn the climate and weather. Many Pacific Rim countries are prone to severe earthquakes, tsunamis, and typhoons. As a result some countries have advanced warning systems and sophisticated mapping designating danger zones; but many countries do not. Don't design and install a beautiful landscape, resort, or restoration on a site likely to be inundated or destroyed.
13. Be wary of bribes and corruption. The West is hardly immune to such problems, but incidents can be more blatant in some foreign countries, in which bribes are considered standard practice.
14. Know what the skills and capabilities of the labor force will be. Skilled, available masons, carpenters, tile layers, pavers, earth

movers and excavators, and utility specialists may be readily available or they may be in short supply, so develop details of installation that are a comfortable fit for who will install them.
15. Find at least a few skilled native foremen who can supervise and inspect the work and direct the native workforce. His or her skills are paramount, particularly at times when the American or foreign firm's key people may not be present or especially knowledgeable about current techniques and customs.
16. Partnership with a local, native firm is usually essential. Often they can develop detailed construction documents based on design renderings and preliminary designs originating from the foreign or non-native guests. Since they are based in the other culture, they usually have long experience in successfully implementing complex designs in that country.
17. Investigate and verify ways to protect the originality and authenticity of whatever you and your collaborators create and develop against anyone who would try to copy it for a quick profit and take credit for it as their own work.
18. Encourage the staff as much as possible to embrace living in a foreign country and participate in local culture and activities: buy groceries in local markets and eat at local restaurants, eschew the fast food franchises; learn to cook (if you haven't already).

Problem 1

INTERNATIONAL WORK: RESORT PLANNING, BOTANICAL GARDENS, PUBLIC TRANSPORTATION SYSTEMS

Consider either or both of the following problems. If you do both, compare results.

(In the following problems, any *terms in italics* may be changed to suit the particular instructor or team leader, or your own personal, work or studio setting. Typical changes might be specific people, such as clients or contractors, plant materials, or the locations of sites, people, institutions or companies.)

1. You have a landscape architecture firm in *location, USA or OTHER* of ten to fifteen professional employees with design degrees and several clerical support staff, including men and women, people of different and diverse races, sexual orientations, ethnic identities, and religions.

Because of your firm's expertise in resort planning or botanical gardens (pick one), you have been sought by a government agency in the country named below and invited to direct the design of a new resort or botanical garden (pick one) in or within 50 miles (c. 81 km) of one of the following locations:

a. Antananarivo, Madagascar
b. Buenos Aires, Argentina
c. Cali, Colombia
d. Cape Town, South Africa
e. Hamilton, Bermuda
f. Istanbul, Turkey
g. Jakarta, Indonesia
h. Kingston, Jamaica
i. Kuala Lumpur, Malaysia
j. Lagos, Nigeria
k. Manila, Philippines
l. Marrakesh, Morocco
m. San José, Costa Rica
n. Santo Domingo, Dominican Republic
o. Tunis, Tunisia

Prepare a document in which you spell out personnel policies that you would incorporate for your staff working on the project. Spell out the following:

a. Who would be on the design team abroad, and who would stay in the local office?
b. What pre-travel examinations might be required of your staff?
c. How will you facilitate collaboration with the leaders of the foreign team who will work with you?
d. What sort of accommodations will you request be provided for your staff?
e. How long will your firm participate in this project, and for how long will a "tour of duty" last for the employees whom you select to travel to this exotic setting?
f. With a firm which is small what is especially challenging about such a project?
g. What do you do if one of the principal's of your firm is a member of a group who is illegal or at best unwelcome in the location where the work is proposed?

2. You have a landscape architecture firm *in location USA or OTHER* of twenty five to thirty professional employees with design degrees

and several clerical support staff, including men and women, people of different and diverse races, sexual orientations, ethnic identities, and religions. Because of your firm's expertise in transportation and infrastructure planning, you have been sought out and invited to help design or upgrade a public transportation system in one of the following locations:

a. Algiers, Algeria
b. Bangkok, Thailand
c. Beirut, Lebanon
d. Belgrade, Yugoslavia
e. Bogotà, Colombia
f. Buenos Aires, Argentina
g. Caracas, Venezuela
h. Karachi, Pakistan
i. Mumbai, India
j. Nairobi, Kenya
k. Puerto Vallarta, Mexico
l. Riyadh, Saudi Arabia
m. Santiago, Chile
n. Shanghai, China
o. St. Petersburg, Russia

Prepare a document in which you spell out personnel policies that you would incorporate for your staff working on the project. Spell out the following:
a. Who would be on the design team abroad, and who would stay in the local office?
b. What pre-travel examinations might be required of your staff?
c. How will you facilitate collaboration with the leaders of the foreign team who will work with you?
d. What sort of accommodations will you request be provided for your staff?
e. How long will your firm participate in this project, and for how long will a "tour of duty" last for the employees whom you select to travel to this exotic setting?
f. With a firm which is larger, what becomes harder or easier?
g. What do you do if one of the principal's of your firm is a member of a group who is illegal or at best unwelcome in the location where the work is proposed?

NOTES

1. Kevin Lynch and Gary Hack, *Site Planning*, 3rd ed. (Cambridge, MA: MIT Press, 1985), 1.
2. Lynch and Hack, *Site Planning*, 11.
3. Gene Leitermann, *Theater Planning: Facilities for Performing Arts and Live Entertainment* (New York: Routledge, 2017), 81.
4. Eugene Raskin, *Architecturally Speaking* (New York: Bloch, 1966). This is a little guidebook to common architectural terms.
5. "Calvert Vaux," Wikipedia, https://en.wikipedia.org/wiki/Calvert_Vaux; see also "Frederick Law Olmsted," Wikipedia, https://en.wikipedia.org/wiki/Frederick_Law_Olmsted.
6. I am also indebted to criteria stated in Nelva Weber's *How to Design Your Own Home Landscape* (New York: Bobbs-Merrill, 1976).
7. Albert Rutledge, *Anatomy of a Park* (New York: McGraw-Hill, 1971), esp. chap. 6, 91–105.
8. Passive House Institute, https://www.passivehouse.com; Matt A. V. Chaban, "Easy on the Environment but Not Necessarily the Eyes," *New York Times*, August 18, 2014; Lindsay Abrams, "One New York City Rowhouse Takes on a Changing Climate," *Sierra*, February 19, 2015, https://www.sierraclub.org/sierra/2015-2-march-april/comfort-zone/one-new-york-city-rowhouse-takes-changing-climate.
9. https://www.nycgovparks.org/greening/planyc/schoolyards
10. Francis Ching and Steven R. Winkel, FAIA, *Building Codes Illustrated: A Guide to Understanding the 2000 International Building Code* (New York: John Wiley & Sons, 2003), 289–290.
11. Steven L. Cantor, *Green Roofs in Sustainable Landscape Design* (New York: W. W. Norton, 2008), 28.
12. Ching and Winkel, *Building Codes*, 289; Cantor, *Green Roofs*, 28–29.
13. Ching and Winkel, *Building Codes*, 149.
14. Much of this material on green roofs originally was published in Cantor, *Green Roofs*, and is incorporated with grateful permission from the publisher. It is updated herein.
15. Cantor, *Green Roofs*, 33.
16. Ivan Penn, "Solar Power to Be Required For New Homes in California," *New York Times*, May 10, 2018.
17. Cantor, *Green Roofs*, 113–118.
18. Cantor, *Green Roofs*, 93–98.
19. Officially called the National Environmental Policy Act of 1969, although it did not take effect until January 10, 1970. See "National Environmental Policy Act," Wikipedia, https://en.wikipedia.org/wiki/National_Environmental_Policy_Act.
20. "National Environmental Policy Act," Wikipedia.
21. Rushmore, a Residential Development, Town of Woodbury, NY, developed by Rushmore Associates, New York City (Carmel McGill and others). Final Environmental Impact Statement, November 1990, prepared by Clarke + Rapuano, Inc., New York City (Raymond J. Heimbuch, President), and Siegmund & Associates, Providence, RI. As a result of draft EIS comments, a waste assimilation capacity study was done, the impacts of development on timber rattlesnakes were studied, a Phase 1 cultural study was undertaken, and the design was adjusted in response to the results of these studies.

22. Final Environmental Impact Statement for Rushmore. See also Steven L. Cantor, *Contemporary Trends in Landscape Architecture* (New York: John Wiley & Sons, 1996), 20–29.
23. New York City Department of Parks and Recreation, Freshkills Landfill Park General super short May 2016.pdf and https://freshkillspark.org/; phone conversations and emails with Eloise Hirsh, Freshkills Park Administrator, June 2018.
24. "Lucky Colors in China," China Highlights, https://www.chinahighlights.com/travelguide/culture/lucky-numbers-and-colors-in-chinese-culture.htm; "Lucky Numbers and Colors in Chinese Culture," Beijing Tourism, http://english.visitbeijing.com.cn/a1/a-XAHSGGD401AB7EA4033C01
25. "MS 522 Motor Vehicle Use and Insurance," Peace Corps, February 8, 2013, http://files.peacecorps.gov/documents/MS-522-Policy.pdf.

CHAPTER 8

Construction Materials, Details and Site Inspections

Square peg in a round hole (*idiomatic*): Something or someone that does not fit well or at all; something that will not succeed as attempted, except possibly with much force and effort, or alteration of either the peg or the hole or both beyond recognition.[1]
—Wiktionary

As a project moves forward, landscape architects, often in collaboration with engineers and other consultants, must generate a design vocabulary in which they experiment with how to use different pavement patterns in concrete, brick, or stone, or different edge treatments, or walls of different types of stone or brick. How does a fence meet a wall? How does a gate fit together? What are its proportions and how does it comply with the building code? Where might a wood deck be an appropriate accent? How do different materials fit together and harmonize? How is ornamental iron used in just the right way as a graceful yet strong element? As these questions are answered, one moves forward and makes final design, dimension, cost, and engineering decisions. Throughout this chapter examples of such details are offered to stimulate discussion and as a way of thinking and imagining. They are usually preliminary, illustrative drawings not to scale. These might become part of your library of options to consider for particular projects. Several details and images are as-built examples from designers and contractors, which the author gratefully acknowledges.

Figure 8.0 Dry laid stone wall at Emory Knoll Farms, a nursery for green roofs plants, in Street, MD solidly anchors the landscape.

Photo by Steven L. Cantor, with permission by Ed Snodgrass

PART 1: PAVEMENTS, STEPS, AND WALLS

Pavements

Pavements are perhaps the most varied item in the site vocabulary of landscape architects, with a myriad array of materials, finishes, textures, joints, edgings, scales, and complexities. The basic starting point with a pavement is who is going to be using it. If the answer is pedestrians and their assorted children, pets, and small vehicles such as bicycles, tricycles, and toy wagons, then one set of requirements comes into play. If the answer includes cars, trucks, buses, and other large vehicles, then construction detailing must include many other considerations, as the pavement cross section must be strong enough to support the weight and vibrations from many moving vehicles. If a pavement is *ever* going to be used by heavy-duty vehicles, then it must be designed as if this is always the case. It only takes one accident in which a car makes a wrong turn and drives on a pedestrian pavement not specifically designed and engineered for vehicles to destroy or damage large sections of pavement.

Asphalt

Although asphalt is ubiquitous for parking lots, driveways, and road surfaces, it has advantages and disadvantages. Because it naturally expands and contracts, expansion joints and other joints are not essential, as they are with rigid pavements. Therefore, it's a useful material for paving irregularly shaped areas as long as there is ample room for rollers and other machinery. Primarily black in color, it absorbs heat, so in settings and climates where excessive heat buildup is a problem, other pavements may be a preferred option. Asphaltic sealants or joints should be avoided, as in hot weather—say, in the upper 80s or warmer—they may start to turn liquid and ooze from joints. It's best to avoid combining it with other materials.

Concrete

Concrete is manufactured or created as a mixture, according to precise recipes, of Portland cement, aggregates (gravel and sand), and water. The wet mixture, when poured or placed into forms, becomes concrete as it sets, cures, and hardens. Adding too much water may weaken the concrete. Too much fine aggregate may result in cracking, and too much large aggregate may result in a weak and porous concrete. If the project for which concrete is being specified is large enough (requiring a few cubic yards) to use a wet mix delivered by a reputable company, then as long as the concrete being delivered is tested a few times, the process of pouring the concrete is straightforward, without the landscape architect having to be constantly on guard for a wet mix that is too liquid, too dry, or with the wrong proportions of aggregates. Poured-in-place concrete is often the pavement of choice when asphalt is not considered, because concrete has so many applications. Many finishes are available as well. Steel mesh as reinforcement allows it to be used for heavy-duty pavements supporting a great deal of weight.

Unit Pavers

Unit pavers abound, from brick in all of its many colors and sizes to manufactured systems of interlocking pavers. The designer must be aware of the advantages and disadvantages of each. Bricks are usually manufactured with set dimensions, including thickness, so if it's to be used as a pavement surface or wall veneer, the designer must take into account its predetermined thickness. For large applications and specialty needs, unit pavers, often designed as thinner blocks, gain strength not from their own nature but as a result of being placed on a reinforced concrete base and may be less expensive while still providing the variety of texture and color of bricks. Unit pavers, whether brick or other designs, often may reinforce direction of movement. For example, running bond or basketweave patterns often suggest movement in one direction, whereas herringbone designs reinforce multidirectional traffic movements.

Brick

Because brick has been used for pavement and walls for centuries, there are long traditions of standard methods and treatments governing how the courses overlap and interlock, strengthening the pattern or the wall. Similarly, there is a vocabulary of mortar joints, which achieve a range of visual effects, but the emphasis should be on resistance to wind and rain as well as aesthetic appearance. Mortar must be of high quality to achieve strong bonds. Lime and well-graded sand (not very fine) are incorporated to keep the mortar mix plastic so that the mason can work with it easily to create the joint patterns and bonds desired. Mortar mixes for brick masonry are rated as Type M, S, N, and O, in decreasing order of strength from high to low. Type M requires 1 part Portland cement, ¼ part hydrated lime, and 3¾ parts sand. Type S is 1 part Portland cement, ½ part hydrated lime, and 4½ parts sand. Type N is 1 part Portland cement, 1 part hydrated lime, and 6 parts sand. Finally, Type O is 1 part Portland cement, 2 parts hydrated lime, and 9 parts sand. The water used should be the quality of drinking water, and enough should be used to bring the mix to a suitably plastic and workable state.[2]

Traditional and Contemporary Materials

Many time-tested materials used or invented decades ago are still in use in the modern world. Since traditional construction materials are often durable, attractive, and available, it often makes sense to continue to use them. But the size of pavers, like the number of ounces of our favorite cereal or crackers in a box, is on a downward trend. For example, Hastings asphalt pavers are ubiquitous in New York City sidewalks, but the current ones manufactured are considerably smaller than the original design. Therefore, if a landscape designer wants to repair an area of the old pavers, either a source must be found of recycled original pavers or large areas of the old pavers must be entirely removed and replaced with new pavers or an alternative design element. Hastings manufactures rectilinear pavers that can be placed in rows to form a boundary separating one area of old pavers from newer pavers of different dimensions.

This same issue of the decrease in size over time of many construction elements must be studied carefully whenever such uses are present. (See the discussion of wood in Part 3 of this chapter.) Brick, for example, comes in myriad dimensions and colors, so if a designer intends to match an existing pattern on a site, it's imperative to check the dimensions of the bricks in the pattern and verify what is available that can match or harmonize with that pattern.

DOS AND DONT'S FOR BRICK PAVEMENTS

1. Know the vocabulary of the mason: header, rowlock, sailor, shiner, soldier, and stretcher.
2. Use stronger (Type M or S) rather than medium- or low-strength mortar (Type N or O) unless you have specific clearance from site manager or engineer that the weaker mortar will suffice. The difference in the costs of the two major categories is minimal compared to the cost of the labor by the masons and any repair work that might be necessary should the mortar fail.
3. Anticipate where expansion (control) joints and construction joints will be needed.
4. Do not work in freezing weather, as the mortar may fail if it freezes before it is fully set. In urgent situations where the calendar is critical, there are ways to manage by the use of plastic enclosures or heating elements.
5. If there is a lot of rain, provide protection so that the brick does not become saturated.

(See Figures 8.1 and 8.2).

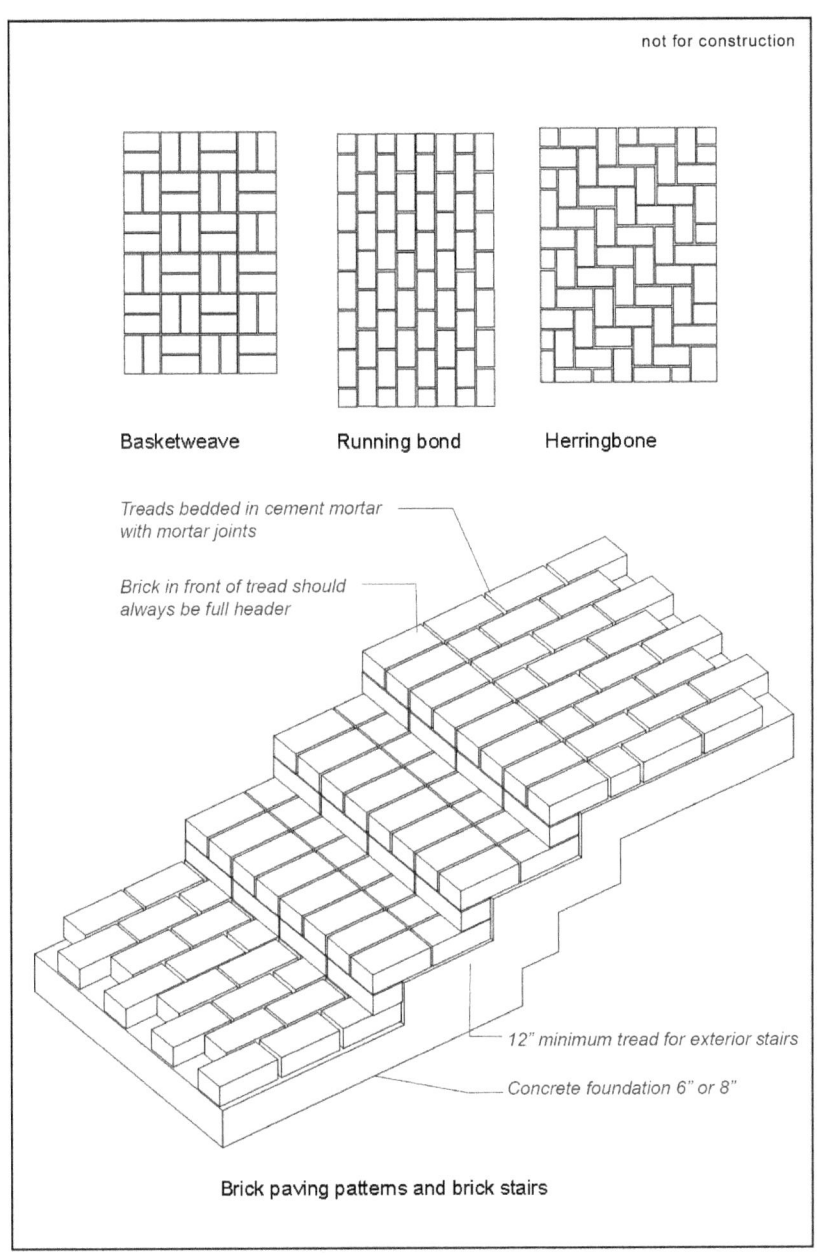

Figure 8.1 Brick paving patterns and brick stairs.
Richard Alomar and Vicky Chan

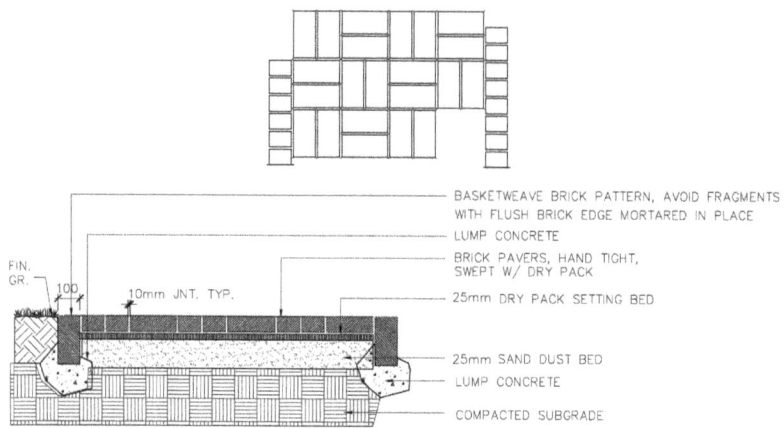

Basketweave Brick Pattern

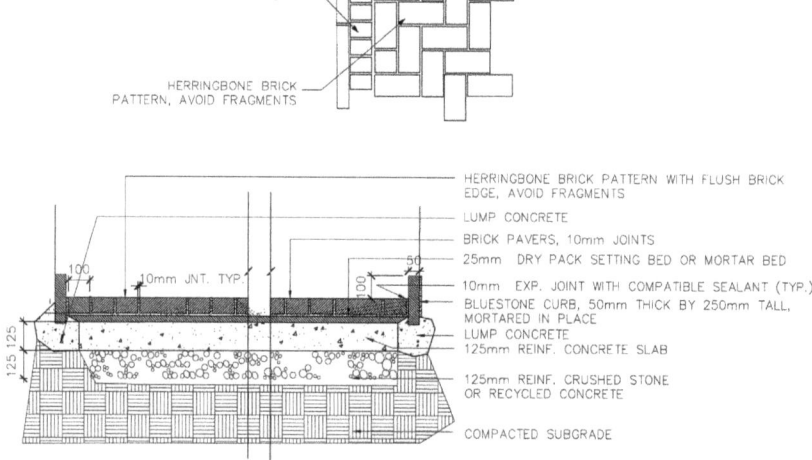

Herringbone Brick Pattern

Figure 8.2 Basketweave and herringbone brick pavements.
Steven L. Cantor and Vicky Chan

Stone Pavements

Stone pavements, such as bluestone and granite, often reveal regional differences. Shipping costs are high due to the weight, so regional preferences emerge based on the reasonably close proximity of the quarry or distributor to the intended destination for installation. This regional quality also applies to gravels and other materials rolled into asphalt or concrete for variable texture or color. Shipping costs over long distance add up quickly, so macadam pavements, although they may be uniform in thickness, durability, and structural design, may show a lot of difference based on the gravel rolled into the bituminous layers.

Reinforcement

Reinforcement for concrete pavement is wire mesh, which is usually a requirement if the pavement is to carry a vehicular load or the underlying soil conditions are weak. Sometimes building codes require mesh for sidewalks, as they may occasionally need to support a vehicular load. For walls and footings, reinforcement is primarily steel bars of varying diameters (often referred to as rebar, for reinforcing bar), which are placed continuously at an on-center spacing based on the needs for increased strength. In locations where strength is particularly critical, the bars are overlapped and tied together with #15 or #16 gauge wire.

Joints

Mistakes are often made in the design and placement of joints in pavement. Control (or contraction) joints allow for controlled cracking where they are placed across concrete slabs. Construction joints, placed where new concrete meets existing concrete, act as control joints. Expansion (or isolation) joints occur approximately every 20 to 25 feet in an expanse of concrete. They consist of an expandable and compressible material, such as foam, with a compatible sealant placed over it to prevent leakage. The purpose of the expansion joint is to allow pavements to expand and contract in response to variations in temperature. Expansion joints also must be placed along the edges where a concrete or other rigid pavement meets a vertical element, such as a wall, so that the pavement is locked in place. The expansion joint allows the pavement to expand and contract independently of the wall or other vertical element or contrasting material. If the expansion joint is not installed, over time a pattern of cracking will occur in the pavement as it responds to the pressure created when it heats up and has no edge against which to expand. All sorts of sealants are now manufactured. Care should be taken to verify that the sealant chosen is compatible with the compressed material and the pavements the sealant abuts. It's often practical to use a sealant with a color matching the pavement adjacent to it. Then the eye overlooks it and the expansion joint recedes from view. (See Figure 8.3).

Finishes

Pavements may have many finishes. Perhaps, the most common finish for concrete is a broom finish, in which a special concrete broom is dragged across partially hardened concrete, to create a texture perpendicular to the direction of traffic, so that there will be better traction in wet conditions. Another common treatment for concrete is a pebble finish in which the material is rolled into the surface of the wet concrete. If the pebble/gravel mixture is not fully integrated into the concrete and if there are too many gaps in the mixture, there is the risk that over time some of the finish will weather and erode, leaving an uneven, unattractive, and hazardous pavement. Yet when executed skillfully, pebble concrete is a nice contrast to standard broom-finish concrete. Also, if the gravel/pebble mixture has an attractive color (tan, for example), it lends itself to the overall character of the concrete and provides a contrasting color.

Avoid Freezing Weather

As freshly poured concrete sets, chemical reactions occur so that the various materials in the mixture bond, and the concrete gains strength as it cures. Freezing temperatures limit this process and may result in weaker, substandard concrete. Therefore, it is urgent *not* to pour/place concrete in freezing weather, nor when there is expectation that freezing temperatures may occur by the evening in which the concrete is placed.

Stamped Pavements

One method of concrete design is stamped pavement, in which a mold is repeatedly placed over the wet concrete and used like a template to impress a design. The most well-established manufacturer is Bowmanite, although many other companies are now in business. Some stamped patterns resemble bricks, others cobblestones. If the concrete is tinted

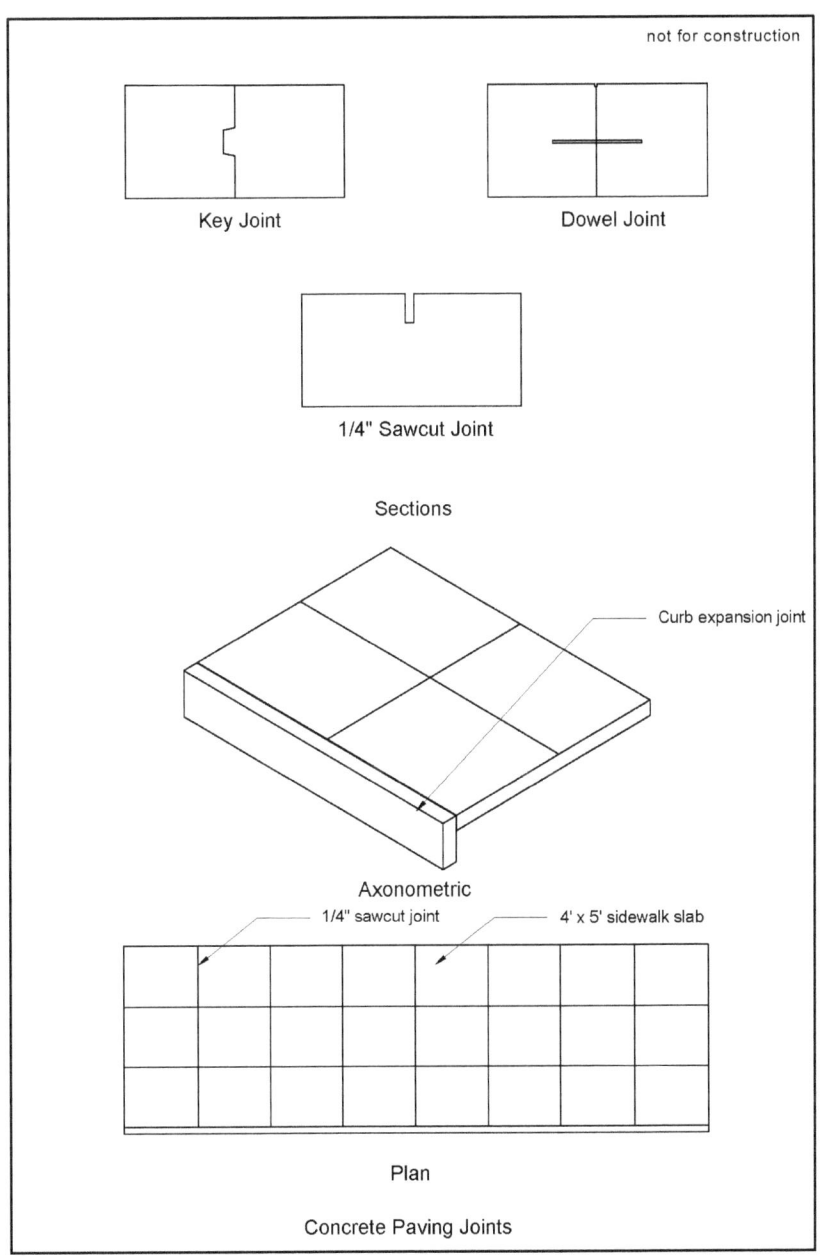

Figure 8.3 Concrete paving joints.
Richard Alomar and Vicky Chan

a brick or cobblestone color, the effect is more convincing. The designer must think through carefully the geometry of the pavement. For example, the more rectilinear the pavement expanse that is to receive a stamp of a running bond brick pattern, the fewer the kinks or leftover areas will be. The designer must also study carefully how and where the stamped pattern terminates and transitions to standard concrete or another pavement material and pattern.

DOS AND DONT'S FOR PAVEMENTS (INCLUDING CONCRETE)

1. Design the full depth of the pavement cross section, not just the surface treatment.
2. Align the depth of the pavement with the depth of other elements, such as curbs or adjacent pavements, so that expansion, contraction, and heaving are controlled as a result of a uniform relationship with the frost line.
3. Determine whether or not the pavement application is directional—that is, is it useful to think of the pattern of joints or some other aspect of the pavement design as helping to direct a pedestrian or other user on the site?
4. Design for a basic minimum PSI. Within reason, an increase in compressive strength of a practical nature to 3,000 PSI from 2,000 or 2,500 does not involve a great deal of additional material or thickness to the cross section, while ensuring a much longer and safer duration.
5. Study and incorporate expansion joints and contraction joints carefully, as their integration into the design ensures a pavement that is both longer-lasting and more attractive.
6. Develop unit paver patterns that minimize the amount of required cutting and use of fragments of bricks or unit pavers.
7. Avoid pavements that taper to acute angles, as such points are inherently weak. (The same principle applies to plantings, as it's difficult to maintain a tapering triangle of sod or perennials because the narrowing of the planting bed reduces the resiliency of the planting medium in that location.)
8. Simple pavement patterns—fewer joints, fewer acute angles, and fewer complex geometries—result in stronger design than complex ones.
9. Consider porous pavements in areas where uniform drainage into the area to be paved is practical and could save on the costs of incorporating a lot of storm drainage lines.

10. Tinting and coloring agents add considerably to the composition of pavement areas, but the agent should be integral to the pavement mixture and not something poured or added only to the surface, which often results in uneven, splotchy results and which may wear off in a short period.
11. Study carefully any areas where old patterns are to be replaced with or merged with newer patterns of the same paver, as the modern version is often smaller.
12. Be careful when specifying a range of colors for a pavement pattern, whether they are bluestone or brick or something else. Require a sample installation so that the color range can be observed and approved prior to installation of the entire area.
13. Edgings may be critical for some pavements. Asphalt, despite the great flexibility in its use since it does not require much jointing, can still bleed out, so wooden or metal edgings often enhance the appearance and create clear adjacent areas for planting or other uses.
14. Joints and sealants now come in a huge range of materials and colors. In those settings where one- or two-step applications of joint material and sealant are used, select a color that matches the pavement or the standard mortar treatment. Verify that the joint and sealant are chemically compatible.
15. White broom-finished concrete is perhaps the most common pavement. Incorporating an integral coloring agent into the mix may result in a nice color effect without a great deal of additional cost and help avoid glare and eye strain in sunny settings. At the same time, since the coloring agents have considerable variation, determine the area that one load from the concrete mixer truck or machine will have, and try to use up the mix entirely and not carry over into the next area for pavement.
16. Be careful in pavements that grates and drains have alignments that will not trap bicycle tires or high heels. This is most easily done by having the grate alignments perpendicular to the direction of the pavement.
17. ADA ramps and landings must be carefully integrated into the overall pavement design. Lay them out first, approve their alignment, and pave them. Then add the surrounding elements, adjusting as necessary to achieve the best dovetailing.
18. Install a sample section of any ramped, crowned, or swaled pavement and be certain it meets all aesthetic and engineering requirements prior to the installation of the balance of the same pavement design throughout the site.
19. Pay close attention to all transitions, where one type or style of pavement meets another. Even with individual stepping-stones, the placement of larger stones at beginnings and ends of walks is often a useful way of accenting and punctuating the design. With concrete,

brick, unit pavers, and other designs, consider letting one form embrace another.
20. Even for walk systems that are primarily for pedestrians, the incorporation of principles of horizontal and vertical alignment results in a much more graceful application.
21. Compact the grade in 3-inch to 4-inch lifts.
22. Slope the subgrade and the subbase under pavements.
23. Allow ample time, at least seventy-two hours, for concrete to cure.
24. Integrate aggregate fully into a concrete mix, rather than just the top surface.
25. Use color additives sparingly and uniformly. Know the precise ratio of additive to standard mix so that uniform results may be obtained.
26. Use welded wire mesh and reinforcing bars that are clean and free of rust and coatings.
27. If there is any uncertainty about the on-center spacing and diameters of reinforcing bars, review with the project engineer. If there is an urgent need to proceed, be on the conservative side: decrease the spacing and/or increase the diameter of the rebars. The steel does have a cost, but it is minimal compared to the cost of the labor, the concrete used to build walls or footings, or the dangerous conditions that might occur if a substandard footing or wall is built.

Stepping-stones

As a strong alternative to pavements, stepping-stones are excellent linkages between two features of a landscape, such as a terrace and the main entrance to a house, or the driveway and a recreational area. Often steppingstones can be placed in a graceful, curving geometry to contrast with rigid rectilinear pavements or other patterns. Individual stepping-stones should be a minimum of 2 inches in thickness (sometimes more for weaker materials) and be set on a compacted base. (See Figures 8.4, 8.5 and 8.6).

not for construction

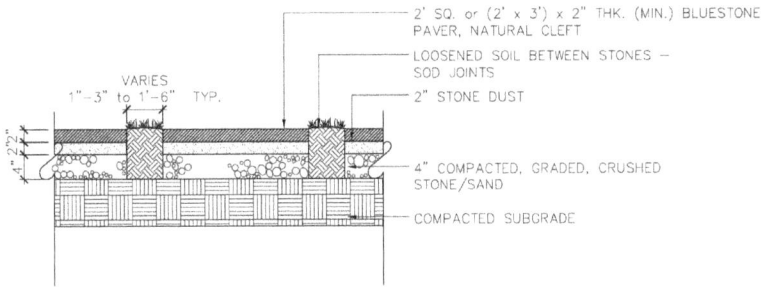

A: Bluestone Steppingstones

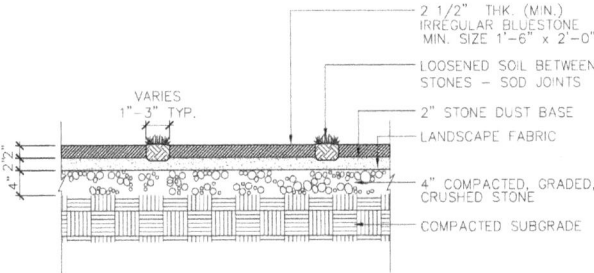

B: Bluestone Steppingstones Path

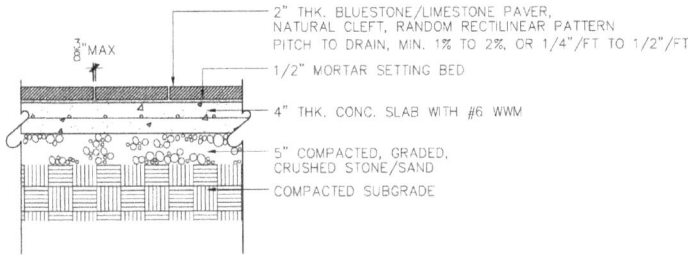

C: Bluestone Pavement on Concrete

Figure 8.4 Bluestone steppingstone and bluestone pavements.
Steven L. Cantor and Vicky Chan

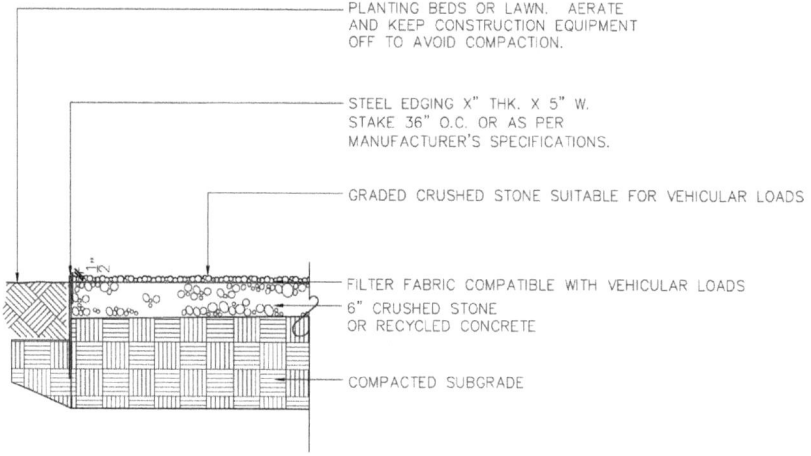

A: Crushed Stone Pavement With Steel Edge

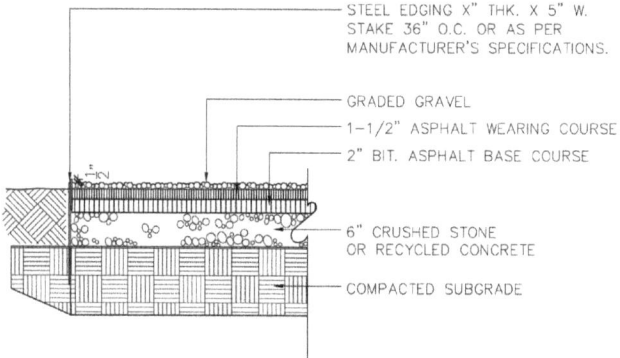

B: Asphalt Pavement With Steel Edge

Figure 8.5 Steel edge with asphalt or crushed stone pavement.
Steven L. Cantor and Vicky Chan

not for construction

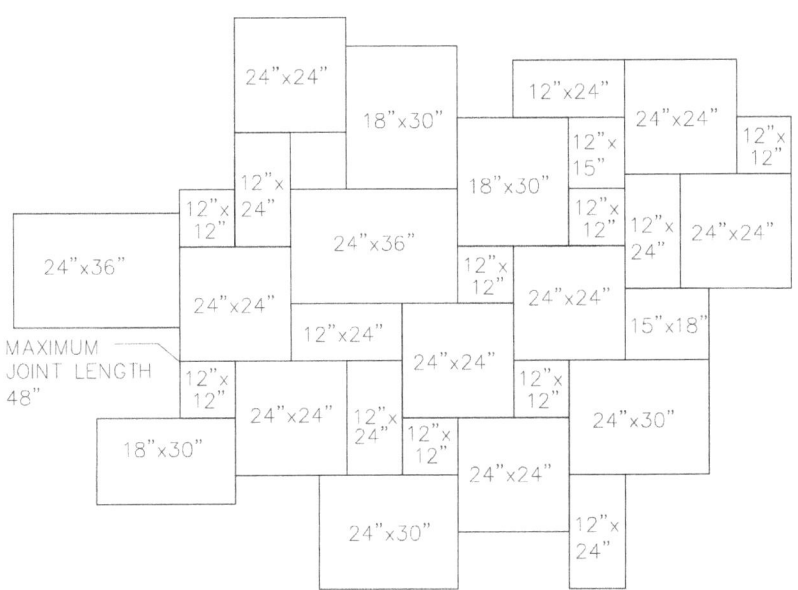

Random Rectilinear Bluestone Paving Pattern

Figure 8.6 Random rectilinear bluestone paving pattern.
Richard Alomar and Vicky Chan

DOS AND DON'TS FOR STEPPING-STONES

1. Select the material carefully. For example, bluestone, sandstone, limestone, marble, and slate all have different qualities in terms of color and texture.
2. Consider a larger stepping-stone or a pair of stepping-stones at the beginning and end of walks as an accent at the transition from the walk to another feature, such as a terrace.
3. Be careful that the finish specified and approved for the stepping-stone gives enough traction for safe walking and prevents slipping.
4. Irregular and geometric shapes are both suitable for stepping-stones.
5. Avoid stepping-stones with acute angles because over time they become weak points and break off.
6. Lay out a curve for alignment the same way that a horizontal curve in a roadway would be set and anchored in the landscape.
7. Straight lines and right angles for layout are also suitable, particularly when formal rectilinear patterns of stepping-stones are used.
8. Plant sod or suitable perennials such as thyme or vinca between the stepping-stones to link them into a continuous design element.
9. Use gravel as a base for a system of steppingstones as an alternative or complement to plantings.
10. Don't mix and match; one stone material in primarily one geometric shape works best, with occasionally some variation at the beginning and end of a walk.

Steps

Steps are as ubiquitous in the landscape as pavements, yet because they occur at changes in elevation where the designer directs people to climb or descend, steps take on an importance greater than the sum of their parts. Precisely because they occur at changes in grade, the design must be precise and safe, sometimes with adjacent or built-in lighting included. There are long-established ratios of risers to treads that are widely accepted and still applicable. Only in unusual circumstances, such as designing steps just for children, should landscape architects depart from these standards. (See Figure 8.7. You may wish to compare it with Figure 6.2).

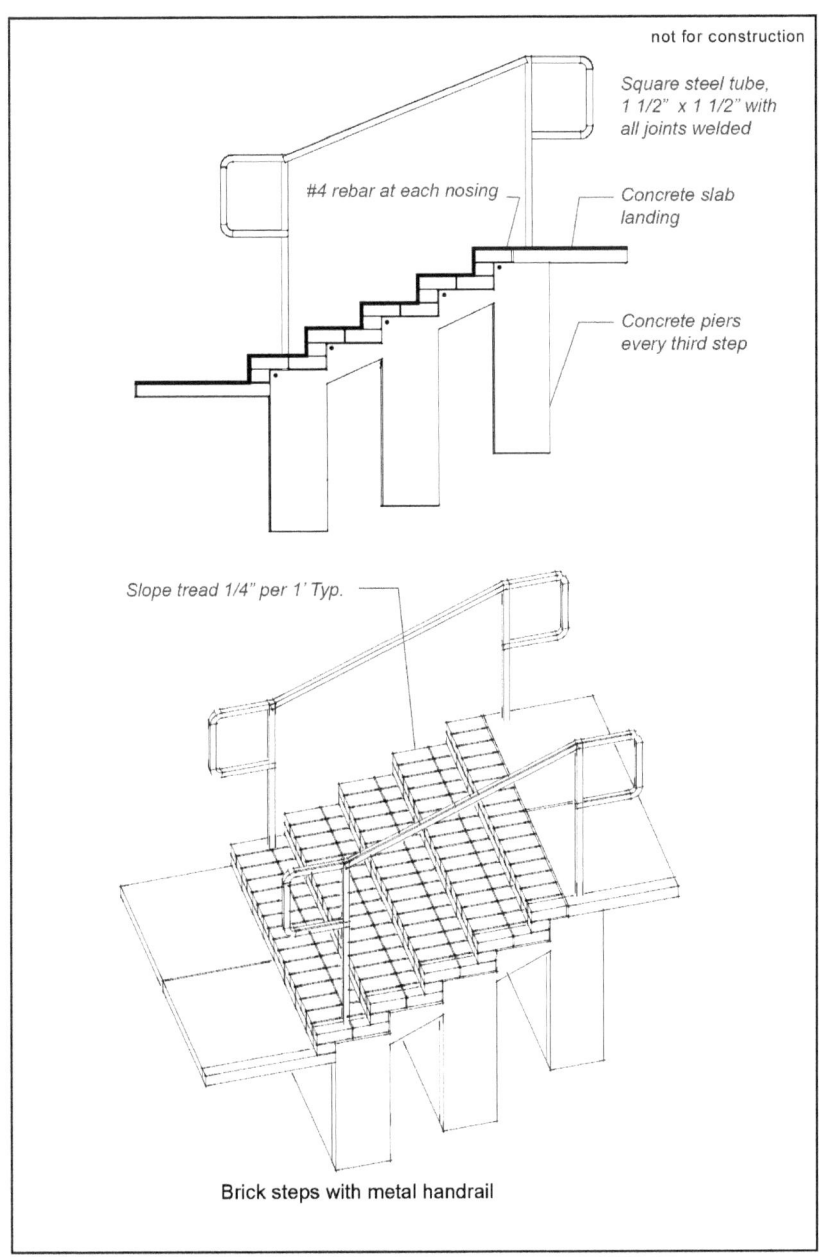

Figure 8.7 Brick steps with metal handrail.
Richard Alomar and Vicky Chan

DOS AND DON'TS FOR STEPS

1. When a series of steps is required, it is appropriate to have a repetition of the same number of risers and treads—for example, four sets of steps, each with ten risers of 6 inches each and treads of 14-inch depth.
2. The maximum number of steps in one run may be determined by local building codes. However, visually it's best if a single set of steps is no higher than a person's eye level, say 5 feet to 5½ feet. If a person can see to the top of the landing, it doesn't seem quite as imposing as if the landing is well above eye level and out of sight.
3. When the designer must have sets of steps with some difference in total height, it's better to put the larger set at the base. People climbing up the stairs will appreciate that the landing of the first set will be slightly past the halfway point.
4. If lighting is not included, the landscape architect must be certain that the steps nonetheless remain highly visible.
5. Consider using a contrasting material for the steps compared to the pavements or walks above and below the steps, or at least some texture or finish on the steps so that they contrast. It's also effective and visually compelling to use a contrasting material for the steps compared to the walls adjacent to them.
6. Even with fairly narrow steps, it's appropriate to stagger the pavement pattern so that there's not a series of continuous joints running the length of the steps. (This is comparable to seating in an auditorium, in which staggered seating is more effective than having every seat align directly with the one in front of and behind it.)
7. Pay attention to the footings, provide adequate steel reinforcement, and consult an engineer to verify that there is adequate structural strength.
8. Place expansion joints or control joints at the top and bottom of each set of steps. It's helpful visually to use a sealant of a color that matches the pavement for the steps or the adjacent pavement so that its appearance disappears in relation to the expanse of pavement or steps below or above it.
9. Treads should have a slight pitch, say 2%, so that water doesn't puddle on them. At the base of a major set of steps, the designer should provide drains either in the pavement or in planting beds or lawn areas. If the pavement is crowned or pitched, the drains may be on both sides or one side only, respectively.
10. Exercise great care in the design of curving sets of steps, as the layout becomes quite complicated to achieve a comfortable riser/tread ratio across the whole set. The risk is that the riser/tread ratio

may be too low, at the inside edge of a curving set of steps and too high at the outer edge, so that walking up and down the steps is only comfortable at the middle of the steps.
11. When steps are built into wall systems, such as cheek walls, expansion joints are required when the steps are locked in between two walls. If there are no walls on one side of a set of steps, it's still advisable to have expansion joints on the continuous edge along the length of the wall that abuts the steps.
12. Other materials such as granite curbstone make beautiful and elegant steps when they are stacked one above the other. Often only minimal footings are required, because the weight of the stone slabs is enough to anchor a set of steps, and only a minimal mortar joint, if any, is needed between adjacent slabs.
13. Handrails should be provided as per code requirements, which may allow for no handrails when only two or three steps are installed.
14. For larger sets of steps, it is better to anticipate the design of the handrail from the beginning and determine the location of the holes to be drilled into the steps. If there is adequate space, it's often preferable to place the handrail immediately adjacent to steps constructed of brick or other unit pavers, because the drilled holes for handrails may be unsightly within the pavements.
15. Similarly, lighting of steps can be independent path lights or built-in systems incorporated into the adjacent cheek walls or the face of the risers. For large sets of steps, it's helpful to have a consistent pattern of lighting, as people walking on the steps (particularly if they have design sense) will sense the rhythm of the spacing and feel safer as they recognize the pattern.
16. The finish on the steps must be textured and abrasive enough to create good traction for pedestrians. Smooth finishes such as polished slate or marble should be avoided for exterior steps.
17. Steps need not be a uniform width. Sometimes alternating widths give the impression that the steps are built into the landscape.
18. Irregular designs of steps in rock gardens and naturalistic settings often result in beautiful and effective results. Execute them with care, and require the construction of a mock-up or sample of at least one set before proceeding over an entire garden or other setting.
19. To avoid a sense of climbing a ladder, consider varying the width of sets of steps, with the wider treads being at the bottom of a large set of steps and the narrower ones at the top or final arrival points.
20. Integrate the placement of the steps within the overall pavement designs so that the steps both accentuate the design and recede in character where and when necessary.

Walls

Aside from groups of specimen plant materials, walls are perhaps the most important elements in a landscape design. They separate one use from another, accentuate the drama of the setting, reinforce major elevation changes, and hold the site together (whether actually or figuratively). Per unit cost, they are among the most expensive of landscape elements, so care should be used in proposing them, designing them, and installing them. Walls may be massive or thin, bold or recessive, straight or serpentine, sculptural or sedate, simple or complex, hidden or accented, yet must always be integrated into the landscape design. A rule of thumb remains that landscape architects should be able to design walls up to the eye level of a person, say 5 to 6 feet tall, without needing the consultative review of a civil or structural engineer; yet having an engineer review the design of any wall is a necessity, and must be considered for walls over a certain minimum height. Even a wall under 5 to 6 feet tall may be a massive construction element, and the study of its layout and detailing by an engineer is warranted.

Skilled Masons

Because walls are such visible elements, finding the right masons and contractors to build them are major requirements. It's useful, where permissible in design drawings and specifications, to require that any contractor proposing to build a wall for a particular project include as references in the bid form the names of two or three clients, and verify the permanence, durability, and aesthetics of walls already built by this contractor using the materials specified in the project to be constructed. Since they may be designed to be visible on two sides, freestanding walls may be as important as retaining walls in a landscape. Don't let a freestanding wall be the first wall that you ask a contractor to build for you; let that be one that's recessive in the landscape.

Critical Aspects of Capstones

Capstone designs are critical to almost any wall design, as those installations lacking them may often be perceived as a headless horseman, a birthday cake without icing, an exclamation point without the vertical line, a window without a lintel. The capstones resolve the geometry of the design into the landscape, providing a horizontal element atop a major vertical one. They appear to anchor the design in place and connect it to its surroundings.

Dry-Laid Walls

Perhaps the most traditional walls seen in a landscape are dry-laid walls, for which a long tradition exists (see Figure 8.0). In rural farming areas, a skillfully built, perhaps undulating ha-ha wall is among the most graceful and elegant structures in a landscape. They are also the most massive, as the taller the wall, the wider the base, so huge stacks of stones are necessary to achieve significant height. Where skilled masons are available and large quantities of stone are present, they can create beautiful impacts.

Stone Veneer

Reinforced concrete walls with or without stone veneer can achieve the same retainage in a much narrower cross section. The landscape architect chooses and develops a design solution based on the available space and materials as well as the skill and experience of those doing the construction.

Footings

Footings for most walls are concrete slabs of different depths, lengths, and widths, reinforced with steel bars according to the requirements of the site engineer. The footing connects to the vertical structure of the wall, which themselves may be poured-in-place concrete, concrete block, or other materials. Often the face of the concrete is veneered with stone, brick, or other materials. So the vocabulary of choices is wide.

Concrete Block Walls

Concrete blocks are manufactured in a range of sizes and shapes, and lend themselves to stacking in overlapping rows, so that vertical rebars may be placed as connectors and reinforcement, and filled with a strong concrete mortar mix. Concrete blocks may be preferred to poured-in-place concrete for many reasons:

1. Access to the site for a concrete mixing truck is not possible, but a small portable mixer may be used.
2. Access to the site for elaborate formwork for poured-in-place concrete is not possible.
3. Concrete block for walls is often less expensive than poured-in-place concrete since the only formwork is for footings.

4. Brick veneer lends itself to concrete block, as wall ties and other fasteners can be set within the courses of the concrete block to align with the brick courses.
5. The dimensions of standard concrete block (8"H × 8"W × 16"L) align with three courses of standard-size bricks (actual size 2¼"H × 3⅝"W × 7⅝"L with ⅜" joints). There are metric equivalents as well.
6. Concrete blocks are available not only in a wide range of sizes but also in decorative shapes.
7. Concrete blocks can be heavy or light, depending on the aggregates used in manufacturing them. The lightweight ones sometimes have sound-insulating properties due to the porosity of the aggregate.
8. Control joints can be provided with relative ease by stacking alternately half-length and full-length blocks.
9. Parge coats, stucco, or waterproof paints can be applied as finishes.
10. Many capstones of concrete block may be used if the wall is not to be veneered.

Review by Engineer

Retaining wall designs over about 5 to 6 feet in height must be reviewed by an engineer, to verify that the footing is adequate in depth, width, and height; that there is adequate reinforcing based on the type of retaining wall; that the design is suitable based on the subsoil conditions; and that drainage details are correct, including the connection to on-site drainage systems.

Weepholes and Footing Drains

Most retaining walls require both weepholes and footing drains. Both types prevent hydrostatic pressure from building up to a point where the wall starts to tip over. Weepholes are made through the full depth of the wall perpendicular to its length and allow water that collects behind the upper levels of the retainage to seep out. Footing drains are placed behind the footing parallel to the length of the wall and collect any water that percolates down to that depth. The gradient of the drain can be almost flat, as the hydrostatic pressure will create a flow. The footing drain is released at some outlet, such as a swale or a relatively flat planted area that can absorb runoff. Finally, some waterproofing layer may be applied to the rear of the wall above the footing for perhaps 2 to 3 feet in height. This prevents water from seeping into the lower levels of the masonry and concrete, and deflects water toward the footing drain.

Corners Are Crucial

As with pavements and other design elements, corners are crucial. They are natural focal points, so the pattern of masonry units, whether brick or stone, should be carefully studied to achieve the best design.

DOS AND DON'TS FOR WALLS (GENERAL)

1. Review information about the native soil, its drainage, and how this may impact the size and type of footing required.
2. Verify with the engineer, based on site information, what the depth of the footing should be so that it can be situated below the frost line, and its length and width so that it can adequately support the masonry wall.
3. For all but minor walls, require the construction of a sample of each type of wall. Give the contractor the option to build the sample so that, if approved, it becomes part of the final design, or to demolish the sample upon completion of the review.
4. Include drainage design, including weepholes and/or footing drains, which are critical elements to prevent hydrostatic water pressure from building up behind the wall and to allow water to drain from retained soils.
5. Clearly define on the drawings exactly where the wall begins and ends, and verify that there is a strong connection at both points to the surrounding site or site construction elements, including plantings.
6. Consider the way the courses of stone, veneer, or other elements align with the landscape itself. If possible, align parallel to the proposed final grades of the site. Consider other alignments for contrast, but the design must be strong if you plan to use alternative patterns.
7. Be clear where expansion joints will occur in the face of the wall, and how they will be disguised if necessary within the pattern of veneer or other finish treatment.
8. If color additives are incorporated in order to give the wall a visual accent, use chemical coloring agents that are integral to the use of the material and not something placed only at or on the surface.
9. Always anticipate where people will sit in relation to a wall. Is it at an ideal height above proposed final grade, 16 to 18 inches, to be a natural seating area for people? If it is shorter or taller, is it clear what its functions are and that seating is provided elsewhere on the site in convenient and appropriate locations?
10. Study how the wall will be lit, if at all, and what fixtures will be used to illuminate it.

11. What material will the capstone be, and what will its dimensions be in relation to the rest of the wall, whether veneer or poured-in-place material like concrete?

DOS AND DON'TS FOR DRY-LAID WALLS

1. What material will be used, and what is its source? Often a large volume of stone is required, so it should be verified that ample material is available on-site or can be shipped at reasonable cost to the construction location.
2. Who will the masons be? Dry-laid walls require a great deal of highly skilled labor. Be certain that the project can support the cost of a skilled mason to execute the design detailed in the construction drawings.
3. Anticipate the thickness at its base, which is buried, and know how wide it will be at the point it first becomes visible above grade. Be certain that the site can absorb its mass gracefully.
4. Most dry-laid walls appear stronger if larger stones or blocks are incorporated in the base courses and smaller ones toward the top.
5. The face of the wall should lean back into the existing grade and area to be retained at a consistent rate.

(See Figures 8.8, 8.9 and 8.10).

Problem 1
SET OF CONSTRUCTION DETAILS

Develop a set of details for a site of 1 acre (0.4 ha) which may be at a location and of a nature of your choosing that includes two walls, two contrasting types of brick pavements, a set of steps, and stepping-stones. Drawing and details may be in imperial or metric scale. Site must have a gradient north to south of 5% and east to west of 8%. Start at the entrance to a building and end at a street or other building. Two walls and two brick pavements must show contrasting construction techniques but must harmonize in design.

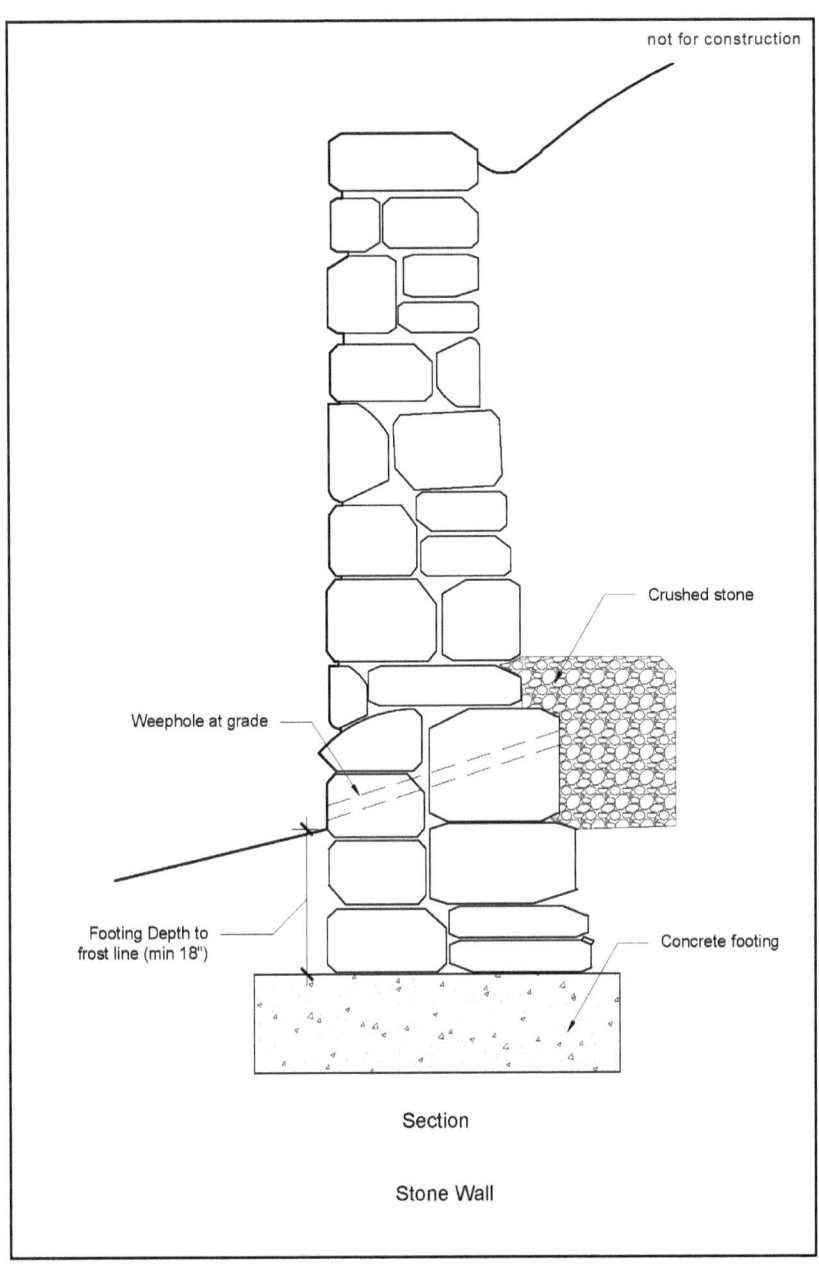

Figure 8.8 Rubble stone wall.
Richard Alomar and Vicky Chan

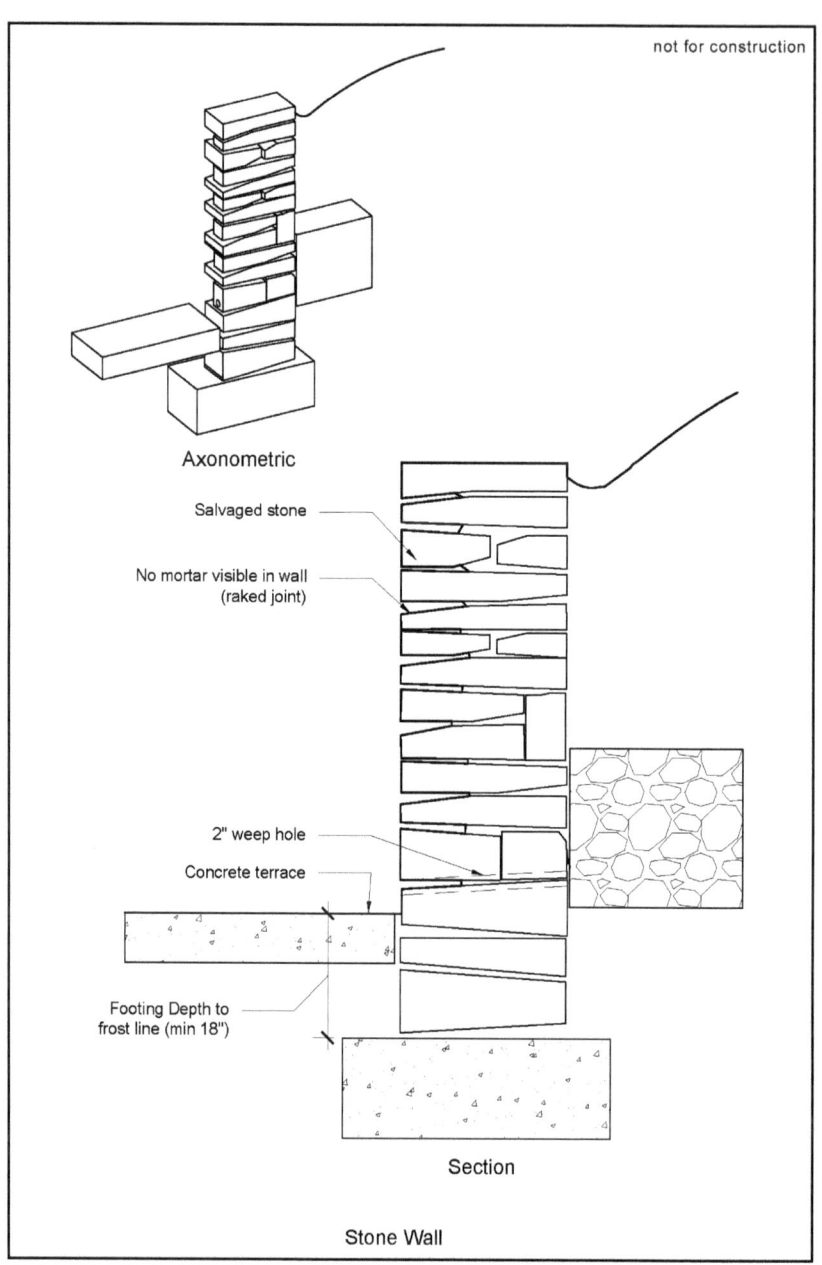

Figure 8.9 Fieldstone wall.
Richard Alomar and Vicky Chan

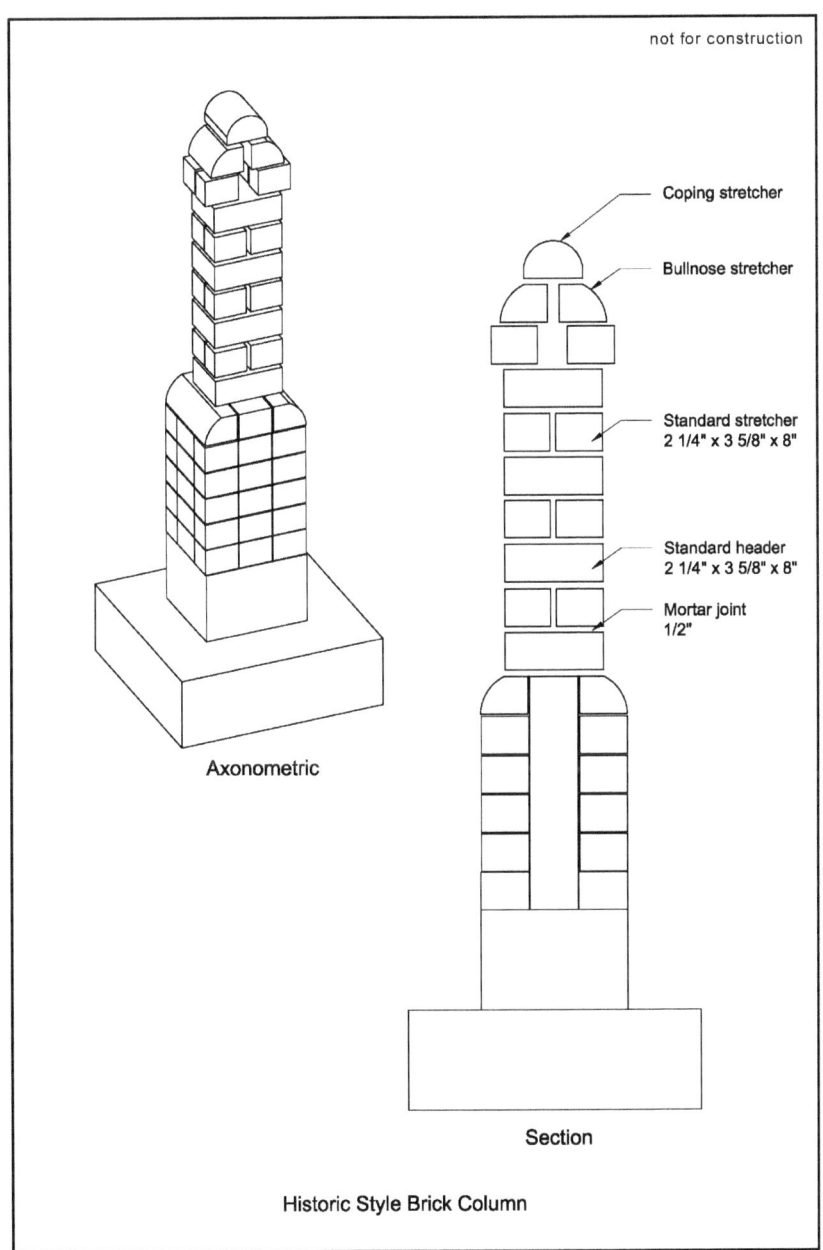

Figure 8.10 Historic style brick column.
Richard Alomar and Vicky Chan

PART 2: LIGHTING, IRRIGATION, AND WATER FEATURES
Lighting

It was not that long ago that lighting for landscape architects consisted of specifying and adjusting in the field various types of outdoor lighting fixtures manufactured or supplied by a small number of companies. In the last few decades, there has been a proliferation of companies that provide a wide array of fixtures for all sorts of outdoor purposes: street and driveway illumination; uplighting and downlighting for garden features and specimen plants; sconces and other outdoor fixtures attached to exterior walls, entrances, or other architectural elements; waterproof fixtures for pool lights and water features; all sorts of path lights and lighting for steps and walls. An interest in historic and cultural landscapes has resulted in the development of many lighting companies specializing in historic adaptations or imitations of old-time gaslights or other fixtures. In general, lighting should be shown on a separate plan, as the general contractor or landscape architect may need to give this drawing and related specifications to an electrician or subcontractor responsible for lighting and other electrical requirements for a site installation. Other items might include security systems, stereo system and speakers, electrical controllers, and so on.

The bulbs or lamps for lighting have also evolved considerably. In addition to standard incandescent bulbs, which achieve a balanced coloring outdoors but are quite expensive per lumen generated, there are fluorescent, mercury vapor, sodium, halogen, metal halide, and light-emitting diode (LED) fixtures in an array of colors and shapes and with vastly different life cycles. It used to be that a mixture outdoors of incandescent and fluorescent light fixtures would approximate outdoor conditions under natural light. Now there are many other combinations or individual fixtures that suffice.

Bamboo Light Walk, Taiwan

A particularly effective example of lighting design comes from Taiwan, Bamboo Light Walk by Ching-Yu Lin. *Sai ge ling* (pigeon racing) is a local game and recreational activity that brings the community together in Yanshuei and Yijhu. Lin transformed bamboo pigeon cages into outdoor lanterns symbolizing prosperity. The lanterns are poetic and dynamic, like racing pigeons, to celebrate peace and the harvest season for local residents, but they also energize the quiet landscape along Yuejin Port. (See Figure 8.11).

Figure 8.11 Bamboo Light Walk, Taiwan.
Design by Ching-Yu Lin; collage by Vicky Chan

DOS AND DON'TS FOR LIGHTING

1. Include a series of notes on the drawing that spell out contractor responsibilities.
2. Verify that all electrical conduits, cables, and other electrical components comply with building codes and electrical codes for outdoor use.
3. Include a range of parallel circuits so that different combination of light fixtures may be turned on in sequence or all at once. For example, for a residential garden space, it may be desirable to turn on all path lights for enjoyment of the garden at night, but only turn on the pool lights if the pool is being used in the evening, whether a pool party or a romantic midnight swim.
4. Reference the location of all utility systems on the drawing so that the contractor and/or electrician responsible for lighting, irrigation, storm drainage, gas grill, and so on may see clearly where the electrical utility system is routed and that there are no conflicts with other systems.
5. Comply with all local and national code systems.
6. As a rule of thumb, place all utilities that carry water in any capacity (water supply, storm drainage, sanitary sewage, etc.) at the proper depth (as per code requirements) well underneath all electrical utilities. This is a way to avoid water spills or leaks shorting out the electrical system.
7. Carefully evaluate the type of lamp used in terms of cost per unit of power needed to generate light, life expectancy, brightness, color, and how outdoor elements (people, plantings, furnishings, pavements) appear when illuminated by that particular source. LED fixtures are becoming more and more favored due to their long life span and relatively low cost, since they need to be replaced much less frequently than many other fixtures.
8. Be careful in choice of the color of the light pole, element, or fixture. Earth tones or black, to blend with or disappear into the landscape, is best.
9. Place uplights and downlights in locations and positions such that light doesn't shine into the faces of people taking the predominant pedestrian and/or vehicular route.
10. Just as with staking out of any major landscape hardscape features or plantings, include language that emphasizes that locations shown on the plans for lighting are approximate and final positioning and adjustment must be made in the field with the approval of the owner/client and/or landscape architect.
11. High-voltage and low-voltage lighting are both available. High-voltage lighting is sometimes needed—for example, for the illumination of a major road, perhaps even a driveway. However, a myriad

of low-voltage fixtures and systems are widely available and much safer to install, use, and maintain.
12. Swimming pools and water features usually have lighting. Be certain that the lighting specified is compatible with the proposed design features, waterproof, and vandal proof. If possible, use low-voltage fixtures in these applications as well.
13. Be careful in the use of colored lights outside. Just as a shade of green paint can look dull, washed out, or even obtrusive, bright colored lights outside can spoil the view of a landscape or its signature elements rather than dramatize them.
14. Lighting for steps is ubiquitous and necessary where any level of regular use at night is anticipated. For steps, a uniform effect with some accents is preferred. Space fixtures at a uniform distance—say, every other or every third riser (whether built into the step or in the adjacent cheek wall)—with an accent at each landing. It's important to illuminate the top and bottom risers.
15. Comply with minimum standards for foot-candles of illumination for parking lots, public walks, and so on. The level required is often lower than one might expect. The goal should be to create a safe environment but one that is still appreciated for its nighttime allure.
16. As with steps, lights may be built into the face of walls or focused onto them as uplights. It is more effective to vary the illumination along a substantial wall than to create a uniform effect, in the same way that a varied planting design avoids monotony.
17. Gaslight is used in rare settings, and is a wonderful atmospheric effect where affordable. Voltage reduction controls applied to incandescent bulbs can result in a glowing effect that simulates gaslight quite well.
18. Specialty lights for specific purposes are common, such as strings of lights for decorating trees in the holiday season. Some municipalities leave these lights in place too long, wrapped tightly around the trunks or twigs of trees, where the cable or wire may become embedded. Either wrap loosely or remove and replace each holiday season.
19. In climates with severe winters with potential for accumulation of heavy snowfall, anticipate path lights that might consist of downlighting onto the paths from trees, structures, or buildings, rather than standard path lights, which may be buried in snowdrifts for long periods.
20. Incorporate into the required task or design process for a contract for outdoor lighting at least one trip to a major lighting manufacturer or supplier to observe the manufacturing of light poles and fixtures, and evaluate a wide range of selections, in the same way that you might go to visit a plant nursery to evaluate what's available and what new products might be worth trying for a special effect.
21. Avoid placements of lights that shine directly into the faces of people.

Irrigation

Landscape architects either design and specify irrigation themselves or provide requirements that a contractor may use as a performance specification to have an irrigation system designed, detailed, priced, and installed. Increasingly, modern practice includes drip systems with emitters designed to minimize water loss from evaporation or runoff. Large spray heads and impulse sprinklers for extensive areas of lawn may still be used for athletic fields and estates. Most zoning codes require a water meter, which can measure the amount of water used for irrigation, and also a backflow preventer, so that water containing dissolved fertilizer or contaminants cannot back up in the potable water supply. Another requirement is to have a certain minimum level of water pressure for a system to function efficiently. Modern systems include controllers that activate the system so that it starts and stops according to a precise weekly or biweekly schedule—for example, Monday, Wednesday, and Friday for 25 minutes, or every other day for 20 minutes (Monday, Wednesday, Friday, Sunday, Tuesday, Thursday, Saturday). Depending on the climate and the nature of the plantings, it may still be preferred to have the system operate for a shorter period of time every day, as it's possible that in very hot and exposed areas plantings can dry up too much in a single day of exposure. Controllers have become more and more sophisticated. One will typically operate for a full season on one battery, and watering can be increased or decreased simply by punching into a keyboard a percentage of increase or decrease in watering. In modern homes and apartments, some controllers can be operated and adjusted from one's smart phone. Rain sensors are useful, particularly in climates with irregular periods of precipitation, so that if a certain level of rainfall occurs, the system will skip a day or period of watering.

It is still common to group into zones plants that have the same watering requirements. However, in simple landscape designs where there is not a need for zones, two species of trees in the same zone with different watering needs can be accommodated by increasing the number of emitters (or sprinkler heads) for the species that has greater need of water. It is usually preferred to operate irrigation systems early in the morning, as once the system turns off, any stray moisture evaporates off the surface of leaves during the course of the day, and leaves don't stay wet overnight, which might encourage fungus growth or other problems. Irrigation systems are activated typically in the spring after the threat of frost has passed and they are winterized in mid- to late fall, before the first hard frosts. Activation involves installing a new battery in the controller and checking all the lines, valves, and connections. Winterization includes pumping out the lines to remove water and storing the controllers. Given the increasing variability in weather patterns, it's prudent to have an alternative manual watering system that can be used when the irrigation system is not operating—for instance, during a spate of warm weather following the winterization of the system, or during an early warm period prior to activating the system. Dissolved fertilizers are often incorporated into an irrigation system, and their use makes a backflow preventer imperative. Due to their increasingly efficient and cost-effective innovations, many landscape architects will tend to want a reliable irrigation system for the first year or so after new plantings are installed, and then expect that, barring droughts or other extreme weather events, that the plantings will adapt and thrive. Yet, even if it's not in constant use, the irrigation system should be tested a few times a year to verify that it's fully operational when needed—much the way a homeowner who has a dishwasher but only uses it after large gatherings must still test it periodically to keep it functional and verify that it works when needed.

Major irrigation manufacturers, such as Rainbird and Toro, often provide comprehensive design services for free or at low cost in return for the exclusive use of their products. Irrigation consultants may improve on these plans by specifying the specific valves, sprinkler heads, and controllers that have been determined to be the best based on years of experience in the industry; such a system will likely incorporate components from various different irrigation companies. Valves, emitters, and controllers manufactured by different companies can be used in the same system.

Irrigation systems and practices have become more efficient in response to growing demand—locally, nationally, and internationally—to conserve water and reduce waste. In the United States, both Atlanta, Georgia, and Tucson, Arizona, which are in vastly different climates, have experienced water shortages. Cape Town, South Africa, has recently been in the news for its water shortages. One way irrigation usage can respond to such situations is to use gray water wherever possible for irrigation usage. Similarly, drip systems use much less water than spray systems. Many green roof, living roof, or ecoroof systems use minimal irrigation after initial establishment of the plantings. Xeriscape plantings may become a mandated, legislated requirement in desert climates. A ban on irrigation for lawns or an outright ban on lawns is to be expected in some regions where water usage continues to exceed water supply.

DOS AND DON'TS FOR IRRIGATION

1. For complex systems, use a consultant with expert knowledge of your region and climate and a knowledge of the best components by different manufacturers.
2. Have a backup manual system—that is, a long garden hose and bucket or a spray nozzle—in case watering of stressed plants is needed during periods when the system is turned off. Attach the hose to a frost-free quick coupling valve.
3. Verify that the water pressure is adequate.
4. Find a location for the water meter, backflow preventer, and controller such that each can easily be checked and serviced. Often local codes have requirements for the location of the water meter and backflow preventer so that they are accessible.
5. Consider a rain sensor to prevent overwatering, particularly in climates with irregular precipitation patterns.
6. Emphasize that maintenance and regular monitoring of the irrigation system must occur. Consider giving the client a maintenance manual that includes directions for the irrigation system.
7. Have the system activated and winterized each year. In subtropical or tropical climates, thoroughly check and monitor the system twice a year.
8. For simpler systems, use components from a single manufacturer whose warranty is strong. If a component fails, the failure cannot be blamed on someone else.
9. Where possible, observe well-established systems in operation prior to having a new one designed.
10. If some spray components are used, verify that the extent of watering does not extend over important walks and pavements, and be certain there is not overwatering.
11. Avoid watering at night, as wet foliage may be prone to fungal growth.
12. Check the main water lines regularly to be certain there are no major leaks.
13. For roof gardens and small courtyards, verify that any emitter lines or other pipes are kept out of view, and that drainage is carefully monitored.
14. Be certain that any contractors working on the site are aware of the layout of the irrigation system (have a set of as-built plans) and know where all major and minor water lines are placed, so as to avoid potential conflicts with gas, electric, or other utilities.
15. Use a different style of switches for irrigation than for other systems like lighting or security.

Water Features

Increasingly, landscape architecture projects include water features, dramatic elements in which water jets or other special effects shoot or spray water to soaring heights, often with illumination and animated effects. Elaborate, computer-animated menus of features coordinated among myriad arrays are possible, like a precision chorus of angels. As with irrigation, consultants whose specialty is water features often must be hired to develop the engineering and details, write the specifications, coordinate the design, check on the installation, and develop maintenance manuals for long-term use of such specialized products. Sometimes such consultants often manufacture specialty components of such systems. Often such features include a basin or storage area from which the water is recirculated when the feature is in active use and which lies fallow when the feature is quiet. The recirculation is a way to conserve water, at least once the system is operational, so that a relatively minimal volume of additional water is needed to keep the water feature operational. At the same time, by incorporating decorative and sculptural elements into the water features, so that they can be attractive during periods when they are not in use—such as during the winter or during droughts—landscape architects and designers make these elements more useful as focal points. (See Figures 8.12 and 8.13).

Figure 8.12 Residential garden water feature, Little Rock, Arkansas. In a corner of a residential garden a small, pump purchased from a local hardware store provides enough flow and sound to mask street noise and create an intimate atmosphere. (See also Figure 1-0.)

Design by Gary Evans and William Gabello; photo by Steven L. Cantor

Figure 8.13 Shepuss has impressed this elegant sink into service as her own personal water feature. Its cool, rigid geometry accommodates her feline contours.
Photo by Steven L. Cantor

DOS AND DONT'S FOR WATER FEATURES

1. Anticipate how the feature will look both when it is activated and when it is not activated.
2. In winter climates, engineer a design, if possible, that will not require draining during freezing weather. If this is not possible, aim for a design that will result in an attractive sculptural appearance.
3. In climates that have only occasional below-freezing periods, consider the use of some sort of heater that might be adequate to ensure year-round use. (This same consideration applies to devices as simple as drinking fountains)
4. Determine the weight limitation for the main water tank, basin, or pool from which the water jets or special effects will erupt.
5. Protect the pool or basin from the public to prevent accidents or drownings.
6. Provide lighting effects.
7. Disguise the source of lighting effects so that they are not readily visible (as with uplights on plantings).
8. Limit spray effects in areas exposed to winds.
9. In areas adjacent to the water feature, avoid pavements that are prone to being slippery when wet or that chip easily when frozen or when they freeze and thaw frequently. Use thicker slabs of pavement to protect against breakage.
10. Chlorinate the water.
11. Design the water feature to mask traffic noise and yet allow comfortable conversations with neighbors sitting or walking nearby.
12. Create plenty of variable seating areas near any major water feature to integrate it into its setting.
13. Locate controls for the water feature in a secure, waterproof, easily accessible, vandal-proof closet or at-grade box or cabinet with a lockable gate or door.
14. Provide a separate, secure water meter and backflow preventer so that the water feature can be turned off, if necessary, during drought emergencies.
15. Design the water feature to recirculate as much water as possible.
16. Include an overflow that would enable the entire system to drain in an emergency.
17. Resolve with the design team whether people, including children, will be allowed to use the water feature, how their safety will be ensured, and how the design will encourage or prevent whatever interactions are anticipated.
18. Be certain that the design team plans with the client/owner for adequate insurance for accidents once the water feature becomes operational.

> *Problem 2*
> ## WATER FEATURE
>
> Design and specify a water feature for a small urban park setting in a temperate climate in which the weather extremes are 10 to 15 degrees Fahrenheit (-26 to -24 degrees C) in winter and 90–95 degrees Fahrenheit (82 to 85 degrees C) in summer. Some or most components may be manufactured. It must include the following:
>
> 1. Year-round use
> 2. Cooling effect in which children will be able to play
> 3. Recirculating systems
> 4. Filtering effect
> 5. Chlorination
> 6. Masking of street noise
> 7. Optional uplighting for decorative effects
> 8. Installed cost less than $125,000–$150,000
> 9. Affordable annual maintenance
> 10. Twenty-five-year guarantee of all construction
>
> Prepare a design drawing to scale. Prepare a construction drawing showing major components and systems, minimum plan, with two sections and elevation or equivalent.

PART 3: METALS, WOOD, PLASTICS, AND SITE FURNITURE

Metals

Metals are widely used and specified by landscape architects. The most important are cast iron and wrought iron, steel and stainless steel, aluminum, brass, and copper. Know the difference between cast iron (also called gray iron) and wrought iron. Cast iron is high in compressive strength, heavy, rigid, and durable. It must be cast from a pattern and cannot be changed through deformation processes. By contrast, wrought iron is a combination of iron alloyed with a very small amount of carbon that results in a much more flexible workable material that is, weldable, bendable, and malleable.[3]

Metal Weights

Perhaps one of the easiest determinations in deciding which metal to use is its weight. The following lists metals by ascending order of weight per cubic inch in pounds:

Aluminum	0.10
Cast iron	0.27
Steel	0.29
Brass, bronze, copper	0.30
Stainless steel, mild steel	0.29

These weights are important not only for roof loading calculations but also for transportation and installation costs.

If its dense weight can be tolerated, stainless steel is an excellent material for planters and water features, as it can hold soil and/or water without any risk of rusting or gradual deterioration. By contrast, aluminum is lightweight, only about a third the weight of steel, but lacks the strength of iron or steel, so its main use is for decorative elements where weight is a concern, such as roof gardens, as well as some nonstructural configurations such as lattices and trellises.

Metal Finishes

Almost any type of metal fencing, from chain link to decorative and ornate forged and cast elements, should have a finish to deter oxidation. Most common are paints over a zinc-rich primer; another possibility is galvanized finishes for simple designs.

Anodized coatings used for aluminum are attractive and can be manufactured in a variety of colors, but they are not as durable as a galvanized finish.

Brass and copper are commonly used for specialty elements, such as light fixtures and decorative piping. Weathered copper develops a greenish patina with age but is best given some protective coating for longer usage. Copper is very malleable and easily worked over a form, and it takes soldering readily.

Most metals can be painted. Sometimes a primer coat is required. In the landscape, a matte black color is effective at receding into the landscape of earth tones, natural browns, grays, and color accents. Of course, some designers want a particular section of fence or a gate to stand out within the landscape, and all sorts of bold colors are possible.

Fastening

Abutting metal elements can be joining by welding or by mechanical fastening. The location of the weld may need to be considered; they are usually located in an area that can be blended with the adjoining pieces by grinding, or otherwise disguised so as to recede into the background of the detail and not be highly visible. Bolting overlapping elements together with galvanized hardware, as is typically done with fences and gates, is more practical.

Processes

Each metal is manufactured differently, and available in different sizes and shapes. There are two distinct classes of metalwork: hot work, where the use of heat allows changes in shape and volume of the piece, and cold work, which can be used to roll or bend straight stock to create a decorative element or component.

Shapes

The most common wrought iron and steel shapes available are bars, rods, tubes, and sheets as well as structural shapes such as channel, angle, and tube. Bar stock can be round, square, or rectangular, and can be worked with heat and easily welded into any number of shapes. For example, the channel iron, an upside-down U-shape, is basic to many fences and gates as both reinforcement of the railing cap and a place to fasten pickets and posts without visible welds. At the top and bottom of a railing, the pickets can be welded from below, or rods can pierce a channel iron at regular intervals to achieve fencing (with welding on the underside, or under the railing cap, where it's not visible). Tubes are another useful manufactured shape that provides stiffness and light weight over their solid counterparts.

A further addition to the material's inherent strength can be achieved by bending an otherwise flat sheet material to create a channel shape, which adds to the stiffness and rigidity without adding as much in mass. This is a standard method on the top of open planters and containers made of metal.

Since gray iron and wrought iron are used for so many ornate elements, their differences in application should be discussed. Grey iron is cast in sand molds and can be used for single or multiple elements that are identical. It is a hard, inflexible material that has great compressive strength but is not malleable and cannot be bent without breaking. Wrought iron is malleable, ductile, and weldable, so it's easier to form, roll, or forge into decorative elements with the right application of heat, but it has little compressive strength.

PLASTICS

Fiberglass and plastics are increasingly in use. Perhaps the most common use for the former is as liners for wood planters or as free-standing planters. Since fiberglass is molded, it can be formed in myriad shapes and sizes. Fitting a liner in a wood planter is an ideal marriage: the two materials reinforce each other. The exterior of wood weathers, it is decorative, and it lasts well as long as it's not exposed to water or other wet materials. The fiberglass is not visible and gives protection to the wood, yet readily holds all the topsoil and plant materials. (See Figures 8.14 and 8.15).

Figure 8.14 These are images of pre-aged Cor-Ten planters with structural fiberglass interior liners, (minimum ¼" or 6.35 mm thick), which are very sturdy and allow a thinner steel sleeve when critical weight savings are desired.
Photo by Dean Anderson, Super Square Corporation; collage by Vicky Chan

Figure 8.15 A close-up showing a fiberglass liner fitting into a wood planter. In this case, the fiberglass insulates the plants, protects the wood, and saves on weight. The return lip is about ¼ inch (6.35 mm). Emitters for irrigation are often preinstalled, hidden behind the planters.

Courtesy Mark Davies of Higher Ground; planter design by Michael Rubin, liner manufactured by Fiberglass Engineering

DOS AND DON'TS FOR METALS AND PLASTICS

1. If designing unique elements, such as a particular handrail or light post, go to the foundry or manufacturer and discuss the details that you are developing with one of the fabricators or other staff. They know how to build such elements and can likely help you improve on your design with possible savings in the long run.
2. Hardware for garden areas should be galvanized or stainless steel, water-resistant and rust-resistant.
3. Use a finish that is water- and rust-resistant.
4. If developing an element such as a light post that which will eventually carry electricity, comply with the electrical code for the size and placement of conduits and fixtures.
5. Calculate the weight of any major metal element and be certain that the design installation detail, such as the pedestal or base on which it will rest, is appropriate for that weight. Consider wind loads as well, especially on rooftops or terrace gardens.
6. Use matte or flat black paint colors for painted metal elements if you want them to recede into the landscape. Use bright colors for bold accents. Be careful with green colors in landscape settings as with few exceptions they usually look poor in comparison to the natural greens of plant materials.
7. Be careful in the designs that there will be no way for water to penetrate a post, column, or other metal element. If there is any possibility for water penetration, provide an exit point for water to drain out.
8. Require shop drawings for all complex metal elements such as planters, lampposts, gates, and decorative fences, so that the contractor has the opportunity to measure precisely the existing conditions and provide a design element that responds to and fits exactly those dimensions without complications or errors.
9. Know why you use one metal compared to another: cost, strength, weight, specific properties, aesthetics.
10. Require a sample to be built of any metal or plastic item that will be used widely in a design: fence section, planter element, lamppost, gate section, etc.
11. Despite the use of the metric system in many detailing systems, for most American offerings dimensioning is still in the imperial system, with the exception of some steel shapes. Yet with the continued globalization of the supply system, training should include learning the metric system, as it will gain in importance.
12. Consider fiberglass liners for wood planters. An even more durable combination is planters of Cor-Ten steel with fiberglass liners. (See Figures 8.14 and 8.15). Stainless steel planters are strong and rigid enough to incorporate removable panels for maintenance.

13. Keep abreast of advances in the industry. For example, plasma cutting services allow manufacturers to precisely cut metals to almost any shape, so that a fabricator can have precise pieces available for welding. This has made complex metal designs more affordable, more practical, and safer to construct.
14. Rolled steel shapes such as channel, angle, tee, and flat bar can be transformed remarkably by plasma cutting to create decorative shapes not possible a few decades ago. Steel tubing can also be cut to provide decorative end shapes.
15. If a new design attaches to a building or must carry a significant load safely, have your design reviewed by a structural engineer.
16. The starting point for any custom design is as-built measurements, so measure things precisely, whether in shop drawings or in existing conditions. Document everything carefully at the site where something you are designing is going to be built.

(See Figures 8.16, 8.17, 8.18, 8.19 and 8.20).

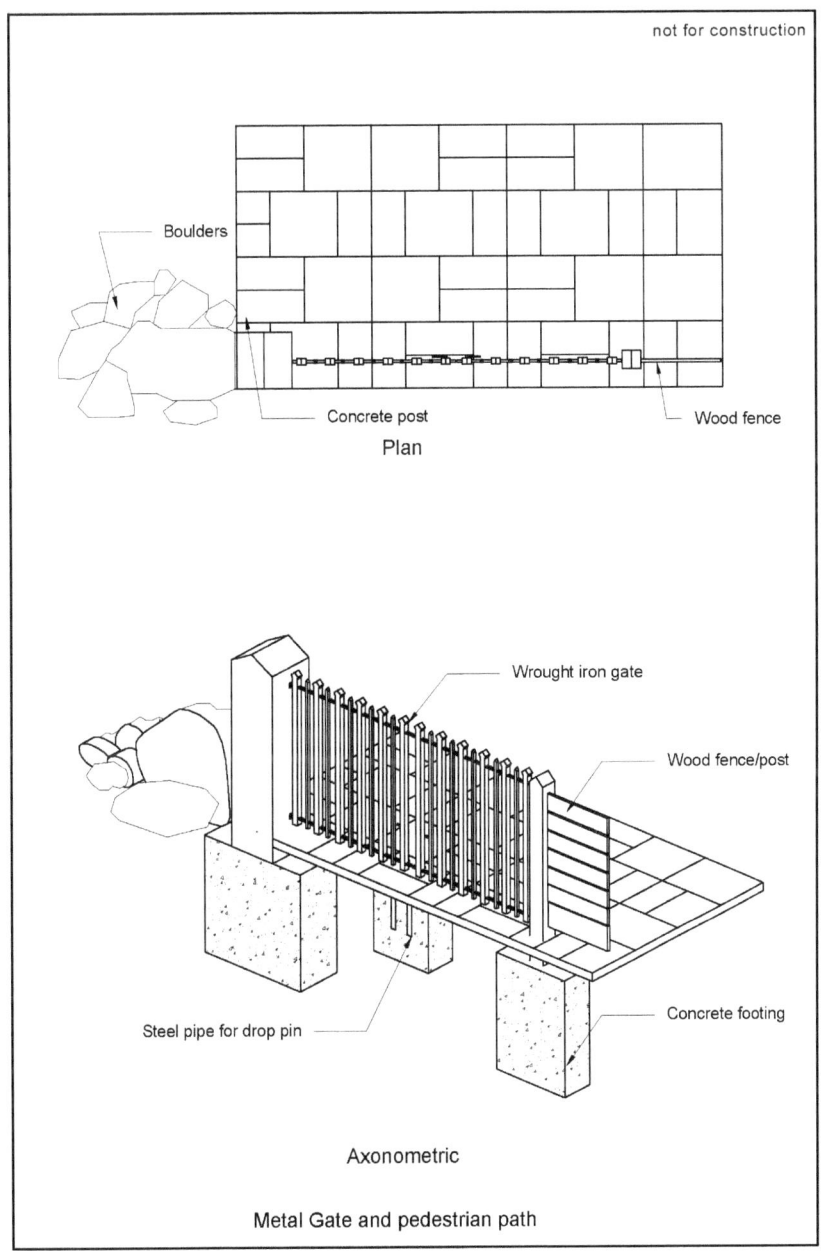

Figure 8.16 Ornamental metal gate.
Richard Alomar and Vicky Chan

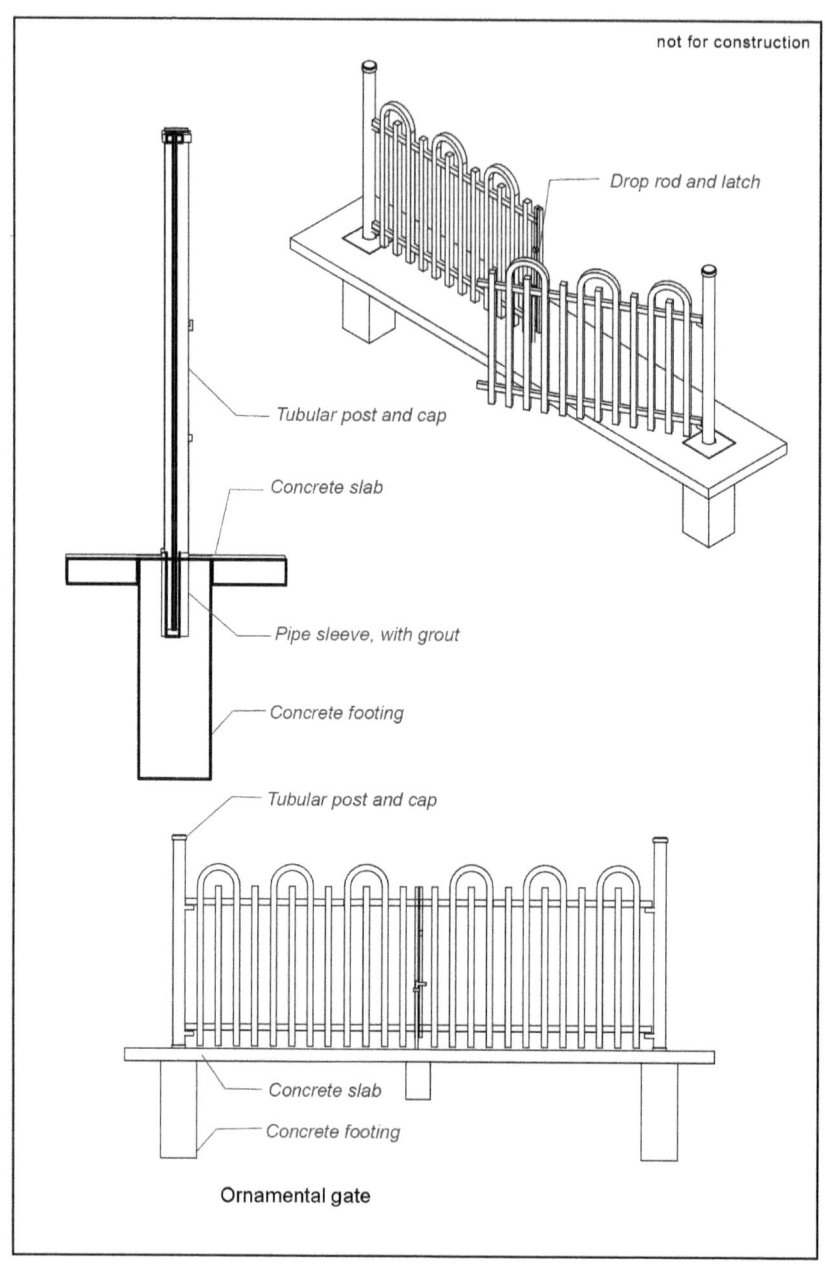

Figure 8.17 Metal gate and pedestrian path.
Richard Alomar and Vicky Chan

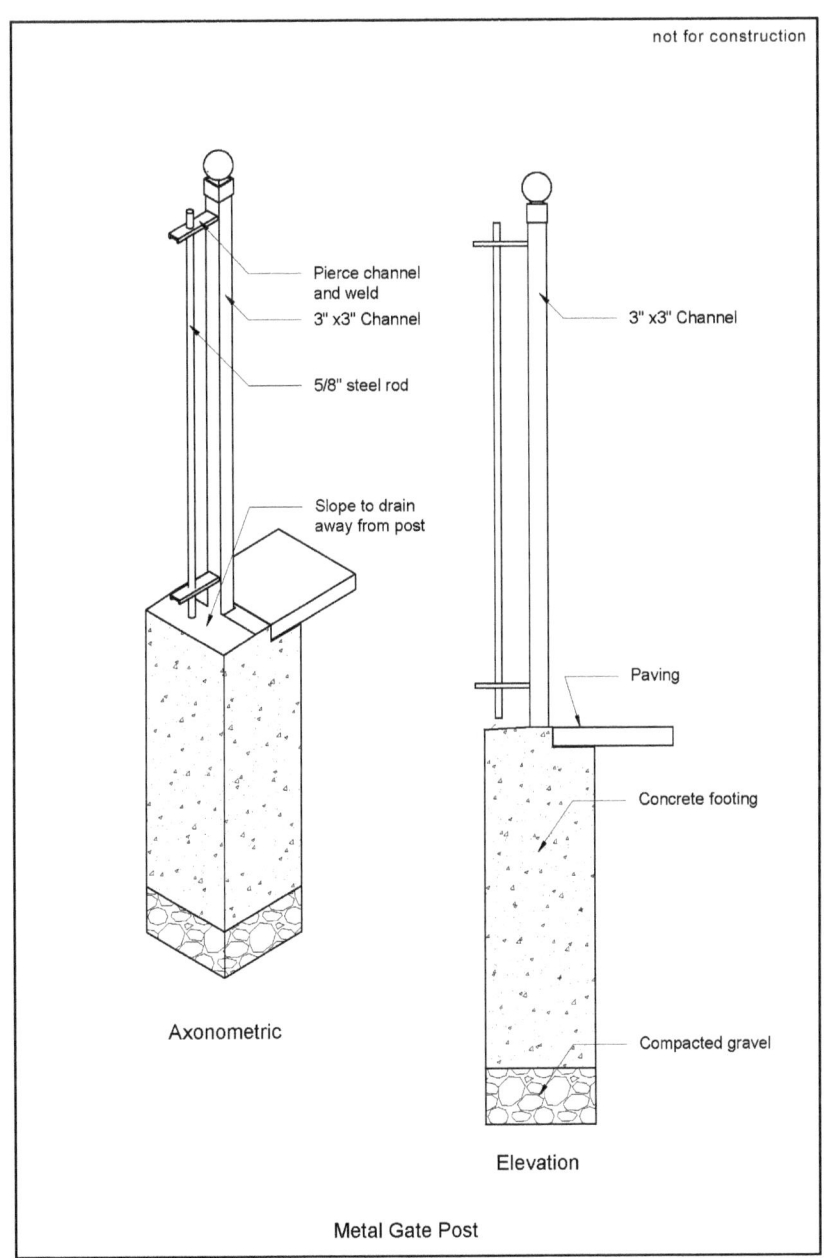

Figure 8.18 Metal gate post.
Richard Alomar and Vicky Chan

Figure 8.19 Collage showing steps in restoration of ornamental iron gate.
Design, and individual digital images by Dean Anderson and Amy Lahey of Super Square Incorporated; collage by Vicky Chan

Figure 8.20 New ornamental iron gate for church.
Design by Dean Anderson and Amy Lahey of Super Square Incorporated

Wood

Wood in its myriad forms, species, components, textures, and colors, with various stains, finishes and paint selections, is a staple of landscape architecture vocabulary. As sustainability and conservation of natural resources have emerged as major themes, there are often restrictions on the use of endangered or threatened species, such as redwood or mahogany or ipe, and the search for wood substitutes has greatly increased. Some vinyl products now used as wood substitutes for decking, lattices, and arbors are so sophisticated as to actually feel like wood in texture and weight. An advantage of vinyl coloring systems is that they are permanent. There is no need to restain or repaint it after a number of years, whereas wood systems often require this, and it can be quite challenging to carry out. For example, just as they are maturing and reaching their most attractive appearance, vines and other plants growing on wood structures may need to be severely pruned or restrained in order to move forward with this process. If vinyl is considered as an alternative, the costs of using wood must be compared to the manufacturing process creating the vinyl substitutes. Although some manufactured products, such a vinyl, are quite attractive and useful, the amount of water and various chemical elements used in creating them must be evaluated before it's decided that vinyl components are a preferred alternative to wood.

Lumber is usually sold according to its compliance with various ASTM standards as to the particular species, percentage of water, degree or extent of imperfections, seasoning, and other qualities. Different species vary in weight, in structural and other properties, in tendency to warp, in durability and so on. The landscape architect must select and specify lumber based on the combination of performance standards that it meets, based on the requirements of the design. Sustainability, cost, and shipping must be major factors. Local products are, of course, preferred to ones that must be shipped long distances. A species that is scarce in its native environment, such as mahogany and teak, is hardly one to be selected if other alternatives are reasonable—and perhaps even if no other alternatives are available. Even within the United States some designers eschew redwood, which is highly valued for its strength, durability, and appearance but is harder and harder to find since so many redwood forests have been depleted. Sometimes the design should be adjusted to meet the available type, size, and quality of lumber.

As with brick, unit pavers, and some other materials, the size of lumber has changed over the years. A nominal 2 × 4 framing stud used to be fully 2" × 4" but is now 1½" × 3½". In drawing complex details in which specific sizes of wood are specified, it is more accurate to draw the *actual* dimensions of the lumber than its nominal size. Just as with pavers, if repairs or additions are being made to a wooden structure over a certain age, the designer must be careful to find a source of planks, beams, and joists, and other members that match the size of those being replaced or added, or at least verify by drawing to scale that the new wood elements will be compatible with the existing ones.

Avoid Tropical Woods

Tropical woods, although still available today, should generally be avoided. As tropical forests are being further explored, such as the vast Amazon rainforest, new species are being discovered. Foreign governments often have fewer resources than the United States and Canada to properly police vast forest areas and prevent poaching of lumber and animals. Although there are undoubtedly some privately owned forests and nurseries in which tropical woods are sustainably harvested, so that the amount of wood logged each year is no more than the additional height and girth of existing trees or the potential of newly planted (or germinated) seedlings, it is very difficult to verify. Few designers have the budget to fly to a particular site to verify that forestry practices comply with standards of sustainability. White oak, red cedar, and longleaf pine are all suitable alternatives to tropical woods.

Think again about forestry practices in the United States. From the founding of the country through the mid-twentieth century, vast acreages of virgin forest were logged. At least in some cases, the climate and environmental conditions were favorable enough so that the forest regenerated, although it's difficult to imagine that trees will reach the size of those originally logged. Advanced scientific procedures and conservation techniques are now routinely implemented, yet the quantity and sometimes the quality of select species that can be logged, harvested, dressed, and marketed is limited. Other species, such as the American chestnut, have been almost wiped out due to a fungus that kills the trees. Although efforts by the American Chestnut Foundation to develop a replacement American chestnut with 99% American chestnut genes and 1% or less Asian genes are far along, nearing fruition, it's hard to imagine how long it will take to replenish the species throughout its former range. American chestnut (*Castanea dentata*) is so rot resistant that wood from huge hulks of chestnut trees that died a century ago can still occasionally be harvested and used. Fortunately, Eastern red cedar (*Juniperus virginiana*) and black locust (*Robinia pseudoacacia*) have been used for fence posts and are widely available. Greatly depleted species such as redwood and bald cypress are highly prized but available only in limited quantities at high cost. It's better to find substitutes, or at least use the prized and rare woods sparingly. If tropical woods must be used for a particular project, perhaps their use can be limited to posts and beams that support and carry the most weight, with joists and planks of local woods.

As with other industries, landscape architects should think globally. The attitude "Not in my backyard, but in someone else's is OK" still prevails. For example, the Patagonian toothfish (*Dissostichus eleginoides*) was discovered by fishermen off the coast of Chile, was renamed Chilean sea bass (although it's not a bass at all), and became a culinary sensation. Huge levels of overfishing have depleted it drastically, although there have been some efforts at conservation. Although the species can reach 100 pounds and live fifty years, it is rare to catch a mature fish. Perhaps, through concerted efforts at limiting the catch in its prime habitat locations, it will be possible for the species to recover. It's not that different with mahogany, teak, ipe, or other tropical woods. Most of the largest specimens have been harvested. They do not grow in large thickets, like aspen or other species, but instead may grow as individual trees or small groups scattered throughout a forest, so harvesting them potentially damages many adjacent trees. It's possible that the species will still survive in protected areas, but if landscape architects keep specifying such species, the demand so increases that people will be willing to harvest them illegally in order to make a large profit at the expense of the native country in which these trees grow. Cumaru (*Dipteryx odorata*, Brazilian or golden teak) is now marketed as a less expensive substitute for ipe, perhaps in the same way that substitutes are offered for redwood in Canada and America now that the prime timber is gone, but would it not be better to err on the side of caution and be certain that significant mature stands of major species remain in tropical forests and can be carefully harvested and sustained? Brazilians elected as President Jair Bolsonaro who took office January 1, 2019 and pledged to open up more of the Amazon rain forest for agriculture and mining. In the period between August 17, 2017 and July 18, 2018 the deforestation of the Amazon rainforest in Brazil reached its highest rate in a decade, 3,050 square miles (7,900 square km) about 5 times the size of London, primarily as a result of illegal logging.[4] Although this rate is still significantly less than some previous decades, the starkly upward trend is not encouraging. Although most of the pressure is from mining and agriculture interests, for the design industries to contribute to the demand on resources from the rainforest particularly at this time, seems unfortunate. Annie Proulx's novel *Barkskins* is full of vivid descriptions of how pioneers experienced the vast forests of North America as a seemingly infinite resource of timber of diverse species and uses.[5] Yet this vast forest vanishes over the course of the novel's 700 pages. The even more diverse Amazon rain forest and other regions of the tropics seem at times on the verge of a similar fate. Sadly, swings in the political pendulum on each continent seem to foretell different outcomes, even though the trees are neither blue nor red. Indiscriminate logging is not limited to one location on the globe; it is an international problem. Putin's Russia leads the world in forest depletion: 16.3 million acres (6,600,000 hectares) of logging occurred in Siberia in 2018, much of it paid for by Chinese timber barons, compared to 9.1 million acres (3,700,000 hectares) in the entire Amazon rain forest.[6]

Certification of Forestry Practices

Two organizations, the Forest Stewardship Council (https://www.fsc.org) and the Programme for the Endorsement of Forest Certification (https://www.pefc.org), were founded in the 1990s to promote the management, sound conservation, and sustained forestry practices of the world's forests. The former focused more on large tropical forests, the latter more on smaller private, often European sites, and progress has been made, but there is a long way to go, given the rapacious demand, intense competition for a diminishing product, and intense volatility in the world.[7]

Painting and Staining

Lumber may be painted or stained. Even when pressure-treated and seasoned, the wood absorbs a lot of the paint or stain, so in painting, a primer coat may first be applied and allowed to dry, then a second coat of the correct color is applied. Two coats of stain are usually required as well. There are also clear stains that allow the natural color and grain to stand out. Some species of wood weather to a beautiful patina, so it's better not to paint or stain them. Tung oil is still used for protection of woods left natural. Polyurethane is a common sealant and protective coating.

In America a great deal of lumber is pressure treated, which is a process in which chemical preservatives are forced into the wood, giving it greater protection against fungal decay and insects like termites. Depending on the chemicals used as preservatives, there can be some limitations on their use. In general, the longer period of use that results from the preservatives justifies their use.

There are now myriad metal hardware elements for use to anchor a wood post or pier into concrete, to connect joists with beams, and so on. Often these elements are labor-saving and prolong the life of the wood. Yet they still must be installed correctly. Detail them carefully, so that the specific applications are clear. Anyone bidding the work must have several years of documentable experience in using such components.

Wood is still a common material for building planters, which are placed as strong decorative or structural elements in a landscape, such as a patio or roof deck. It is advisable to use fiberglass or other liners within wood planters to protect the wood from exposure to the moisture in the soil, greatly prolonging the life of the wood planter.

Don't overlook plywood as a construction material. Through its manufacture by gluing layers of thin wood together and pressurizing them into a single dense board, plywood has remarkable strength and versatility. It's useful for cabinets, boxes, doors, and tables, and as a veneer or cap attached to heavier-duty structural elements, perhaps made of steel or heavier wood.

A rather recent substitute for some uses of wood is vinyl. In the last decade, some manufacturers such as Walpole Woodworkers have introduced vinyl products for lattices and arbors that have the texture and character of wood, can be formed and connected like wood structures, and come in a range of preapplied permanent colors. One advantage is that they do not need to be repainted or restained just as the vines reach maturity on the arbor. On the other hand, the process whereby they are manufactured may deplete the environment in ways that the harvesting of wood does not.

DOS AND DON'TS FOR WOOD

1. Select and specify the species of lumber based on the properties needed, on the design, and in consideration of the availability and cost of the species.
2. Avoid tropical woods, or verify that a small amount for a special design element comes from a sustainably managed site.
3. Draw details based on the actual dimensions of the lumber.
4. Design decks, pergolas, and other landscape structures while keeping in mind the available sizes of the wood being specified. For example, if pressure-treated yellow pine 2 × 4 planking is available in 4-, 8-, and 12-foot lengths, don't design a deck 10 feet wide.
5. Consider carefully all of the connection details for wood elements. Often the use of metal hardware, such as post bases, joist hangers, and so on, is not just practical but prolongs the life of the wood, because the metal protects the wood from water damage and also

protects the wood members at their weakest points. All such hardware should be galvanized or otherwise rust-resistant; some are manufactured from stainless steel. Don't scrimp on the quality of all screws, nails, and other miscellaneous fittings and hardware.
6. Notching under planks helps prevent warping over time.
7. Staining is generally preferable to painting because staining can enhance the natural qualities of the grain while opaque layers of paint tend to disguise or cover it.
8. Find a way to use scraps of wood generated at a job site.
9. Be aware of code requirements, as some jurisdictions may forbid the use of flammable materials in certain locations, such as roof gardens. To date, no known treatment renders wood immune to fire.
10. When wood is in contact with water, devise a way to protect it—for example, use fiberglass liners placed inside wood planters.
11. Err on the side of caution. When sizing wood for joists or beams, incorporate a safety factor. In much the same way as it's prudent to increase the hardiness zone for a roof garden by at least one due to the extreme weather conditions the plants will experience, anticipate that on occasions a wood deck or other structure may experience a degree of use a little greater than it was designed for.
12. Be careful in selecting colors of stains and paints. Often wood looks best when it appears to be in its natural state, with clear stains for protection.
13. Based on the type of wood specified and its use, consider application of a sealant.
14. Standard details for handrails, steps, deck patterns are readily available from many sources. If you wish to create a unique element, one way is to brainstorm and start from scratch. Another is to make incremental changes to a standard detail that you know meets code requirements and is also durable in adverse weather, so that your eventual creation may also pass muster.
15. Insist on having samples or mockups fabricated of any major wood element, and require shop drawings for any non-standard elements.
16. Verify local code requirements for load criteria and specific site requirements for wind resistance, handrails, decks, pergolas, and pool fences and gates. Many manufacturers and suppliers have reliable standard details.

(For typical wood details, see Figure 8.21).

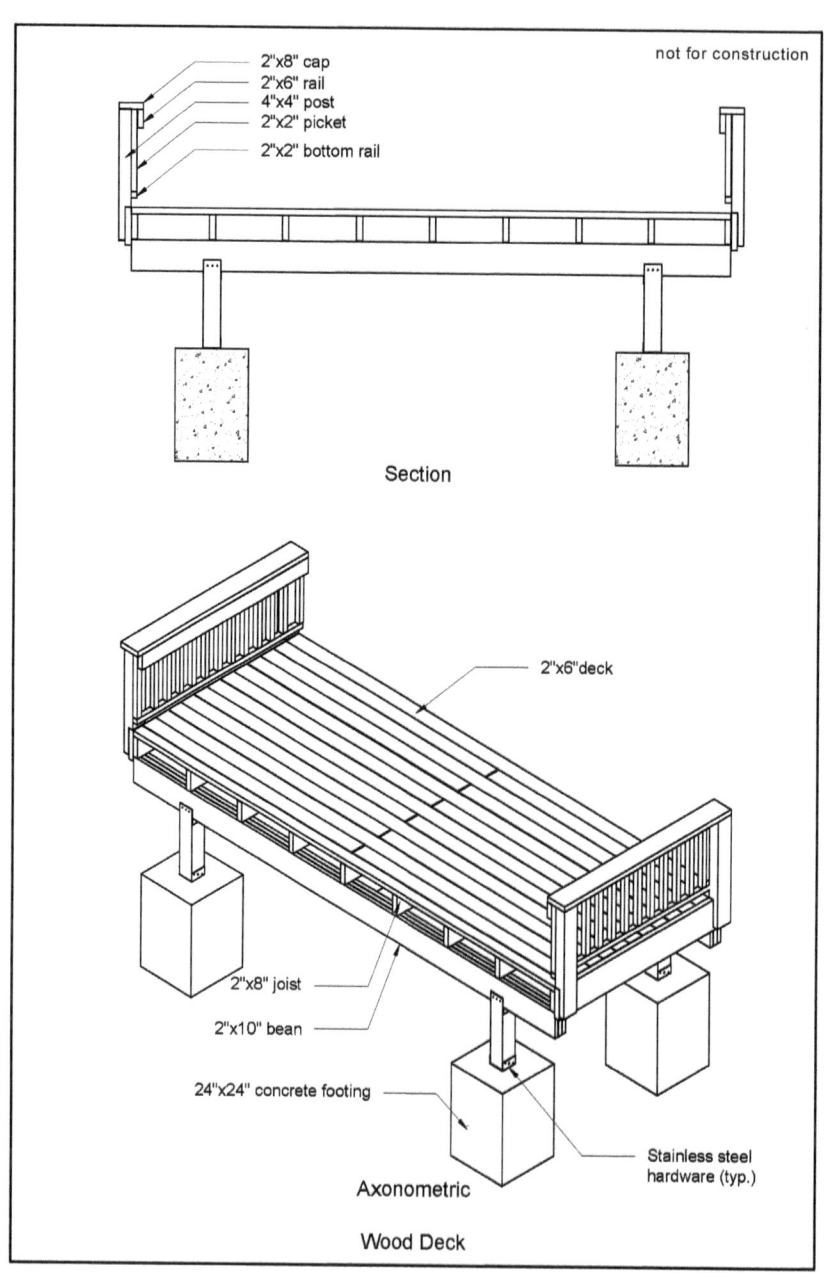

Figure 8.21 Wood deck.
Richard Alomar and Vicky Chan

METALS, WOOD, PLASTICS, AND SITE FURNITURE (407)

The living room of this residence extends outdoors to a large rear deck, comfortable enough for entertaining but intimate enough for a small family dinner, cooked on a built-in BBQ. The sloping back yard was re-graded to provide a flat lawn, allowing open and flexible outdoor space. (See Figures 8.22a and 8.22b).

not for construction

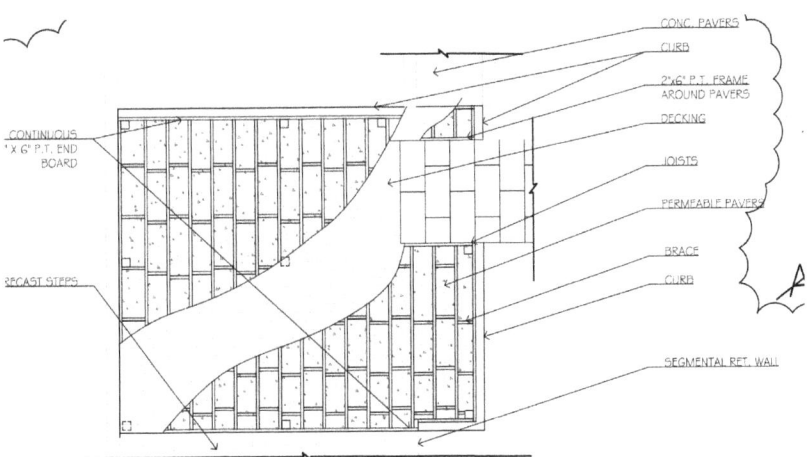

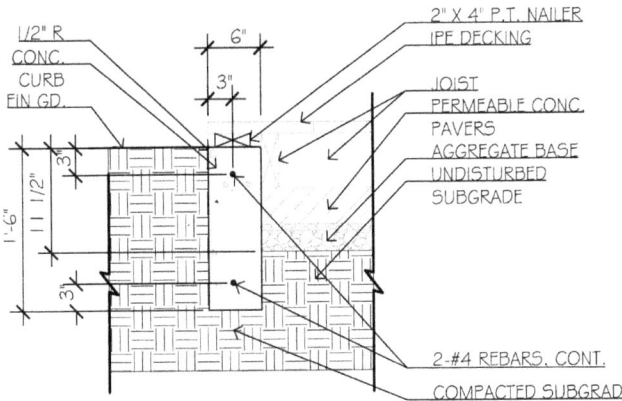

Figure 8.22a, 8.22b Plan view and section of wood deck for residence and as-built photograph.
Jeff Dragan, LDGN Landscape Architects, DPC

Figure 8.22a, 8.22b Continued

Site Furniture

Site furnishings abound and can be found both mass-produced and custom-designed. Benches, chairs and seating systems are perhaps the most common. Trash receptacles and all sorts of planters are common. Water fountains, bicycle racks, and decorative or ornamental drainage grates and tree grates are also widely used. How does a landscape architect take stock of so many possibilities? It helps to take a thematic approach, in which all site furniture is similar in style and materials, locations where groups of site furniture are placed are carefully analyzed and designed, and recessive, natural colors are specified. On the other hand, the occasional accent of bright colors can be an effective focal point. Yet, how many pleasant views of parks, or streetscapes have been badly marred by the grotesque intrusion of an ugly green trash can in a location by itself at a focal point?

DOS AND DON'TS FOR SITE FURNITURE

1. Not everyone is the same size. Therefore, in designing benches and other items of furniture with which people directly interact, include or incorporate a range of sizes. Just as indoor settings include urinals of different heights, outdoor settings should include benches of different dimensions, to accommodate people who are large or small, tall or short.
2. Remember that water features may not be operational all year round due to winter temperatures below freezing. Design water features to look attractive and have some function, at least as sculpture in the landscape, during the winter and other periods when they may be drained and not operating.
3. Increasingly, with winters being milder or at least of shorter duration, consider some drinking fountains that could be operational all year long.
4. Design outdoor spaces for ease of maintenance, such as collecting garbage. Trash receptacles should be vandal-proof but also easily operational for the people who collect the garbage and install a new liner on a regular basis.
5. Find recessive locations on the site for most street furniture, so they are not the dominant features in the landscape.
6. Consider portable furniture, particularly tables and chairs that users of the site can move around to accommodate their particular preferences. Such furniture should be lightweight and durable.

7. Provide convenient storage locations so that portable furniture may be locked up at night. Sometimes the perimeter of the site is fenced, so all furniture within it may remain in flexible locations.
8. Remember that wide sets of steps are often desirable as sitting areas, so designers must accommodate both functions: seating and walking.
9. Locate bicycle racks near the areas where bicyclists most frequently dismount. Bicyclists typically want to be able to see their bicycles while they are engaged in other activities, such as resting, buying food and refreshment, using restrooms, or participating in entertainment.
10. Place trash receptacles in locations near where people are generating trash, and program regular trash pickup schedules.
11. Anticipate recycling: design trash receptacles to comply with required functions.
12. If night use is anticipated, incorporate lighting fixtures.
13. Include restrooms as permanent elements of the landscape, or at least portable elements during periods of peak use. Odorless composting toilets are widely available.
14. Develop a unified vocabulary of site furnishings.
15. Place site furniture to animate the spaces, not to use up space in leftover areas where people are unlikely to congregate.
16. Be certain that the major site furniture elements are easily replaceable; don't use exotic or unique elements that may be difficult to replace in a few years, when replacement may be necessary.
17. Be wary of integrated systems of elements that require a level of maintenance and repair that may not be available in the future. (The High Line in Manhattan is a good example of a high-maintenance integrated system.)
18. Be certain that any planters are of adequate size to accommodate insulation and enough volume of soil so that plantings will survive the winter.
19. Anticipate how "undesirable" elements, however they are defined, will be discouraged from congregating for long periods.
20. Provide at least a few frost-free and insulated elements to accommodate winter usage.
21. Give clear directions for recycling and provide separate clearly labeled receptacles for plastics, paper, metals, composting, etc.

Problem 1
SITE FURNISHINGS AND LIGHTING

(In the following problem, any *terms in italics* may be changed to suit the particular instructor or team leader, or your own personal, work or studio setting. Typical changes might be specific people, such as clients or contractors, plant materials, or the locations of sites, people, institutions or companies.)

You are working on a 10-acre (2.5 hectares) park for the *City of D.* It features a range of recreational facilities for people of all ages, including one renovated and two new ball fields for baseball, two basketball courts, a handball court, a children's playground, a 1-acre (0.4 hectare) field for open play, some shaded areas with chess tables and other sitting areas, parking, pedestrian paths, restrooms, and a small concession building selling snacks and food. Come up with a complete range of site furniture for this park, including benches, chairs, tables, trash receptacles, and water fountains. You may design one seating area for group sitting. All other proposed furniture must be selected from a manufacturer's available products. All furniture need not be from the same manufacturer.

The park is designed for night use, so the parking lot, the three baseball fields, the basketball courts, and the handball court are illuminated. Propose lighting standards/fixtures for these fields and parking lots and smaller-scaled lighting for the pedestrian areas, including the walk system, sitting areas, and other areas.

Problem 2
WOOD AND METAL GATE

(In the following problem, any *terms in italics* may be changed to suit the particular instructor or team leader, or your own personal, work or studio setting. Typical changes might be specific people, such as clients or contractors, plant materials, or the locations of sites, people, institutions or companies.)

Prepare a design drawing and construction details for a wood and metal gate 4 feet wide by 8 feet tall for the entrance to a garden. One side of the gate may be attached to a *wall of a residence or other brick building,*

> and the other side may be freestanding. By the freestanding side is a *long wall, hedge, or fence.* The pavement underneath the approach to the gate is gravel, and there will be a 30-foot-long walk leading by the side of the *building,* beyond which is the *main garden.* The gate will provide privacy but also have enough openings to give a glimpse of the interior.

PART 4: SITE INSPECTIONS

Site inspections are a crucial aspect of completing a project; they must be carried out in an organized, thorough, and detailed manner and documented clearly in order to ensure that the design defined, described, and detailed in the contract documents is implemented to the satisfaction of all parties: the client, the owner, and the contractor(s). Years ago it was common to refer to this process as "site supervision," but this term is inaccurate. The landscape architect is not directing the work of the contractor, nor is s/he responsible for its implementation. Instead, the landscape architect comes to the site to facilitate the contractor's work, to verify that it is in compliance with the contract drawings, and to emphasize which elements, whether layout, grading, details, or planting, need correction or are incomplete. At each site inspection, the landscape architect compiles a checklist of problems and summarizes them in a set of minutes (or in a list of items to be corrected), which is distributed to all parties to the contract. Responding to this document becomes the first order of business at the next site inspection. A binder kept with these documents in reverse chronological order, with the most recent at the front, then becomes a record of the history of the implementation of the work.

One of the hardest aspects of a project is to estimate the number and duration of site inspections. Therefore, if possible, it is best to reach a contractual agreement with the client on the number and length of site inspections and the fee to be charged *as the contract documents* (contract, drawings, and specifications) *are being completed.* This fee can vary tremendously depending on a number of variables. The number and complexity of drawings, the range and type of specifications, the bid cost of the work to be implemented, and the distance the landscape architect must travel to reach the site where the construction is to occur all become factors in calculating a fee for site inspection services. In many contracts for public agencies, the fee for "construction services," "site administration," or similar term is predetermined by the agency and is defined as a percentage of the total fee, anywhere from 10% to 25%. Sometimes this phase of work includes the solicitation and review of bids and includes assistance required by the client to finalize a contract with the low bidder or other bidder to the satisfaction of the owner and client. In other public contracts, construction services may just be limited to all those aspects of work that occur after the bids are complete and have been reviewed and a contractor has been selected and is ready to begin work on the project.

Most, if not all projects, include the following phases or tasks. The number of meetings or site inspections required for reviewing the entire phase of work, from beginning to end, will help the landscape architect estimate the total number of meetings and site inspections and be able to calculate a fee. If the project has a preset fee, then a careful review of all of these tasks or meetings will give the landscape architect a sense of the amount of his fee that will be available for each item. On projects in which the total fee for these services is limited, the landscape architect must find ways of working at maximum efficiency in order to stay within budget.

Start-up Meeting

This is an organizational meeting in which all the key parties—owner and/or owner's representative, general contractor and some specialized contractors, landscape architect and/or project manager, and key consultants (such as engineers and architects)—come together and decide on protocols and procedures that will apply for the duration of the project. Everyone should have a complete set of the contract drawings (often half-size so that they can be handled easily yet are still clear and legible) and specifications. It is customary to decide how often such meetings will occur and set standard dates so that everyone can reserve time in the schedule for future meetings. By reviewing the drawings and specifications, a preliminary list of permits and approvals for which the contractor should apply and a list of shop drawings the contractor will prepare or request over time can be determined. A location for a contractor's site office, if applicable, is determined. The sources of electrical and water service for the contractor are identified. (The specifications usually indicate if these are provided for free or at a charge.) Larger projects often include the location of portable toilets. A list of attendees, with full name, company, phone number, and e-mail address, is prepared and exchanged so that everyone knows how to contact one another. Any necessary restrictions or taboos are clearly described—for example, where a contractor and his staff may or may not park, if some areas of the site are off-limits, the hours of operation, when the site must be closed, where and how deliveries will occur, who will sign for deliveries, where a temporary contractor's storage area will be set up, and so on. All of these matters are discussed thoroughly and agreed upon. The project manager distributes a first set of minutes that include all of this information, including names and addresses of all participants. Usually a set period of time is permitted for all participants (as well as the occasional person who was unable to attend) to review the minutes and submit corrections, and a corrected set of minutes is issued that includes the date of the next meeting.

Demolition

For small projects, or ones in which there is a very limited amount of demolition, this phase is often combined with layout.

1. Review the site survey. Although it should have been clearly reviewed and verified during the design phase, it's still useful to review it in the field, with a few new sets of eyes, to see if anything is at odds with conditions as they exist just before demolition starts.
2. Verify that all existing utilities have been staked out. Many utility companies do this for free, as it's in their self-interest to avoid major problems.
3. Identify any discrepancies between what is shown on the drawings and what is found in the field. Mark up the set of drawings and correct and reissue them as necessary.
4. Verify disposal methods and that contractor has secured proper permits for this purpose.
5. If dumpsters or other containers for disposal of debris and garbage are to be maintained on the site, determine their location and verify that they present no potential problems or conflicts with existing utilities, vehicular traffic, neighboring properties, or the activities of key personnel.
6. Verify that all elements labeled for demolition are correctly indicated, including the precise limits of the demolition.
7. Focus attention on any conflicts between areas to be demolished and proposed conditions.
8. As demolition proceeds and concludes, verify that as the site is given over to the contractor in charge of layout and grading (even if it's the same company and people as those doing the demolition) there are no loose ends or potential conflicts.

Layout

1. Verify that the contractor has correctly identified starting points for systems of measurements for each key area of a design.
2. Routinely check the dimensions of random elements. Further investigate where there are significant departures from key dimensions shown on the drawings compared to those found in the field.
3. Take in and evaluate both overall sitewide dimensions and localized subsections.
4. Make certain that center points of circles, beginnings and ends of arcs, key corners of geometric figures, and intersections of these elements are consistent with what is shown on the drawings and the goals of the design.

Grading

1. Proceed from the broad brush of landforms to the specifics of detailed grading.
2. Visualize the landforms to be created and gradients to be achieved, and check enough specific locations to verify the results in the field.
3. Approve the location and elevation of all low points and coordinate the placement of drainage structures and pipes.
4. Approve the location and elevations of all proposed high points.

Construction Details

1. Ask to see certificates of delivery of raw materials, such as bluestone, and observe and approve the range of colors and textures.
2. If you have any doubt about the compliance of a material with its specification, stop work while it's being tested. Do not risk allowing a contractor to proceed with the construction of a major element of the design—for example, a pavement, a wall, or a set of steps—if there is any doubt about the materials.
3. Be certain that the contractor is aware of maximum and minimum dimensions, thickness, hardness, porousness, proportion, and other characteristics of each material. Verify compliance at both a number of key locations and some random locations.
4. Pick out important locations where the same detail is being implemented and observe that it is being constructed in a consistent manner.

Planting Design

1. Review tags on trees delivered to the site to verify they are the ones previously selected by you or your staff.
2. Arrange for the client or owner to approve locations of major materials prior to excavation of tree pits or planting.
3. Approve the staking out of major trees and the outline of planting areas that are structural to the design before proceeding to less important areas.
4. Garden hose or thick colored string is useful to mark areas, as it stands out and can easily be adjusted in the field.
5. Request that the contractor client dig a few test holes for major trees to verify drainage prior to planting (this requirement should be incorporated in the specifications). If water is encountered or there is poor drainage, consider use of a French drain or other method for achieving well-drained planting areas.
6. Routinely count quantities of plant materials, particularly large trees, flowering trees, and shrubs, to verify compliance with drawings.
7. For groundcovers and perennials, verify quantity and spacing by counting the quantity within a set area and calculating the number of plants per square unit area to verify the spacing requirements.

Special Elements, Such as Water Features, Lighting, and Irrigation

1. Anticipate the best time for the review and approval of shop drawings and plan with the contractor and required consultants accordingly, as it may take several submissions before the shop drawings are approved.
2. Request that the contractor stake out key elements, and discuss any potential conflicts and adjustments in the field.
3. Let the contractor set priorities and implement the various items of work in any particular area as she sees fit. It's fine to ask for an explanation, but unless the work lags behind or the landscape architect observes problems caused by the order in which the contractor is implementing the work, it is the contractor's responsibility to achieve what is required by the contract documents in whatever order she prefers, unless some specific order is spelled out in the contract documents

Pre-Final Inspection

1. Usually this is requested by the contractor when she feels the work is nearing a threshold of completion as defined in the specifications. However, usually it is appropriate for the landscape architect, in consultation with the contractor, to set a date. Sometimes a nudge is helpful. Since plantings often are one of the last items implemented on a site, sometimes it's appropriate to have inspections for major construction and approval of this work prior to starting the planting implementation. Yet, depending on the nature and complexity of the design, its overall size, and the skills of the contractor, these two phases may overlap. In order not to delay the work, it's useful to defer to the contractor as long as the landscape architect verifies that the work is correct, in compliance with drawings and specifications, and on schedule.
2. Develop a checklist that identifies all items of work that are incomplete or need corrections.
3. Set a date for achieving full compliance with checklist items.

Final Inspection

1. In a well-run project, this task goes quite smoothly, as it's a simple matter of checking off items on the list from the previous inspection.
2. Final requirements may be implemented. For example, it's common to require that the contractor provide the owner with a maintenance manual for all planting and other special items and equipment. This should be reviewed carefully and the contractor should respond to any questions at the landscape architect's or owner's request.

Problem 1
SITE INSPECTION

(In the following problem, any *terms in italics* may be changed to suit the particular instructor or team leader, or your own personal, work or studio setting. Typical changes might be specific people, such as clients or contractors, plant materials, or the locations of sites, people, institutions or companies.) Imagine exactly where the site is, what your role is, pin down your responsibilities, and then proceed.

As a *landscape architect*, you are going out in the field to meet with the *general contractor* first and then the landscape contractor to evaluate the progress of the work on a residential estate. The landscape contractor who has begun planting is a subcontractor (or subconsultant) to the general contractor and has recently begun planting part of the first phase of construction, as per directions from the general contractor. The project has been running smoothly for several months, but a recent slowdown in the progress of work has you concerned, so the timing seems appropriate to meet at the site.

You notice the following issues. How would you resolve them and keep the project on track? What office materials should you bring with you to the site?

1. The concrete sidewalks are well constructed in one location, and the color, texture, and finish appear uniform and in compliance with the contract documents. However, a stretch of newly poured sidewalk adjacent to the first location has a color that doesn't match, an uneven finish, and a different texture altogether, as if a different gravel mix was used.
2. There is a pool of standing water adjacent to a drain in a newly developed lawn area. The drain appears to be in the correct location at the lowest point in the grading, but water is not flowing into it.
3. There is a formal allee of twelve *Acer saccharum* (sugar maple) trees lining the driveway. The planting plan calls for six pairs of matched shade trees to be placed 25 feet (7.62 m) on center and symmetrically flanking the driveway, with each pair perpendicular to the alignment. The driveway is 12 feet (3.7 m) wide with a granite curb. You had previously tagged the twelve matched trees at a nursery, yet one tree does not seem to have the same height and character as the other eleven. Due to a few architectural changes, the garage is about 5 feet (1.5 m) closer to the front property line than shown on the drawings. The trees are set in an even rhythm 24 feet (7.3 m) on center. The drawings call for the center of each tree to be 4 feet back of the curb on either side of the driveway. You find that each tree is uniformly set back 5 feet (1.5 m).
4. In an adjacent part of the site, five *Quercus palustris* (pin oak) are to be planted in an informal grouping at the edge of a large lawn area with a gentle gradient, after which there is a long moderately sloping landscape. You had previously staked the approximate locations for each tree, yet the landscape contractor has stopped the work because the first two tree pits dug to the specified depth of 6 feet (1.8 m) were quickly filling with a few feet of water.

5. The landscape contractor successfully planted masses of ten species of shrubs, as called for and specified in the planting plans. The size, spacing, and character of nine of the species easily comply with the minimum requirements of the specifications; in fact, some are more robust and exceed the specifications. Yet in one prominent location, instead of ten *Ilex verticillata* (winterberry, eight female and two male) are a group of ten handsome *Ilex opaca* (American holly).
6. Adjacent to the concrete sidewalk is a 20-foot-long (6.1 m) by 8-foot-wide (2.4 m) terrace of bluestone pavers on a concrete slab subbase and mortar bed. The 8-foot (2.4 m) edge abuts the concrete sidewalk, but rather than a smooth flush edge, it's irregular and bumpy, a potential tripping hazard.
7. The bluestone terrace is specified as "random rectilinear" according to a pattern shown in the drawings, with a range of colors that was approved by the landscape architect in the shop drawing phase. The pattern executed is consistent with the contract documents, but instead of a random range of colors, the pavers are grouped in large swaths of each color.
8. Some of the new plantings appear to be drying out prematurely. The correct mulch is placed to a depth of 2 to 3 inches (50–75 mm), as specified.
9. There are some existing gardens with large shade trees adjacent to the new construction zone. A heavy-duty backhoe used for grading has been placed underneath the drip line of one of the mature existing trees, and some deep ruts are noticeable, showing the route that the vehicle has followed. Fortunately, it has not come directly against the trunk of any of these trees. The general contractor claims that the landscape contractor is responsible for the backhoe even though it's the general contractor's equipment.
10. There's an elegant 4-foot-tall (1.2 m) cast iron pool fence, which seems to have been built as per shop drawing, with channel irons, pierced with vertical pickets, welded, painted black, and so on, but you notice places near several of the piercings where the paint finish is not uniform.
11. There are teak benches with a clear stain that are attractive, but when you sit on two of the ten on-site, you find that they wobble.
12. The lighting system around the pool looks attractive, and all the poles and lamps are as specified. You elected to use incandescent fixtures because the client didn't like the color of the halide, mercury vapor, or LED options, but when turned on, the lights are very bright, like being at an amusement park. What can be done?
13. The dry-laid wall is being capped, and the capstone is being placed. It does not have a reveal—that is, the face of the first few capstones is flush with the alignment of the face of the wall. The drawing detail shows a 1½-inch (38 mm) overhang.

14. The stepping-stone path between the shade garden and the residence is laid out well, and the plantings are well arranged. There is not a larger, transitional steppingstone at the beginning and end of the walk, as called for on the drawings, as the contractor says he could not find any.
15. In the entrance courtyard, the contractor has finished rough grading and laid out the edges of proposed paving. There is to be a water feature in the center. What should you discuss with him to do next?

Problem 2

PROJECT MANAGEMENT WRITING

(In the following problem, any *terms in italics* may be changed to suit the particular instructor or team leader, or your own personal, work or studio setting. Typical changes might be specific people, such as clients or contractors, plant materials, or the locations of sites, people, institutions or companies.)

Write a letter to a *client* in which you recommend a solution to a *particular problem* relating to design drawings you have prepared. The letter may not exceed two double-spaced pages. You may include one sketch, at 8½ by 11 inches (same size as letter), with the letter.

Address the reason for your site visit and describe the problem. Indicate the drawings and specifications that are relevant to this problem. Explain how what you've encountered on-site is different from what is required by the drawings and specifications. Provide your recommended solution.

Some examples could be:

1. Plant materials are not in compliance with drawings and/or specs.
2. Materials or construction are not in compliance with drawings and/or specs.
3. Some other issue where there is lack of compliance.

If there is a significant change in scope of work that would result in a change order, describe the reasons this is appropriate (and not something for which you should be held responsible) and why it is necessary, and indicate what the anticipated cost increase is going to be. Call it Change Order One and any drawings relating to it SK-1a, SK-1b, et cetera. This change order text should not exceed one to two pages. You may include one additional sketch with it. If you submit this, then the maximum document will be three pages for the initial letter and three pages for the change order. Please do the basics immediately:

> Set up a sheet size with *office name, address, phone, and email*.
> Include the *name and address of the client*.
> Write in business letter format.
> Be certain that all pages are in the same format.
> Sketches must be the same size as the letter, 8½ by 11 inches.
> Number the pages (the better format is "page 1 of 6," "page 2 of 6," etc., so that anyone reviewing it knows the maximum number of pages that are included).

NOTES

1. en.m.wiktionary.org
2. R. J. DeCristoforo, *Handyman's Guide to Concrete and Masonry* (Reston, VA: Prentice-Hall, 1978), 108.
3. The author is appreciative of the review and comments by Dean Anderson of Supersquare Corporation on the section on metals, as he provided much detailed and useful information.
4. https://www.bbc.com/news/world-latin-america-46327634
5. Annie Proulx, *Barkskins* (New York: Scribner, 2016).
6. Andrew E. Kramer, "China's Logging in Siberia is a Source of Jobs and Worry," *New York Times*, July 26, 2019, A10.
7. See "Promoting Sustainable Forest Management Globally," PEFC/16-01-01, https://www.pefc.co.uk/system/resources/W1siZiIsIjIwMTcvMDgvMjEvOXAxY XRnejE2al9QRUZDX0ZTQ19wcmVzZW50YXRpb24ucGRmIl1d/PEFC%20%20 FSC%20presentation.pdf.

CHAPTER 9

Ethics and Sustainability

PART 1: ETHICS

Ethics is the code of behavior we share on the basis of reason, law, honor, and an inborn sense of decency, even as some ascribe it to God's will.[1]
—Edward O. Wilson, *The Creation*

Ethics is, quite simply, distinguishing between right and wrong. For landscape designers this means setting and enforcing standards for employee behavior that promote a safe and productive workplace environment. Ethical concerns also govern how one reacts to and interacts with clients, contractors, consultants, and anyone else. There is an increasing emphasis on environmental ethics, meaning the use of the land and, by extension, the study of the best use of a property or parcel of land that you analyze, design, develop, and implement as part of your role with a particular client and owner. Sustainable design may be thought of as one of the evolving subjects created as a result of ethical concerns about the use of resources in the environment. Legal aspects of ethics are embodied in laws governing contracts, bidding, and construction, as well as in building codes and zoning requirements. Each landscape designer must become familiar with all of the legal aspects of the community, city, state, and region in which work is occurring. Such requirements can vary significantly from place to place, which is one reason that some practitioners choose to concentrate their practice on just a few areas—residential design or health care, for example—and within a limited geographic area. When my own career as a landscape architect began in ancient times, there were many landscape architecture firms with a general

Figure 9.0 A brightly painted drainage structure explains its purpose in Bentonville, Arkansas. An education program utilizes art to communicate the function and importance of storm drains. Art by Laura Neill. The NWA UpStream Art storm drain mural project is managed by the University of Arkansas System Cooperative Extension Service to protect waterways from stormwater pollution.
Photo by Steven L. Cantor (courtesy of https://nwaupstreamart.com)

practice. That is, they would do diverse projects in both the public and private sectors: historic preservation in a suburb, an expansion of a children's hospital garden, residential estates in exclusive neighborhoods, urban design for a community park, visitor parking and gardens for a carpet mill, a landscape for a high-rise retirement tower, a corporate headquarters, playgrounds for a public school, recreational fields for a private school, tree planting for city streets. One modern trend has been the specialization of firms, which now focus on only a few areas of practice. As design requirements and knowledge expand and become more complex, it is more and more difficult to be a jack-of-all-trades. It becomes critical to have a sense of the collective expertise of the firm, be aware of limitations, and know when it is necessary to seek out consultants or other assistance.[2]

One way to consider ethical concerns is to imagine how they might arise in different types of design projects in office settings and how you would respond to them. What are you willing to do for the client, and what seems, feels, or is unethical? What actions advance the profession of landscape architecture, and what actions demean it? What response on your part to a challenging conundrum could be a defining moment in your career, which you would look back upon years later with appreciation that you made the right decision?

Questions

(In the following questions, any *terms in italics* may be changed to suit the particular instructor or the personal, work or studio setting. Typical changes might be specific people, such as clients or contractors, plant materials, or the locations of sites, people, institutions or companies.)

1. As a landscape designer you have an opportunity to meet with a potential client, who wants you to do a master plan for a forested residential property that abuts on two sides a fenced conservation reserve managed by a conservancy. Although there is considerable potential for its use for passive recreation, the reserve is not currently used much, and mainly functions as an aquifer recharge area and ecological study area. The client wants you to provide standard but lavish amenities, such as a swimming pool, gardens, and a few additional parking spaces near the house, and also insists that you provide two private pedestrian gates into the conservation reserve, one on each side. You would like this job, as you need some income and this client is well connected. What do you say or discuss with the client?

2. You work in a small design office and are assigned by a principal to join a colleague to drive to a large estate under construction 20 miles away and perform an inspection. Normally the colleague would do the inspection herself, but there is so much to review that you join her for the trip. While driving there, she confides to you that from the moment she starting taking lithium to treat bipolar disorder, she's had several fender-bender accidents while driving for work, probably because the medication, particularly in early stages of treatment, can sometimes make her groggy. On the way back you insist on driving, and she falls asleep in the passenger seat. Are there additional actions you should take upon returning to the office? Why or why not? Would it make a difference if she confided in you that she were taking other medications that didn't have such obvious side effects?
3. As a project manager, you review an invoice in which your company is billing for time spent by four employees. The average time each has charged over a two-week period is 20 hours. You are able to quickly verify the accuracy of the time charged by three of the employees and their specific tasks accomplished. However, you talk to the fourth employee, as, from your recollection, he has not worked on the project. He tells you that he was directed to charge time to this project by one of the principals, even though he did not work on it. Work has been slow in the office, and the principal indicated that the workload would likely pick up soon. The utilization rate of the employee in questions is 85% (that is, 85% of his time must be charged to billable tasks), and the 20 hours he charged enables him to comply with this requirement. If he moved the block of time to "general office" or another non-billable category, he would fall below the minimum threshold. As project manager, you are required to submit the invoice to the client and verify its accuracy. What do you do?
4. You meet a prospective client at a 3-acre property where you might do some landscape design work. Upon arrival at the site, you notice a spectacular *Liriodendron tulipifera* a considerable distance from the house. It's easily 20 feet in circumference and 150 feet high, and appears to be in excellent condition. The tree is so large that, together with some other specimen hardwoods, it frames a view toward the residence. The client nevertheless wants to cut it down in order to make space for an additional room to the residence, to be used for business activities relating to the client's landscape nursery and for creating extensive planting beds that they wish to build for growing endangered wildflower and perennial species that need full sun. What do you do? Why?

The next five questions apply to the same project, in which you as a *landscape designer* or *landscape architect* are preparing landscape development plans for a residence in a windy, exposed site in *the foothills of Blue Ridge Mountains of North Carolina*.

5. You receive bids for demolition, construction, and planting. You express some discomfort with the low bidder, as this firm does not have nearly as much experience as most of the other bidders. However, the client insists that you negotiate with this low bidder, and you agree on a final price for a contract between this contractor and your client. You previously had started to discuss with the client a fee for construction inspections and observation based on visiting the site about twice a month. What might be a practical way for you to continue to serve the client's needs and also work with the contractor?

6. You have already completed a landscape plan for the same client's residence in *the Research Triangle, NC* for a sheltered site, which has a climate at least one climate zone milder. Due to her busy schedule, she did not pay that much attention when the plantings were being implemented on the current site under development. One day, however, she comes by and is upset to see that instead of the *crape myrtles* and *'Nellie Stevens' hollies* that she loved so much in the *Triangle* location, you have specified Meserve hollies and redbud trees for a similar location at the *Blue Ridge* home. What is your explanation to her for these two changes? In hindsight, what could you have done differently at an earlier point in time to reinforce your point of view?

7. The swimming pool is almost finished and the client likes the way everything looks but is upset at the height of the wooden pool fence, painted white, which complies exactly with the building code requirement. The client's husband swims a lot and feels that the height of the fence will create too much of a visual barrier between the residence and the property, which will discourage him from going swimming. As the landscape designer, what can you do to adjust the design of the fence without creating any legal jeopardy? What else might you propose to assuage the client?

8. The preliminary design calls for the construction of a tennis court in a gently sloping natural valley somewhat lower in elevation than the residence. As you proceed with design studies, you find out that this area is within the floodplain as mapped by the Army Corps of Engineers for the nearby *PDQ River*. Although construction and grading are not forbidden, the requirements are that any filling within the floodplain must be matched by comparable excavation in order to maintain the

same capacity of the swale to carry water. You do a few grading studies and find that although you could engineer a plan that would work, the result would not be aesthetically pleasing, as the tennis court would be bordered by slopes that are relatively steep and subject to erosion. You recommend to the client not to build the tennis court in this location; however, it is not possible to find an alternative location on the site. The client insists on moving ahead with the tennis court. What do you do?

9. The client's husband is a successful hedge fund manager, and although you met him initially when you began work, he is rarely present, as he is quite comfortable with his wife taking on all the responsibilities of meeting with you, responding to questions, reviewing and approving design decisions and installations, paying bills, and so on. Then one day, as a lot of the work is nearing completion, he approaches you at the regularly scheduled bimonthly meeting every two weeks with his wife on the site and requests that you design in a large adjacent meadow, for an additional fee, a "cemetery" to memorialize all of the financiers he has defeated in the course of his financial management. There would be a tombstone, quite realistic in appearance, simulating the "death" of each rival. Around the cemetery, there would be an elegant ornamental iron fence with gates, and there would be specimen plantings and other landscape elements to harmonize with the other aspects of the design. How do you respond?

10. You serve on a committee with several laypeople, an architect, and an engineer in which you are meeting to open bids from five contractors for proposed landscape development for a private residence. The chairman of the committee announces that since one of the bidders, who has previously done a few projects for the homeowner, has had a family emergency, his request to submit a late bid has been approved. The other four bids are discussed; they are competitive, and several are within budget and seem to meet all requirements. The chairman leaves with copies of the four bids and announces that she will review the late bid when it's received and report back to everyone. What is your response?

11. You have been working on the implementation of an exciting urban design project for a public agency for two years. Construction is under way, and you regularly attend project management meetings at the site with all key members of the design team, including the contractor. The last meeting of the year occurs before Thanksgiving, and the site will be closed through Christmas and open again after New Year's. As the last meeting ends, the foreman for the contractor asks you for a

moment of your private time. He says, "Happy holidays," and hands you an envelope that contains a gift certificate for $200 for a popular restaurant. You are appreciative and thank him. Is it all right for you to accept this gift?

12. Repeat question 10 except the bid discussion is occurring for a public works project for a federal agency. Does this change anything?

13. Repeat question 11 except the project management meeting is occurring at a large private estate. Does this change anything?

14. You are a landscape designer directing a complex urban design project for a public agency. A deadline is approaching, by which time the complete package of construction documents must be finished so that the project can move ahead within the current fiscal year. Unfortunately, the agency did not give you final critical review comments until a few weeks past the point at which you'd need to receive them in order to be able to complete the documents in a timely manner. You review the comments and mark up everything that must be revised by the deadline. In previous phases of work on this project, most of the drawings have been revised by Employees A and B, most of the specifications have been written by Employee C, and most of the cost estimate has been developed by Employee D. All four of them are available, but each would need to work at least ten hours of overtime for three consecutive weeks—that is, thirty extra hours by each employee—in order to meet the deadline provided by the client. There is no budget for overtime for this project. How can you ensure that the project will be completed on time to the satisfaction of the client?

15. You are an employee of a small landscape architecture firm and have just completed your first three months of work. You have had a solid review with the principal, who has praised your work and indicated that you have graduated from probationary to permanent status. From this point forward you earn full health insurance benefits and accrue vacation time. You are quite pleased, as the principal has managed to give you a steady flow of work, you find the projects are interesting, you are learning a lot, and you collaborate well with your more experienced colleagues. Three months after your promotion, one of your colleagues must take maternity leave and you must take over her workload for two months until she returns. You are excited about the opportunity to work on three additional projects but also concerned about how much time this might take you. What actions can you take to ensure that you succeed?

16. You are hired in your first job after earning a degree in landscape architecture. The principal welcomes you to the firm. Her husband shows up

one afternoon with their two children in tow. Your boss suggests that you go out for ice cream, a nice treat on a hot summer day. The five of you leave the office and there is a long line of people ahead of you. On cue the children are told to start whining and crying. The husband says apologetically to others ahead of you in line, "I'm so sorry, my children have had a long day. Would you mind if we went ahead of you in line?" This strategy works, as you proceed to the front of the line, get your ice cream, and return to the office. What do you do or say to your boss or her husband?

17. You arrive at a residence to meet with the mason who has been building an elaborate bluestone pavement design. You have reviewed and approved his latest shop drawing for the design detailing of a central, decorative pavement for the main entertainment terrace at the residence. However, when you meet with him and look at the layout he has staked out and started to implement, you notice that although the size of the pavers is correct, the breakdown of the colors is not. The specifications and shop drawings call for random distribution of five tints, each of which is to constitute 20% of the layout. However, you notice that there are three dominant colors, and the pavers in those colors each constitute about 25% of the total distribution. The central medallion shape has the correct distribution of sizes of pavers but is lacking the correct color distribution. You point this out to the contractor, who disagrees at first but gradually acknowledges the lack of compliance. The overall installation is still quite attractive. What do you do?

18. You have designed a white garden for an estate. All the major flowering trees, shrubs, and perennials are white-blooming. In the fall, you approve the layout and planting installation by the contractor. All the major plants that you'd previously tagged in the spring at a specified nursery were delivered and installed properly, as you could verify the nursery's tags on the specimens. Nevertheless, the following spring, you go to carry out a final inspection, and find that three specimen dogwoods in prominent positions within this garden are pink-flowering rather than white. You investigate and find that the nursery switched the tags by mistake, and the three tagged white-blooming dogwoods were mistakenly delivered to another site and planted. The nursery acknowledges the error and apologizes. What do you do?

19. You are nearing the completion of a set of contract documents for a complex project for a public client. You receive the final comments from the client, and one reviewer, who has not previously seen the drawings, recommends a valid change in the design that would improve it. She

proposes to eliminate some steps and substitute some graceful ramps, for which there appears to be adequate space within the site, and to highlight the ramps with different accent plantings and lighting. However, as you consider these changes, you realize that it would take considerable time for your staff to develop these changes and incorporate them into the contract documents. What do you do?

20. You have graduated from a strong BLA/MLA program and have two interviews with firms in a city where you would like to relocate. Firm A specializes in high-end residential design and has an excellent, comfortable work environment and solid benefits. From interviewing you understand that you would work hard but rarely be expected to do much overtime. One reservation you have is that you're offered two weeks vacation (beyond the standard holidays). Firm B focuses on public works; it has major projects with the National Park Service and the Federal Highway Administration, which would provide you with significant travel to some national parks but pays $12,000 to $15,000 less than the other job. You prefer the second firm's type of work, although the atmosphere in their office is more frenetic. The cost of living in this city is high. When you discuss the differences with the principal at Firm B, s/he explains that the contracts for government work are not as lucrative, and they're often locked into lower fee structures, but after a year, if your work is excellent, they will guarantee you a raise of $10,000 and a third week of vacation, instead of two for the first year. Which job do you take? Why?

21. Because she says she's swamped with work, another landscape architect with whom you've been friends for some years refers you to a potential client. You meet the potential client, for whom your friend once designed a fence along the joint property line with a neighbor. On that fence were planted some climbing hydrangeas and clematis which had matured, climbed the lattice, and bloomed beautifully. The neighbor inexplicably cut them way back, exposing a view of some very ugly play equipment used regularly by some noisy children. Apparently the blooms had attracted bees, which stung some of the neighbor's children, so the parents acted abruptly. What do you do?

22. You are hired by the Parks Department of City A to do a redesign of a neighborhood park, roughly rectilinear in shape. It abuts a school playground. Both the park and playground have chain-link fence in poor condition that needs replacement, so at least some of your budget must be spent on new fence, probably with school fence on one property line and park fence on another. You find that each bureaucracy, parks and schools, has a different design and specification for chain-link fence,

and the two don't dovetail well. Ideally, what is the best thing to do as a designer? Practically, what is the best thing?

23. You are a middle-level project manager on a large project, and high enough in the hierarchy that you can see printouts of staffing and workload. From this you realize that the team assigned to the project is way over budget in relation to where they should be as a proportion of the fee spent to date. On several occasions you've brought this to the attention of the senior-level project manager who supervises you and is directing the project, but no changes have been made, so you fear that the situation will only worsen. What do you do?

24. As a landscape architect specializing in native species, you've been excited to work on a planting restoration project. You've been assigned to develop the plant list for the project and have worked diligently to come up with a comprehensive list of plant materials that are available from local nurseries, are appropriate for the site, would harmonize with what exists, and would transition well from the existing palette that will be present at the time of planting to the intended mature succession forest. You are required to work with a preapproved contractor who has considerable experience and has successfully implemented a number of similar projects. Twice you've presented plant schedules to him, and he has rejected them, offering substitutions that you know to be inappropriate, because he prefers plants from other sources that are not part of the native community being established and apparently are on his list only because they are available from some friend of his. What do you do?

25. You have worked with the ownership of a department store in the suburbs to redo the entrance plantings. There was a limited scope of work, primarily involving plantings in two linear planters flanking an entrance plaza. Since the budget was somewhat limited and there were time constraints, the owner's representative agreed to let you work with a landscape contractor you recommended who had a lot of experience in urban landscape. The existing irrigation system had malfunctioned, so all the plantings in the linear planters had died. While being very careful not to dig down any farther than necessary, the contractor carefully removed all the plantings, verified that the waterproofing and root barrier layers were still intact (another 18 inches lower), replenished the soil mix, planted new plantings according to the plant schedule, and repaired and reactivated the irrigation system. Twice while he was doing his work, the contractor complained that a mason who was doing repointing of brickwork was being so sloppy that a lot of loose masonry dust and debris was falling into the planters. You

observed this happening yourself on one occasion, complained to the mason, and reported the problem to the owner's representative. The mason was working under the direction of the engineer under contract to the building for various infrastructure and structural issues. Just as the planting was finished, the engineer wrote a report saying that one of the planters had a leak into the basement, so there must be a puncture of the root barrier caused by the landscape contractor. The engineer therefore recommended removing all the plantings, soil backfill, putting in a new root barrier and waterproofing membrane. When this was investigated by another engineer, it was found that the leak was caused by a small fissure in a vent pipe on the roof, which was easily repaired by an application of tar on the pipe. No remedial work was needed on the planters, although to be certain, the landscape contractor did excavate the planter over the location of the basement leak to its full depth, and no sign of disturbance or damage to the root barrier or waterproofing was found. The landscape contractor meticulously removed all the loose mortar from the surface of the planter, and the mason was directed to use a plastic sheet underneath his work to catch debris. He did not do so, and more mortar spilled into the new plantings. Why would the engineer have written his report after the plantings were finished? Why would the mason have ignored a warning about simple directions? What do you do?

26. You have just finished a contract adding ADA-compliant access ramps to a complex of four 6-story public housing apartment buildings. You receive an inquiry about working on a residential estate for a young couple who have just purchased a home on 6 acres and have as a budget for improvements four times the total budget for the project you just finished. Your fee could be completely open-ended as long as you work fast. Is this a problem?

27. What is the relationship between landscape architecture practice, teaching, and research? Not to coin a phrase, but consider the following: Those landscape architects who can, do. Those who can't, teach. Those who can't teach well focus on research and apply for government grants, which, if funded, lead to tenure, prestige, more students, more lab personnel, and even more faculty.

28. You are a registered landscape architect in your state. A landscape designer approaches you, and you learn that this person has an encyclopedic knowledge of plant materials but is not registered because she has not passed the exam in other areas. This person is working on a set of planting plans for a state agency that requires that the plans be prepared by a licensed engineer or registered landscape architect. You

look at the work and are impressed not only by its technical facility and its compliance with all requirements but also by its graphic strength and aesthetic qualities. All the work has been done, and she offers to pay you to sign and seal them, if you will then reimburse her when you receive payment. What do you do? Why?
29. Based on where you are in your career at the moment, write a summary of what your code of ethics is within landscape architecture in general, and your own particular job and your personal life. How are they similar, and how are they different?

PART 2: SUSTAINABILITY

> A thing is right when it tends to preserve the integrity, stability, and beauty of the biotic community. It is wrong when it tends otherwise.[3]
> —Aldo Leopold, Sand County Almanac and Sketches Here and There

"Sustainability" and "sustainable design" have become catchwords, like "environmental design," "Earth Day," or "green roofs," to signify that whatever the term is being applied to is on the right side of the vast spectrum of public opinions and professional statements dealing with these topics. Some accurate definitions are important. *Sustainable design* is simply design that nurtures itself; it uses and recycles all materials and systems so that there is no net loss to the environment as a result of implementing the design, and once it is completed, it functions without a net drag on the resources and systems on which it depends.[4] *Sustainable design* may be thought of as a large tapestry of processes and design approaches, as well as materials and construction technologies. On a human scale, the renowned Mahatma Gandhi noted, "Sustainability requires an understanding of how our actions affect our environment, economy and community. We must become the change we wish to see in the world."[5] Green roofs are perhaps one small step in this process.

Yet sustainable design must go well beyond the mere easing or amelioration of environmental problems and be an integrative process through which people and the built environment are linked in an overarching concept of nature. Perhaps the most far-reaching definition of sustainable design is by the noted social ecologist Stephen R. Kellert, who has developed the concept of restorative environmental design, which enriches the human spirit by creating connections between nature and the built environment. He explains, "Restorative environmental designs avoid and minimize harmful impacts on the natural environment and human health, while also providing positive opportunities for beneficial contact with nature in places of cultural and ecological significance. The major difference from conventional approaches to sustainable design is the tendency only to focus on the first part of the definition (minimizing harmful environmental impacts) while ignoring the inherent human need to affiliate with natural systems, referred to as biophilia by myself and colleagues."[6] Some of the projects presented in this book could be categorized as restorative environmental designs.

A somewhat more technical definition of sustainability comes from the American Society of Landscape Architects: "Sustainability is the capability of natural and cultural systems to maintain themselves over time. It is impacted by: (a) individual and collective actions; (b) the amount and the rate of consumption; and (c) the intrinsic properties and carrying capacity of each system."[7] As the science and technology have advanced, there is more sophistication to the measurement: the carbon footprints of different projects may be compared, and the potential amounts and rates of key variables and resources can be estimated for a proposed project. Such methodology has major implications in environmental impact assessment.

A correlation to sustainable design is the concept of whole building design, which aims to create a successful high-performance building or project in which all components contribute to the whole. In 1926, Jan Christiaan Smuts, a South African statesman and philosopher, wrote *Holism and Evolution*, and pioneered the term "holism" as "the tendency in nature to form wholes that are greater than the sum of the parts through creative evolution."[8] He believed that there are no individual parts in nature, only patterns and arrangements that contribute to the whole. While working on the space program in 1969, the visionary Buckminster Fuller said, "Synergy is the only word in our language that means behavior of whole systems, unpredicted by the separately observed behaviors of the system's parts or any subassembly of the parts. . . . The whole is greater than the sum of its parts."[9] Decades later, that approach is still essential.

Although sustainable design suggests a process, it is like other types of design in that a thorough, site-specific approach is required when it is applied to a landscape architecture project. The landscape architect must ask and find answers to an array of questions having to do with myriad processes and design approaches, as well as materials and construction technologies. For example, what sort of design brainstorming and programming started the long trajectory of coming up with a design concept and turning it into reality? What criteria were applied to develop a sustainable design process and a final design ready to be implemented? What traditional design approaches and techniques might need to be deleted from the beginning in order to honor the spirit of sustainability? For example, where do the plant materials originate? The closer they are located to the site where they will be planted, the less energy that is likely to be expended in transporting them to their final destinations. At the same time, one must determine that the nursery or other site from which they are extracted itself follows sustainable practices so that it could providing those particular plant materials for installation at a new project. If durable tropical wood products, like mahogany or ipe, are specified for construction, what safeguards are there in the specifications for these materials to prevent wood from being harvested from some far-distant tropical forest in which sustainable practices are not used? What construction techniques will be used in building and installing the project? What sort of fuel or energy will be required for digging plant pits, installing water supply and stormwater drainage pipes, pouring concrete, building walls, and other essential details of the design?

Some sustainable design processes and safeguards are much easier to track than others, so the landscape architect must be careful what is included and specified in a design. For example, when tropical woods are specified, there is often language in the specifications stating that the wood must be certified as being from a sustainably grown and harvested resource. (See the discussion in Chapter 8 under "Wood.") Such language, it is thought, would prevent the purchase of such lumber from entrepreneurs who have simply gone surreptitiously into a private or public forest (or hired others to do so), located and harvested selected tropical wood specimens, and dragged them out of the forest by the shortest possible route, regardless of the damage this does. But what's to prevent such operators from stamping their lumber "sustainable" and selling it for quick profit? It's not often that a project landscape architect would have the resources to go check on sources of tropical wood in Honduras, Brazil, or other distant lands. Some years ago I worked on a project involving the construction of a new boardwalk along a storm-battered section of the coastline for a state park in Connecticut. To limit the use of tropical wood, ipe was specified only for the main piers, beams, and joists, and planking was to be pine or local lumber. There was no way to prove absolutely that the ipe provided was sustainably harvested. Perhaps a better strategy for ensuring sustainability is not to specify tropical woods at all. It's also a matter of degree and intensity of use. There are many tropical foods, such as coffee, in which some sources are certified as sustainable—for example, where the coffee plants are grown under the shade of canopy trees so that the entire forest is not destroyed in order to plant and harvest the crop. Also, the coffee beans are at least a crop that recurs year after year. By contrast, a specimen mahogany or ipe tree hundreds of years old is harvested, it will never be replaced. How much boardwalk structure or outdoor furniture is this worth?

Leadership in Energy and Environmental Design (LEED) is an environmental rating system originally developed by the United States Green Building Council to help identify and rate sustainable designs. (See Chapter 1.) One important aspect of the LEED process is that the design and construction team must document compliance with criteria for sustainability from the beginning to the end of the building process. Therefore, even if a building is constructed with many sustainable features, if these are not documented in a step-by-step way, it might not be possible to receive a LEED certification.[10]

In either the American or Canadian approach, points (or credits) are awarded for beneficial or positively rated environmental accomplishments in six areas: sustainable site development, water efficiency, energy efficiency, materials and resources selection, indoor environmental quality, and innovation in design. Green roofs could typically earn points for the design in five of these categories, all but indoor environmental quality. A particular green roof design, for example, could reduce stormwater runoff; eliminate or reduce site irrigation and water usage; reduce energy usage, use highly rated building materials; and provide innovation in design. Each goal must be carefully documented and verified according to an elaborate process.

DOS AND DON'TS IN SUSTAINABILITY

1. If LEED certification is a goal, start in the initial phases of a project.
2. Focus efforts at sustainability on products, techniques, and processes that can be easily verified, such as plant materials from local nurseries.
3. Hold back on sustainable processes and certifications that might require verification beyond the resources of the project team.
3. Protect existing sites and restore damaged areas. Not only is protection desirable, it's also cost-efficient, because the larger the areas left undisturbed, the less funds required for ameliorative activities.
4. Incorporate green roofs, living walls, bioswales, and rain gardens.
5. Consider methods for efficient water use, such as recycling gray water, slowing down stormwater discharge, and increasing wastewater treatment.

6. Store and collect recyclables.
7. Balance cut and fill on the site, so that there is not a need to bring in or haul off large volumes of material.
8. Incorporate photovoltaics where feasible.
9. Develop construction and planting practices that take less time to fully implement.
10. Expand your vocabulary by including one newly introduced recycled, recyclable, or sustainable product or material in each design.
11. Specify one new sustainably sourced plant material for a new project.
12. Use only drip irrigation.
13. Specify sustainable maintenance practices so that after the landscape architect finishes with a client and a site, other practitioners will apply the same philosophy to ongoing operations.
14. Do not overreach.
15. Consider phasing if budgets are tight.
16. Incorporate sustainable practices into the in-house operations of a project. For example, print legible half-size drawings, rather than full-size; print on both sides of the paper for all but the final versions of specifications, cost estimates, and other documents; minimize the use of color inks; primarily use digital images, which can be reviewed on computer monitors instead of needing to be printed; print only those documents that are essential to have copies of, and minimize the number of copies printed.
17. Strive to keep the project within budget, in terms of both staffing and implementation costs. Beautiful results that exceed the budget are not sustainable.
18. Incorporate quality control processes in the document production stage of a project as well as in evaluating its implementation.
19. With selected projects, try to partner with scientists to monitor some environmental components, such as water usage, temperature moderation, or the impacts of insulation on cooling and heating.
20. Archive the project carefully and systematically so that you don't reinvent the wheel when there is reason to do a similar project for another client or site.
21. Engage with other designers via websites, blogs, and user groups to share ideas, improve strategies, and increase the likelihood of high-quality design products.

Horsetopia, South Korea

Horsetopia is a competition submission for a horse park in Yeongcheon, South Korea. Half of the 1,474,883 m² site (364.4 acres or 147.5 ha) is dedicated to horse racing, and the other half is dedicated to horse-related activities. This proposal is a multidisciplinary approach to creating a sustainable and profitable park. It aims to elevate the status of horse racing in the next ten years. Architecture is designed as an extension of the landscape. Undulating landforms are used to define zones and create a park without an obvious boundary. The racing track is used as a focal point, and various features spin out from the fast track to create slow zones. The master plan is defined by a never-ending figure eight loop, based on the sign of infinity, combining nature and architecture. Horsetopia is a park created with programs of "fast" excitements and "slow" enjoyments. A variety of programs are multi-season, educational, and connected to the local community. The low-carbon design brings ecotourism to customers, while sustainable features will keep the park economical, functional, and aesthetically pleasing. The existing site is largely forested primarily with thick growths of pitch pine, with some acacia and oak, and the program requires the siting of a huge oval racetrack for horseracing within a difficult topography.

As part of their approach to sustainability, the team designed eight on-site wind turbines to power all path lighting. Ninety-five percent of the earth excavated will be reused on-site, in part by tilting the oval of the racetrack at an angle but still situating it within a graceful alignment while maintaining the integrity of a large existing hill. The design team focused also on the application of carbon sequestration on the site as a way of managing timber resources. The logic is that a grown tree has sequestered carbon in its wood throughout its lifetime. Although efforts were made to preserve the most pristine and valuable forests, it was inevitable to have to remove some existing trees to create the racing track and ancillary facilities for the horse park, so the design team felt that the most efficient method was to reuse these local woods as source material for site furniture or other required wood products in the design. As a result, the carbon sequestered within these trees not only will stay on-site but also will not be released into the atmosphere.

The design team calculated a rough carbon footprint of the design consisting of the sequestration of carbon by *existing* trees converted into built elements added to the sequestration of carbon dioxide absorbed by the *proposed* trees. For each existing species of trees to be re-used on the site, the number of trees and their area was determined, and multiplied by the unit amount of carbon that particular species can hold. The subtotals were added and the result became the carbon to be re-used on site from existing trees. A second set of calculations encompassed new trees proposed for the site, since they absorb carbon dioxide. As before, for each species a separate subtotal and total were made. The sum of the two totals resulted in the carbon footprint for the entire site.

As Vicky Chan, the principal, indicates, "Obviously, if we want to dig deeper, the carbon calculation can be highly detailed and very sophisticated. In a life cycle assessment, there are calculations for the machinery and labor required to cut the old trees and to plant the new trees. The carbon footprint of the construction, transportation, and future operation were not taken into account in my case. The full assessment is a three-year process." For instance, the American Hardwood Export Council, https://www.americanhardwood.org, an industry council with funding from the U.S. government, has documented the exact carbon footprint for each wood product their members manufacture. To give one example, there is a specific calculation for red oak flooring from the East Coast of the United States being used in a European project.[11]

Following the illustrations is a spreadsheet showing the calculations used to determine the carbon footprint of the park, the total of which is summarized graphically in Figure 9.7.

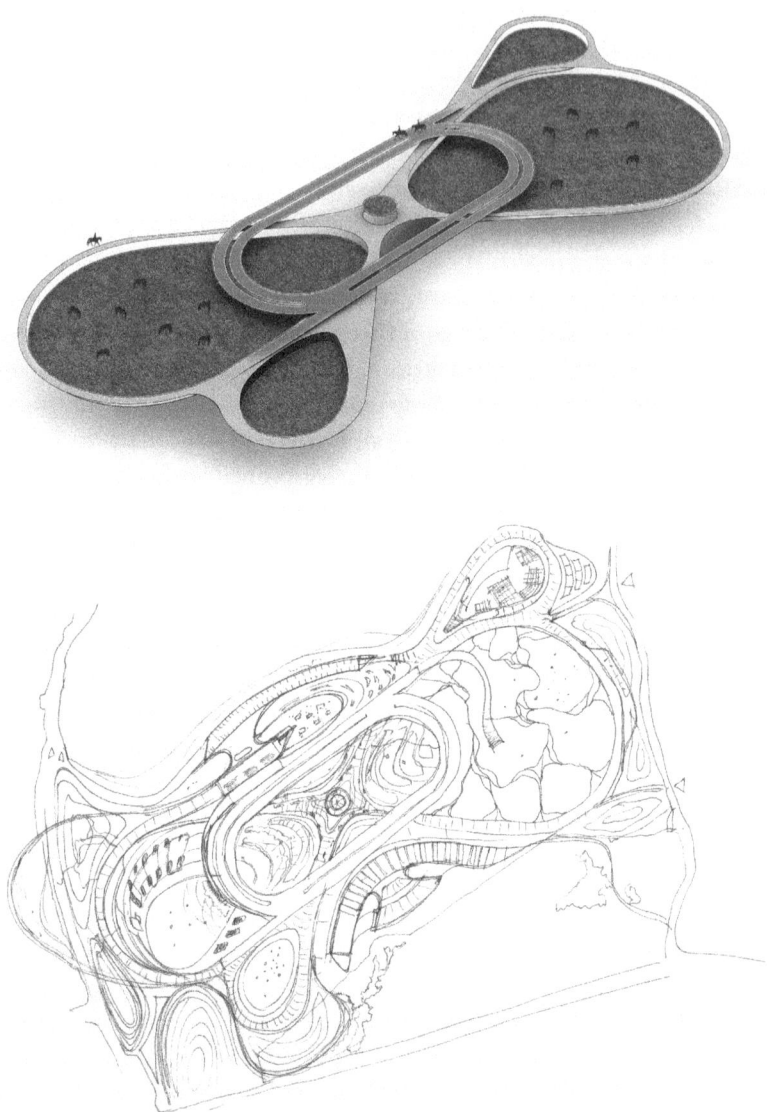

Figure 9.1 Concept Sketch of Horsetopia showing landscape dividing programs into zones.
Daewook Lee, Vicky Chan, Melissa Chan, Krystal Lung, and Leo Le

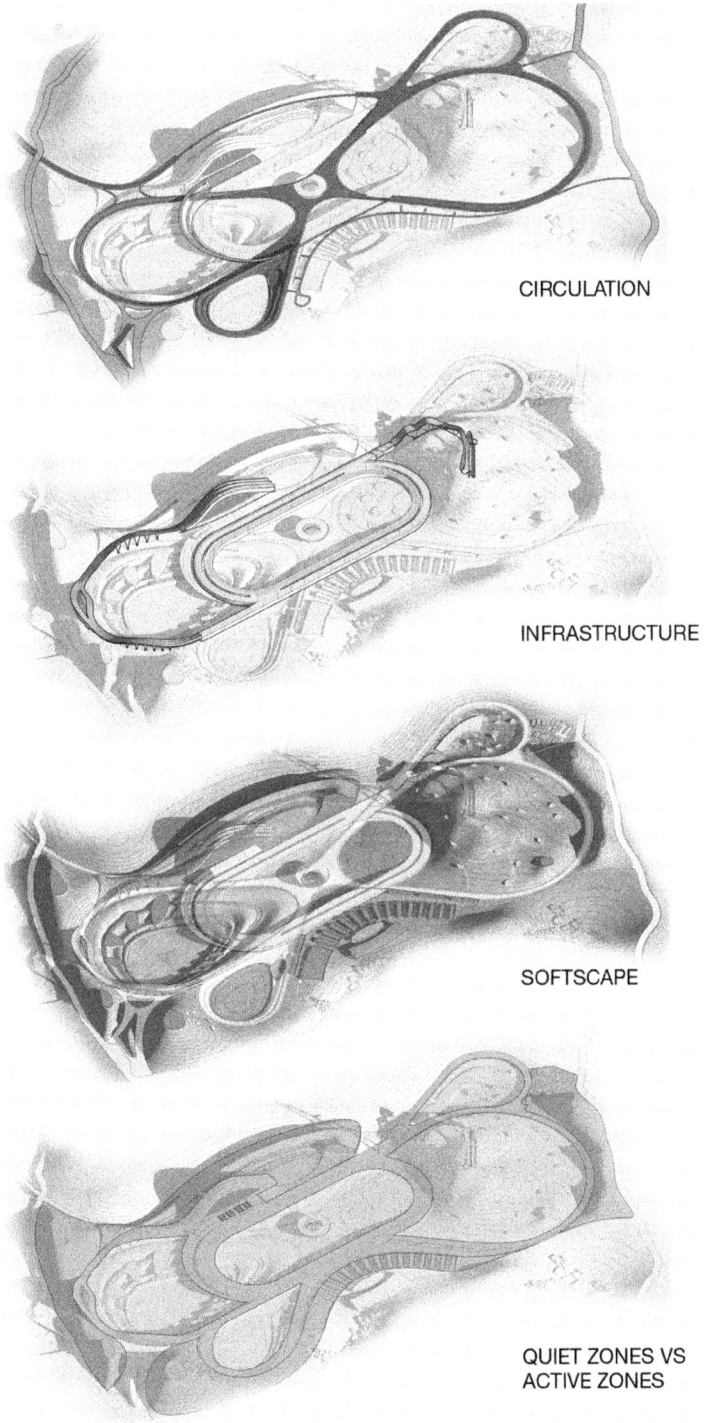

Figure 9.2 Exploded Axonometric Diagrams of Horsetopia showing multiple system. Top to Bottom: 1) Circulation 2) Infrastructure 3) Softscape 4) Quiet Zones vs Active Zones.
Daewook Lee, Vicky Chan, Melissa Chan, Krystal Lung, and Leo Lei

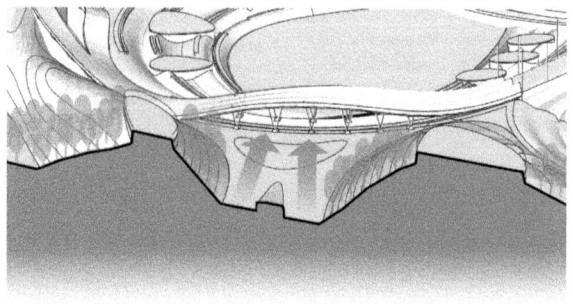

ENTRANCE

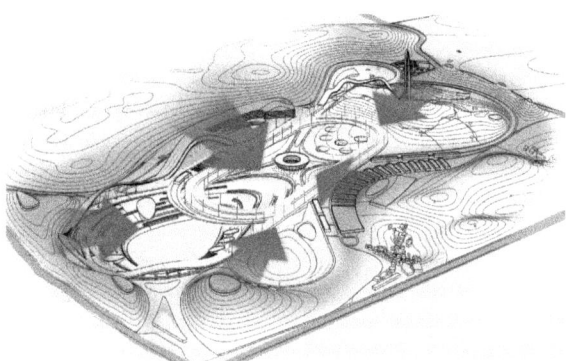

HIGHER ELEVATION
SURROUNDING THE
RACING TRACK

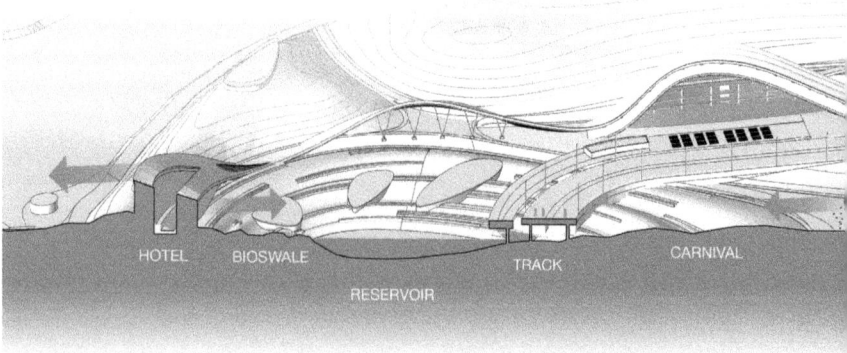

Figure 9.3 Sectional diagrams showing relationship between architecture and landscape.
Daewook Lee, Vicky Chan, Melissa Chan, Krystal Lung, and Leo Lei

HOTEL

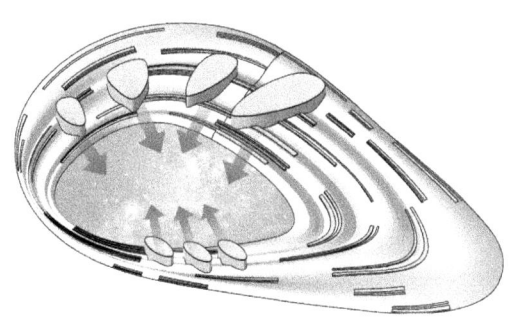

RESTAURANTS AROUND RESERVOIR

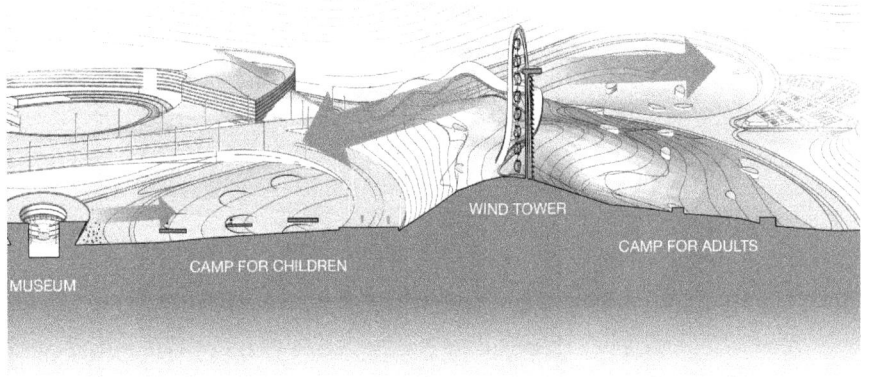

MUSEUM   CAMP FOR CHILDREN   WIND TOWER   CAMP FOR ADULTS

Figure 9.4 Rendered Master Plan of Horsetopia showing integration of the architecture into the landscape.
Daewook Lee, Vicky Chan, Melissa Chan, Krystal Lung, and Leo Lei

Figure 9.5 Night Rendering of Horsetopia showing pockets of activities within the landscape.
Daewook Lee, Vicky Chan, Melissa Chan, Krystal Lung, and Leo Lei

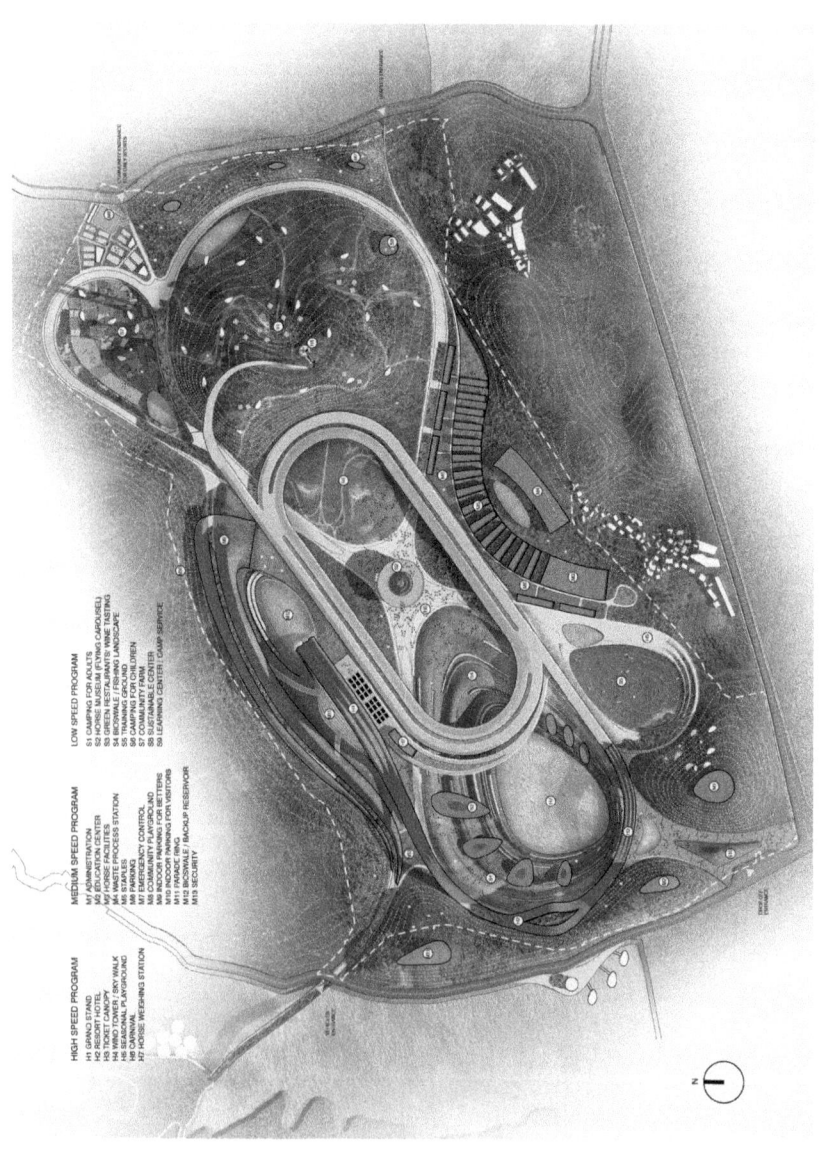

Figure 9.6 Low, Medium, High Speed Program. Daytime rendering of Horsetopia showing integration of the architecture into the landscape organized by the speed at which the various elements are experienced.
Daewook Lee, Vicky Chan, Melissa Chan, Krystal Lung, and Leo Lei

Boa'an G107, China

There was a competition to convert a borough in Shenzhen, China, the size of Manhattan, into a sustainable manufacturing area. The area in focus was the neighborhood on both sides of a highway. The existing site consisted of old factories with lots of pollution. The drainage systems and waste treatment plant were overutilized. The following diagram is meant to show how the sheer volume of rainwater going into the waste treatment plant can be reduced.

Rainwater is collected in different part of the boroughs. Some is absorbed immediately in the landscape, while some goes underground. The rest can be transported to the proposed basins below the G107 highway. The water will then be distributed to other wetlands for a filtering and purification process brought about by the layers of sand and gravel and proposed plant materials. (See Figure 9.8).

K-Farm, Hong Kong

K-Farm is a 1900 m^2 urban farm in the Western Pier next to Victoria Harbor in Hong Kong. The farm will be the only farm in the region to combine aquaponic, hydroponic, and organic farming into a single location. The concept is to show the residents that smart farming is part of the smart city while at the same time giving people a nice relaxing area to enjoy the ocean view. The farm will be run privately on a nonprofit basis. (See Figures 9.9 and 9.10).

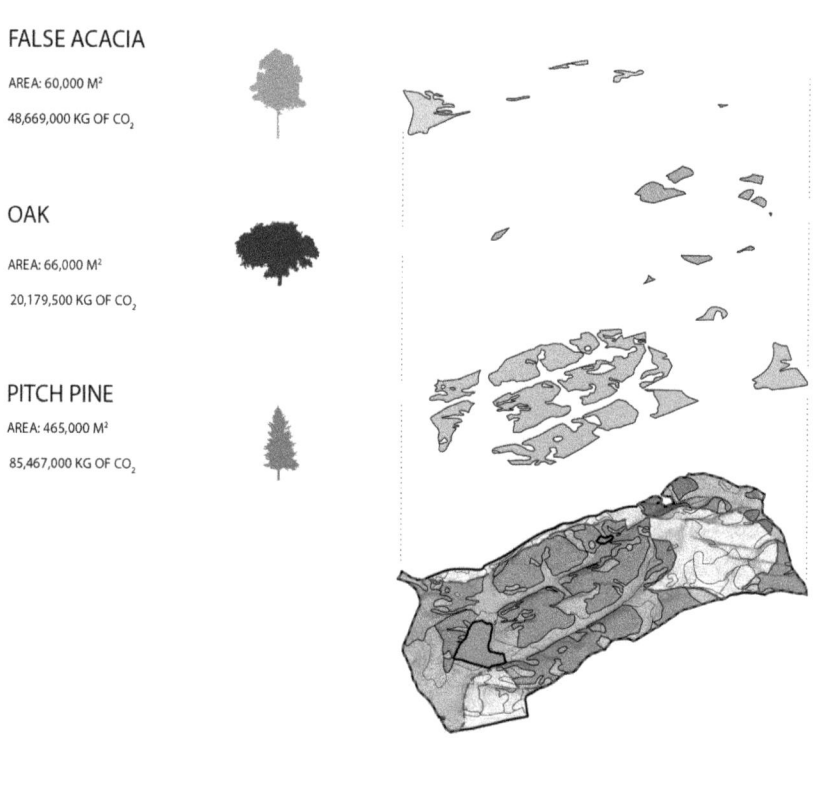

Figure 9.7 Carbon Sequestration
Diagram showing the amount of carbon that can be stored by using woods harvested on the site as part of the proposed architectural elements in the park.
Vicky Chan and other

PROPOSED

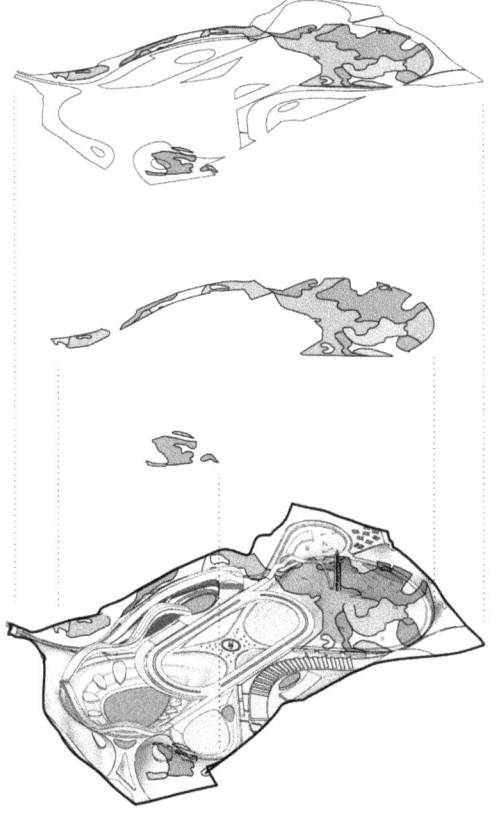

NEW TREES

500,000 M²

PROTECTED FOREST

FALSE ACACIA: AREA: 45,000m²

OAK: AREA: 130,000m²

PITCH PINE: AREA: 145,000m²

TOTAL
PRESERVATION:
   AREA: 320,000m²

320,000 M²

465,327,750 KG
449,120 TONS OF CO_2 STORED AS ARCHITECTURE WHILE NEW TREES ARE PLANTED

(450) Ethics and Sustainability

HORSETOPIA CARBON CALCULATIONS

AREA (m²)	TYPE	# OF TREES	CARBON PER TREE (KG)	TOTAL CARBON (KG)
HARVESTED TREES				
60,000	False acacia	3,000	16,223	48,669,000
66,000	Oak	1,650	12,230	20,179,500
465,000	Pitch pine	46,500	1,838	85,467,000
PROTECTED TREES (SAVED)				
45,000	False acacia	2,250	16,223	36,501,750
130,000	Oak	3,250	12,230	39,747,500
145,000	Pitch pine	14,500	1,838	26,651,000
NEW TREES				
160,000	False acacia	8,000	16,223	129,784,000
160,000	Oak	4,000	12,230	48,920,000
160,000	Pitch pine	16,000	1,838	29,408,000
			Kilograms	465,327,750
			Imperial tons	449,120
In the table above, method for calculating total carbon is shown				
for existing trees only.				
See Figure 9.7.				

CALCULATION METHOD APPLIED TO HARVESTED TREES			
TREE VOLUME			
CONICAL SHAPE: VOLUME(m^3) = HEIGHT(m) × DIAMETER(m^2) × .7854 / 3			
CYLINDRICAL SHAPE: VOLUME(m^3) = HEIGHT(m) × DIAMETER(m^2) × .7854			
TREE MASS			
MASS(kg) = VOLUME(m^3) × DRIED WEIGHT(kg/m^3)			
CARBON SEQUESTRATION PER TREE			
CO_2(kg) = MASS(kg) × .65(DRY MASS) × .5(CARBON %) × 3.67 × 1.2 (ROOT VALUE)			
FALSE ACACIA			
AVG. HEIGHT	20–30 m		
TRUNK DIAMETER	60–90 cm		
DRIED WEIGHT	770 kg/m^3		
VOLUME (m^3)	25m × .75m^2 × .7854 =		14.7
MASS (kg)	14.7m^3 × 770 kg/m^3 =		11,340
CARBON PER TREE (kg)	11340 × .65 × .5 × 3.67 × 1.2 =		**16,223**
AREA PER TREE (m^2)	20		
SITE AREA (m^2)	60,000		
TREES COUNT	3,000		
TOTAL =	**48,669,000**	total carbon kg?	
OAK			
AVG. HEIGHT	20–30 m		
TRUNK DIAMETER	60–100 cm		
DRIED WEIGHT	680 kg/m^3		
VOLUME	25 m × .75 m^2 × .7854		12.5

AREA (m²)	TYPE	# OF TREES	CARBON PER TREE (KG)	TOTAL CARBON (KG)

CALCULATION METHOD APPLIED TO HARVESTED TREES			
MASS	12.5 m³ × 680 kg/m³		8,545
CARBON PER TREE (kg)	8545 × .65 × .5 × 3.67 × 1.2		**12,230**
AREA PER TREE (m²)	40		
SITE AREA (m²)	66,000		
TREES COUNT	1,650		
TOTAL =	20,179,500	total carbon kg?	
PITCH PINE			
AVG. HEIGHT	15–20 m		
TRUNK DIAMETER	30–60 cm		
DRIED WEIGHT	545 kg/m³		
VOLUME	20 m × .45 m² × .7854 / 3		2.4
MASS	2.4 m³ × 545 kg/m³		1,284
CARBON PER TREE (kg)	1284 × .65 × .5 × 3.67 × 1.2		**1,838**
AREA PER TREE (m²)	10		
SITE AREA (m²)	465,000		
TREES COUNT	46,500		
TOTAL =	85,467,000	total carbon kg?	

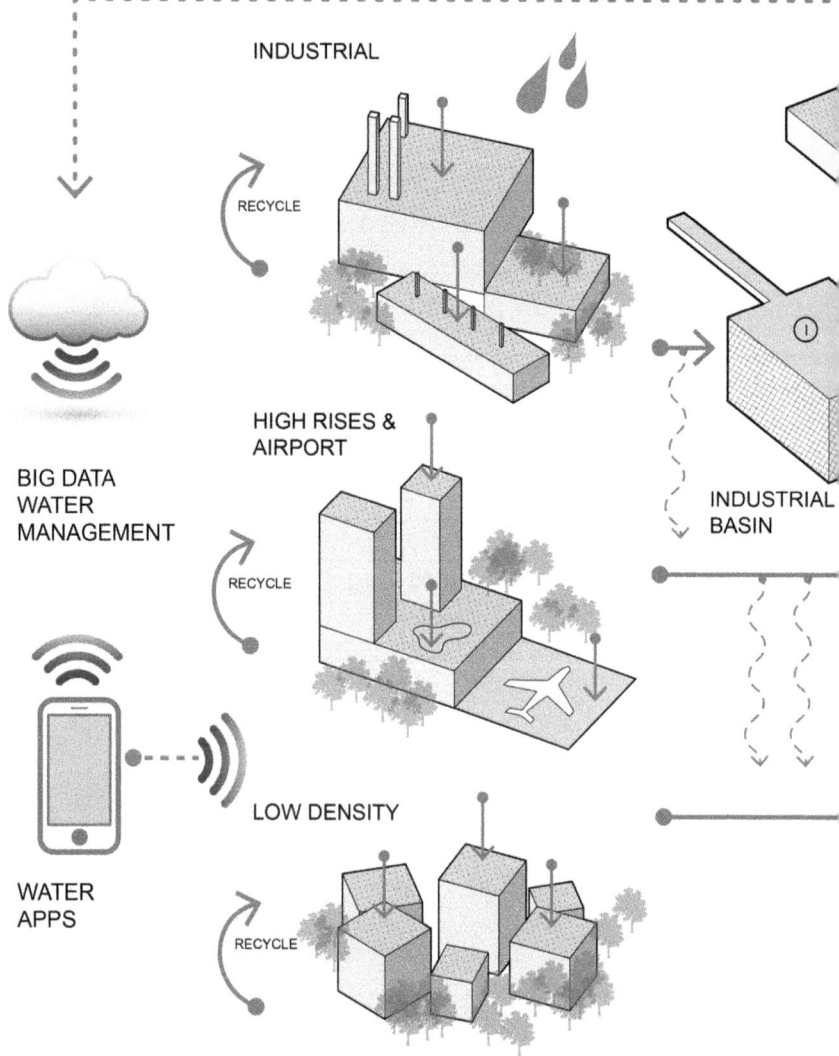

Figure 9.8 Boa'an G107 Highway, Shenzhen, China.
Water Management Diagram showing low density, high density, industrial and regional water treatment, all as part of a master plan to reconnect the scenic mountains and long established waterfront.
Richard Alomar, Vicky Chan, and Eve Hocheng

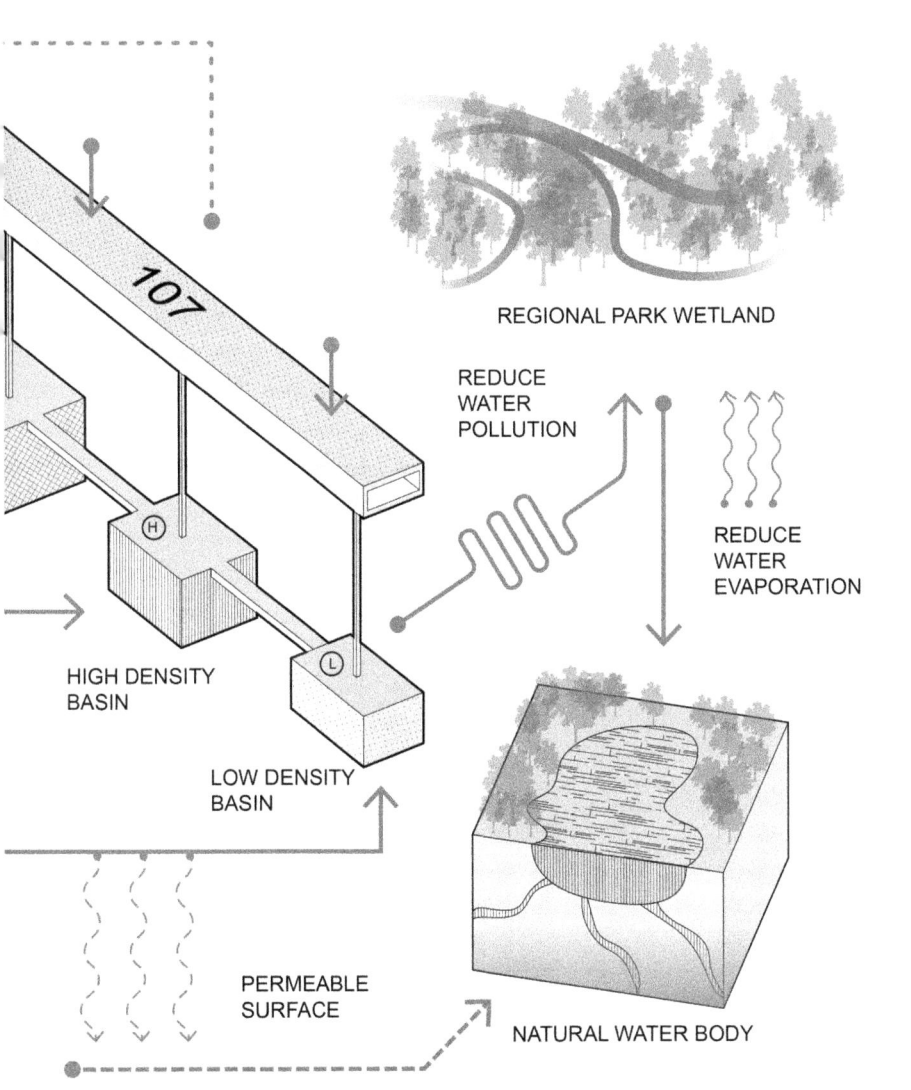

Figure 9.9 K-farm, Urban Farming, Hong Kong, illustrative perspective.
Vicky Chan

Figure 9.10 K-farm, Urban farming design in Hong Kong. Vicky Chan.

Problem 1
SUSTAINABLE DESIGN

(In the following problem, any *terms in italics* may be changed to suit the particular instructor, or the personal, work or studio setting. Typical changes might be specific people, such as clients or contractors, plant materials, or the locations of sites, people, institutions or companies.)

It is proposed to build a park primarily for passive recreation on a former 25-acre (62.5 ha) landfill that was gradually phased out as a refuse dump and capped as a landfill site ten years ago in the *city or town of Beverly, XY*. Roughly rectilinear in shape, it is now in the midst of a residential neighborhood and surrounded by single-family housing on two sides, multifamily housing on one side, and a mixed-use business development on the fourth side. There are engineered landfill caps, and some sparse vegetation has become established. Some small animals such as birds and mice have been seen by people wandering into the site. Effluent samples show acceptable water quality. Methane is captured from exhaust pipes at several locations at high points of hilltops.

The town council proposes the following recreational activities:

1. Bird watching
2. Walking
3. Bicycling
4. Skateboarding
5. Kite-flying
6. Frisbee and/or Frisbee golf
7. Open lawn activities
8. Sledding
9. Cross-country skiing
10. Several basketball and/or tennis courts
11. Parking for 50 cars and 200 bicycles

Develop a preliminary site plan and propose materials and details for each of the above activities (as well as any other necessary proposed activities) as follows:

1. All runoff must continue to be contained or absorbed within the site.
2. Assign a sustainable design footprint for each activity based on the degree of impact it would have on the site.

3. Propose plantings or other mitigations to compensate for the negative impacts caused by each proposed activity or detail.
4. Summarize your results in a chart.
5. Achieve a net result of zero impact or a positive impact on the site.

NOTES

1. E. O. Wilson. *The Creation*. New York: W.W. Norton, 2006, p. 4
2. Please note that I have chosen to discuss social media in Chapter 2, on human resources, although there are many ethical challenges that derive from it. The employer, as embodied by the head of HR or the principal(s) of a firm, must set ethical and practical standards on the use of social media
3. Aldo Leopold. *Sand County Almanac and Sketches Here and There*. New York: Oxford University Press, 1949
4. Some of this text first appeared in *Green Roofs in Sustainable Landscape Design* by Steven L. Cantor (New York: W. W. Norton, 2008), which the author incorporates herein with permission from the publisher, and has updated.
5. Gandhi, quoted from exhibit at Brooklyn Museum. See also www. www. gandhiinstitute.org.
6. Stephen R. Kellert, email to author on May 14, 2007, provided the definition quoted. See also his book, *Building for Life: Designing and Understanding the Human-Nature Connection* (Washington, DC: Island Press, 2005) for a range of relevant projects.
7. American Society of Landscape Architects, Sustainable Design and Development Professional Practice Network, newsletter, April 5, 2018
8. Jan Christiaan Smuts. *Holism and Evolution: The Original Approach to the Holistic Approach to Life*, edited by Sanford Holst (Sherman Oaks, CA: Sierra Sunrise, 1999), 95.
9. Don Prowler, "Whole Building Design," revised and updated by Stephanie Vierra, https://www.wbdg.org/resources/whole-building-design.
10. U.S. Green Building Council, www.USGBC.org; Colorado Green Building Guild, www.CGBG.org.
11. American Hardwood's Life Cycle Assessment Tool, https://www.americanhardwood.org/en/environmental-profile/american-hardwoods-life-cycle-assessment-tool.

CHAPTER 10

Resources

This section provides resources, including books, websites, and other documents, useful to students and emerging professionals in landscape architecture. As each reader of this book pursues a career, she or he will undoubtedly want to acquire a library of books and other resources that are most suitable for the specific setting. As you expand your library, don't overlook books that have been reprinted or issued in revised editions: either is a sign that the book has remained useful over a long period of time. Most major and many small publishers have websites, such as McGraw-Hill (www.mcgraw-hill.com), W. W. Norton (www.wwnorton.com), and Oxford University Press (www.oup.com). Amazon (www.amazon.com) is one source for books at a discount, although keep in mind that authors earn better royalties from sales at bookstores or from the websites of their publishers. Website addresses, at the end of this section, multiply rapidly but also change very frequently.

DOS AND DON'TS FOR RESOURCES

1. If you find some *unusual* chestnuts to be the centerpiece of a major essay, try to dig up some corroborating acorns.
2. Provide your sources through a bibliography, notes, and footnotes.
3. Be careful with humor, particularly in serious writing. (See number 1.) It is exceedingly rare for every reader to appreciate your sense of wit consistently.
4. After you've made what you feel is a good start on your writing, ask a friend—or, better, a respected writer or scholar in the appropriate

Figure 10.0 Little Rock Main Street Water Quality Demonstration Project: The downtown of this central Arkansas city features bioretention areas and bioswales, protected by metal grate boardwalks, part of a water quality demonstration project designed by Crafton Tull, in coordination with the City of Little Rock and the Arkansas Natural Resources Commission in which low impact development techniques are used as an option for stormwater management. See http://littlerock.org/lid and https://www.facebook.com/creativecorridorlittlerock.
Photo by Steven L. Cantor

field who has available time—to review and comment on a sample of your writing.
5. Outline what you propose to do. An outline organizes your major thoughts and gives you perspective on where you are within a project. The worst that can happen is that you don't follow the outline, in which case revise it and make it better.
6. Write in a comfortable setting and if at all possible have an automatic backup. If you insist on keeping a handwritten journal in backcountry conditions that inspire you, know the risks.
7. In the outdoors, whether urban or rural, polluted or pristine, learn to identify a dozen trees, shrubs, and perennials in each season wherever you happen to live, and keep expanding your knowledge. When you visit a new place, seek out an oldtimer experienced in the regional flora and fauna long enough to buy coffee or tea or the local brew and learn a few new ones.
8. Learn how to use a camera. Learn how to save, store, and edit digital images.
9. Be a frequent user of your nearest public library, the library of the university or college from which you graduated, or a nearby university library that grants you visiting rights.
10. Explore one step at a time.
11. Don't overreach.
12. Enjoy the process, however you set it up for yourself.

Abbett, Robert W. *Engineering Contracts and Specifications*. 4th ed. New York: John Wiley & Sons, 1963.

Abbey, Edward. *Beyond the Wall: Essays from the Outside*. New York: Holt, Rinehart and Winston, 1971.

Abbey, Edward. *Desert Solitaire: A Season in the Wilderness*. New York: Ballantine Books, 1968.

Ambrose, James E., and Patrick Tripeny. *Simplified Engineering for Architects and Builders*. 12th ed. Hoboken, NJ: John Wiley & Sons, 2016. Good guidebook.

Amidon, Jane. *Radical Landscapes: Reinventing Outdoor Space*. New York: Thames & Hudson, 2001.

Baggs, Sydney, and Joan Baggs. *The Healthy House: Creating a Safe, Healthy and Environmentally Friendly Home*. New York: Harper Collins, 1996.

Beery, William. *Business Law for Landscape Architects*. Boston, MA: Pearson Custom Publishing, 2002.

Beveridge, Charles E., and Paul Rocheleau. *Frederick Law Olmsted: Designing the American Landscape*. Edited by David Larkin. New York: Rizzoli International, 1995.

Birnbaum, Charles, et al. *Pioneers of American Landscape Design*. National Park Service Historic Landscape Initiative Project. New York: McGraw-Hill, 2000.

Brickell, Christopher, and Trevor Cole, editors. *American Horticultural Society Encyclopedia of Plants and Flowers*. New York: DK Publishing, 2002.
Brookes, John. *The Small Garden*. New York: Van Nostrand Reinhold, 1983.
Bye, Arthur. *Art into Landscape, Landscape into Art*. Mesa, AZ: PDA Publishers, 1983.
Cantor, Steven L. *Contemporary Trends in Landscape Architecture*. New York: John Wiley & Sons, 1997.
Cantor, Steven L. *Green Roofs in Sustainable Landscape Design*. New York: W. W. Norton, 2008.
Cantor, Steven L. *Innovative Design Solutions in Landscape Architecture*. New York: John Wiley & Sons, 1997.
Carpenter, Philip L., Theodore D. Walker, and Frederick O. Lamphear. *Plants in the Landscape*. San Francisco: W. H. Freeman, 1975.
Carson, Rachel. *Silent Spring*. Boston: Houghton Mifflin, 1962. A classic study by a passionate scientist on the effects of pesticides on the environment
Ching, Frances D. K. *Architecture: Form, Space and Order*. New York, Van Nostrand Reinhold, 1979.
Ching, Francis D. K. *Building Construction Illustrated*. Hoboken, NJ: John Wiley & Sons, 2014.
Ching, Francis D. K., and Steven R. Winkel. *Building Codes Illustrated: A Guide to Understanding the 2000 International Building Code*. New York: John Wiley & Sons, 2003.
Clark, Kenneth. *Landscape into Art*. Boston: Beacon Press, 1961.
Cole, George M. *Water Boundaries*. New York: John Wiley & Sons, 1997.
Cooke, Michael, and Brigid Arnott. *Disobedient Gardens: Landscapes of Contrast and Contradiction*. Sydney: Murdoch Books, 2017.
Crane, Robin, and Malcolm Dixon. *The Shape of Space: Indoor Sports Spaces*. New York: Van Nostrand Reinhold, 1991.
Crowe, Sylvia. *Garden Design*. Cambridge, UK: Packard Publishing, 1981.
Davis, David A., and Theodore Walker. *Plan Graphics*. 5th ed. New York: John Wiley & Sons, 2000.
DeCristoforo, R. J. *Handyman's Guide to Concrete and Masonry*. Reston, VA: Prentice-Hall, 1978.
Dickey, Page. *Breaking Ground: Portraits of Ten Garden Designers*. New York: Artisan, 1997.
Diekelmann, John, and Robert Schuster. *Natural Landscaping: Designing with Native Plant Communities*. New York: McGraw-Hill, 1982.
Dines, Nicholas. *Landscape Architect's Portable Handbook*. New York: McGraw-Hill, 2001.
Dines, Nicholas, and Charles W. Harris. *Time-Saver Standards for Landscape Architects*. New York: McGraw-Hill, 1997.
Dingwall-Main, Alex. *The Angel Tree*. London: Ebury Press, 2004.
Dingwall-Main, Alex. *The Luberon Garden*. London: Ebury Press, 2002.
Dirr, Michael. *Manual of Woody Landscape Plants: Their Identification, Ornamental Characteristic, Culture, Propagation and Use*. Champaign, IL: Stipes Publishing, 1977.
Dube, Richard L. *Natural Pattern Forms—A Practical Sourcebook for Landscape Designs*. New York: Van Nostrand Reinhold, 1997.
Duffield, Mary Rose, and Warren D. Jones. *Plants for Dry Climates—How to Select, Grow and Enjoy*. Tucson, AZ: HP Books, 1981.

Dunnett, Nigel, and Noel Kingsbury. *Planting Green Roofs and Living Walls.* Portland, OR: Timber Press, 2004.

Eckbo, Garrett. *Home Landscape: The Art of Home Landscaping.* Rev. ed. New York: McGraw-Hill, 1978.

Ellis, Barbara W., editor. *Burpee Complete Gardener: A Comprehensive, Up-to-Date, Fully Illustrated Reference for Gardeners at All Levels.* New York: Macmillan, 1995.

Fisher, Thomas. *Ethics for Architects: 50 Dilemmas of Professional Practice.* New York: Princeton Architectural Press, 2012.

Fitch, James Marston. *American Building: The Historic Forces That Shaped It.* Boston: Houghton, Mifflin, 1966.

Fleming, John, et al. *The Penguin Dictionary of Architecture and Landscape Architecture.* London: Penguin, 1998.

Frederick, William H., Jr. *100 Great Garden Plants.* New York: Alfred A. Knopf, 1975.

Galantay, Ervin Y. *New Towns: Antiquity to the Present.* New York: George Braziller, 1975.

Girardet, Herbert. *Cities, People, Planet: Liveable Cities for a Sustainable World.* Chichester, UK: John Wiley & Sons, 2004.

Gold, Seymour M. *Recreation Planning and Design.* New York: McGraw-Hill, 1980.

Halprin, Lawrence. *RSVP Cycles: Creative Processes in the Human Environment.* New York: George Braziller, 1969.

Harlow, William H., and Ellwood S. Harrar. *Textbook of Dendrology, Covering the Important Forest Trees of the United States and Canada.* 5th ed. New York: McGraw-Hill, 1969.

Hayward, Robert. *The Brick Book.* New York: Thomas Y. Crowell, 1977.

Hightshoe, Gary L. *Native Trees, Shrubs and Vines for Urban and Rural America.* New York: Van Nostrand Reinhold, 1987.

Hiss, Tony. *The Experience of Place.* New York: Alfred A. Knopf, 1990.

Hopper, Leonard. *Landscape Architectural Graphic Standards.* New York: John Wiley, 2007.

Howard, Love Albrecht. *So You Want to Be a Garden Designer.* Portland, OR: Timber Press, 2010.

Huxley, Aldous. *The Painted Garden: The Garden Through the Artist's Eye.* London: Quarto, 1988.

Irons, Karen Phillips, editor. *Trees and Shrubs, Ground Covers, Vines.* Birmingham, AL: Oxmoor House, 1980.

Jacobs, Jane. *The Death and Life of Great American Cities.* New York: Modern Library, 1993.

Jefferson, Thomas. *The Garden and Farm Books of Thomas Jefferson.* Edited by Robert C. Baron. Golden, CO: Fulcrum, 1987.

Johnson, Hugh. *The Principles of Gardening: A Guide to the Art, History, Science and Practice of Gardening.* New York: Simon and Schuster, 1979.

Joyce, David. *The Perfect Plant for Every Site, Habitat and Garden Style.* New York: Stewart, Tabori & Chang, 1998.

Karner, Gary E. *Contracting Design Services.* Washington, DC: American Society of Landscape Architects, 1989.

Kassler, Elizabeth A. *Modern Gardens and the Landscape.* Rev. ed. New York: Museum of Modern Art, 1984.

Kellert, Stephen R. *Building for Life-Designing and Understanding the Human Nature Connection.* Washington, DC: Island Press, 2005.

Kellert, Stephen R. *People and Nature in the Modern World*. New Haven, CT: Yale University Press, 2012.
Kellert, Stephen R., Judith Heerwage, and Martin Mador. *Biophilic Design: The Theory, Science and Practice of Bringing Buildings to Life*. New York: John Wiley & Sons, 2008.
Kimball, Theodora, and Frederick Law Olmsted Jr. *Forty Years of Landscape Architecture: Central Park*. Cambridge, MA: MIT Press, 1973.
Koberg, Don, and Jim Bagnall. *Universal Traveler: A Soft-Systems Guide to Creativity, Problem Solving and the Process or Reaching Goals*. Los Altos, CA: William Kaufmann, 1991.
Lane, Kenneth F., and John M. Robert. *Fundamental Land*. Ames: Iowa State University Research Foundations, 1970.
Laurie, Michael. *An Introduction to Landscape Architecture*. New York: Elsevier, 1976.
Leopold, Aldo. *A Sand County Almanac: And Sketches Here and There*. New York: Oxford University Press, 1949.
Lewis, Jack R. *Construction Specifications*. New Jersey: Prentice-Hall, 1975.
Lyall, Sutherland. *Designing the New Landscape*. New York: Van Nostrand Reinhold, 1991.
Lynch, Kevin. *Managing the Sense of a Region*. Cambridge, MA: MIT Press, 1980.
Lynch, Kevin, and Gary Hack. *Site Planning*. 3rd ed. Cambridge, MA: MIT Press, 1985.
Maclean, Norman. *Young Men and Fire*. Chicago: University of Chicago Press, 1992.
Marshall, Lane L. *Landscape Architecture: Guidelines to Professional Practice*. Illustrated by James E. Hiss. Washington, DC: American Society of Landscape Architects, 1981.
McHarg, Ian. *Design with Nature*. New York: Doubleday, 1971.
Melby, Pete. *Simplified Irrigation Design*. New York: Van Nostrand Reinhold, 1995.
Molnar, Donald J. *Anatomy of a Park*. 4th ed. Long Grove, IL: Waveland, 2015.
Moyer, Jan. *The Landscape Lighting Book*. 2nd ed. New York: John Wiley & Sons, 1992.
Munson, Albe E. *Construction Design for Landscape Architects*. New York: McGraw Hill, 1974.
Nelischer, Maurice. *Handbook of Landscape Architectural Construction*. Washington, DC: Landscape Architecture Foundation, 1985.
Newton, Norman T. *Design on the Land; The Development of Landscape Architecture*. Cambridge, MA: Belknap Press of Harvard University Press, 1971.
Odom, Eugene. *Ecology*. New York: Holt, Rinehart & Winston, 1963. A brief summary of ecological theory.
Osler, Mirabel. *A Gentle Plea for Chaos*. New York: Arcade, 1989.
Osmundson, Theodore. *Roof Gardens: History, Design and Construction*. New York: W. W. Norton, 1999.
Oudolf, Piet, with Noel Kingsbury. *Designing with Plants*. Portland, OR: Timber Press, 1999.
Paul, Anthony, and Yvonne Rees. *The Water Garden*. New York: Penguin, 1986.
Pevsner, Nikolaus. *Pioneers of Modern Design from William Morris to Walter Gropius*. New Haven, CT: Yale University Press, 2011.
Pevsner, Nikolaus. *The Sources of Modern Architecture and Design*. New York: Fredrick A. Praeger, 1968.
Proulx, Annie. *Barkskins*. New York: Scribner, 2016.
Ramsey, George, and Harold Sleeper. *Architectural Graphic Standards*. New York: John Wiley & Sons, 2000.

Rapuano, Michael, P. P. Pirone, and Brooks Wigginton. *Open Space in Urban Design.* New York: Spiral Press, 1964.

Raskin, Eugene. *Architecturally Speaking.* New York: Bloch, 1966.

Robinette, Gary O. *Plants, People and Environmental Quality.* Washington, DC: U.S. Department of the Interior, 1973.

Rogers, Walter. *Professional Practice of Landscape Architecture: A Complete Guide to Starting and Running Your Own Firm.* 2nd ed. Hoboken, NJ: John Wiley & Sons, 2010.

Rolston, Holmes, III. *Environmental Ethics: Duties to and Values in the Natural World.* Philadelphia: Temple University Press, 1988.

Rosen, Harold J., and Tom Heineman. *Construction Specifications Writing: Principles and Procedures.* 3rd ed. New York: John Wiley & Sons, 1990.

Rubinstein, Harvey M. *A Guide to Site and Environmental Planning.* 2nd ed. New York: John Wiley & Sons, 1980.

Russ, Thomas. *Redeveloping Brownfields.* New York: McGraw-Hill, 1999. A reference on reclamation.

Schlechtl, Hugo. *Bioengineering for Land Reclamation and Conservation.* Translation co-ordinated by N. K. Horstmann. Edmonton: University of Alberta Press, 1980.

Schuler, Stanley. *The Gardener's Basic Book of Trees and Shrubs.* New York: Simon & Schuster, 1973.

Schuler, Stanley. *How to Build Fences, Gates and Walls.* New York: Macmillan, 1976.

Segal, Paul. *Professional Practice: A Guide to Turning Designs into Buildings.* New York: W. W. Norton, 2006.

Sharky, Bruce. *Ready, Set, Practice: Elements of Landscape Architecture Professional Practice.* New York: John Wiley & Sons, 1994.

Snodgrass, Edmund C., and Lucie L. Snodgrass. *Green Roof Plants: A Resource and Planting Guide.* Portland, OR: Timber Press, 2009. Fully illustrated in color

Spiegler, Jennifer C., and Paul M. Gaykowski. *The Bridges of Central Park.* Charleston, SC: Arcadia, 2006.

Spreiregen, Paul D. *Urban Design: The Architecture of Towns and Cities.* Malabar, FL: Robert E. Krieger, 1981.

Starke, Barry W., and John Ormsbee Simonds. *Landscape Architecture: A Manual of Environmental Planning and Design.* 5th ed. New York: McGraw-Hill, 2013.

Stevenson, Elizabeth. *Park Maker: A Life of Frederick Law Olmsted.* New York: Macmillan, 1977.

Strom, Steven, and Kurt Nathan. *Site Engineering for Landscape Architects.* 3rd ed. New York: John Wiley & Sons, 1998.

Strunk, William, Jr., and E. B. White. *The Elements of Style.* 50th anniversary ed. New York: Pearson Longman, 2009.

Tallamy, Douglas W. *Bringing Nature Home: How Native Plants Sustain Wildlife in Our Gardens.* Portland, OR: Timber Press, 2007.

Tandy, Cliff, editor. *Handbook of Urban Landscape.* New York: Whitney Library of Design, 1974.

Tanner, Ogden. *Garden Construction.* Alexandria, VA: Time-Life Books, 1979.

Taylor, John S. *Commonsense Architecture: A Cross-Cultural Survey of Practical Design Principles.* New York: W. W. Norton, 1983.

Taylor, Lisa, editor. *Urban Open Spaces.* New York: Rizzoli, 1979.

Taylor's Guide to Garden Design. Rev. ed. Boston: Houghton Mifflin, 1988. Many excellent color illustrations.

Thoreau, Henry David. *Thoreau's Wildflowers.* Edited by Geoff Wisner. Illustrated by Barry Moser. New Haven, CT: Yale University Press, 2016.
Tourbier, J. Toby, and Richard Westmacott. *A Handbook for Small Surface Coal Mine Operators.* Washington, DC: Department of the Interior, 1980.
Tyler, Norman. *Historic Preservation: An Introduction to Its History, Principles and Practice.* 2nd ed. New York: W. W. Norton, 2009.
Untermann, Richard K. *Grade Easy.* Washington, DC: Landscape Architecture Foundation, 1973. Introduction to principles of grading and drainage.
Van Pelt Wilson, Helen. *Helen Van Pelt Wilson's Own Garden and Landscape Book.* New York: Weathervane Books, 1973.
Weber, Nelva M. *How to Plan Your Own Home Landscape.* New York: Bobbs-Merrill, 1976, paperback.
Whitehill, Walter Muir. *Boston, A Topographical History.* 2nd ed. Cambridge, MA: Harvard University Press, 1968.
Whyte, William Hollingsworth. *The Social Life of Small Urban Spaces.* Washington, DC: American Psychological Association, 1979.
Wigginton, Brooks. *Trees and Shrubs for the Southeast.* Athens: University of Georgia Press, 1963.
Wilson, Edward O. *Anthill.* New York: W. W. Norton, 2010.
Wilson, Edward O. *Biophilia.* Cambridge, MA: Harvard University Press, 1984.
Wilson, Edward O. *The Creation: An Appeal to Save Life on Earth.* New York: W. W. Norton, 2006.
Wilson, Edward O. *The Diversity of Life.* Cambridge, MA: Harvard University Press, 1992.
Wilson, Edward O. *Half-Earth: Our Planet's Fight for Life.* New York: Liveright, 2016.

MISCELLANEOUS

The Brooklyn Botanic Garden issues publications on many topics, such as *Japanese Inspired Gardens, Hummingbird Gardens, Landscaping Indoors, Old-Fashioned Flowers, Spring Blooming Bulbs, Gourmet Herbs,* etc.

Sunset, Ortho, Southern Living, and Better Homes and Gardens publish handbooks on many topics.

Google is a great search engine and can help you find all sorts of resources, including photographs of plant materials.

The Dummy series, despite the self-deprecating names, is a collection of thorough, excellent references, and very affordable. See the series website, www.dummies.com.

> *Container Gardens for Dummies*
> *Roses for Dummies*
> *Perennials for Dummies*
> *Gardening for Canadians for Dummies*
> *Herb Gardening for Dummies*

Photoshop for Dummies (many versions based on the various editions of Photoshop)

PowerPoint for Dummies (similar to Photoshop based on the various editions)

TRADE ASSOCIATIONS AND ORGANIZATIONS

The following organizations, even many with "American" in their titles, have strong Canadian, Mexican, European, Asian, and other foreign representation and membership. Sometimes useful information such as specifications, details, and documents can be downloaded for free; in other cases there are charges for members and nonmembers. Many of these organizations sponsor courses, webinars, annual conferences, and so on. Some are clearinghouses for major publications. Particularly with the internet such a source of information it's important to take advantage of what particular sites have to offer.

AAPQ, Association des architectes paysagistes du Quebec, https://aapq.org

Alberta Association of Landscape Architects, https://www.aala.ab.ca

Agricultural extension services by state or university—for example,
 www.extension.umass.edu
 www.extension.uga.edu
 www.cce.cornell.edu

American Bird Conservancy, www.abcbirds.org

American Fence Association, www.americanfenceassociation.com

American Hardwood Export Council, www.americanhardwood.org

American Wood Council, www.awc.org (represents over 75% of North American wood product manufacturers)

American Arbitration Association, www.adr.org (although based in the US, it functions internationally, hosting conferences, providing arbitration services, resolving disputes, and offering webinars, courses, and publications)

AmericanHort, www.americanhort.org/page/our_story (group founded in 2014 through the consolidation of the American Nursery and Landscape Association and the Association of Horticultural Professionals; advocacy, resources, research)

American Association of Nurserymen, https://www.amerinursery.com

American Association of State Highway and Transportation Officials, www.transportation.org (vast array of data, covering both national and state criteria for roads and highways, engineering and design criteria for roads)

American Coatings Association, https://www.paint.org
American Concrete Institute, www.concrete.org/standards (technical resources, educational programs, certification programs)
American Forest & Paper Association, www.afandpa.org
American Galvanizers Association, www.galvanizeit.org (nonprofit trade association providing technical support on today's innovative applications and state-of-the-art technological developments)
American Institute of Architects, www.aia.org
American Institute of Timber Construction, www.aitc-glulam.org (national technical trade association of the structural glued/laminated timber industry, representing a majority of the glued/laminated timber manufacturers in the United States in addition to a number of installers, suppliers, sales representatives, engineers, and other professionals)
American Insurance Association, www.aiadc.org (represents about 300 firms that provide property and casualty insurance throughout the United States)
American National Standards Institute, www.ansi.org (fosters the development of technology standards throughout the United States)
American Nursery and Landscape Association, www.anla.org
American Standard for Nursery Stock, www.jerseygrown.nj.gov/jgstandards.pdf (download free copy of American Standard for Nursery Stock)
ASTM International, www.astm.org (over 12,000 standards, many of which were first developed in the United States but now applied internationally; 140 member countries; standards available in many languages)
American Society for Landscape Architects, www.asla.org
American Plywood Association, www.apawood.org
American Wood Council, www.awc.org (including www.awc.org/codes-standards/publications/dca6, a free download of DCA6: Prescriptive Residential Deck Construction Guide, which shows compliance with the 2012 international residential code for single-level residential wood deck)
American Wood Protection Association, www.awpa.com (formerly Wood Preservers Association)
Americans with Disabilities Act of 1990, https://www.ada.gov (U.S. Department of Justice website with wide array of technical information and resources)
Asphalt Institute, www.asphaltinstitute.org
Associated General Contractors of America, www.agc.org
Association of Landscape Architecture of Quebec, https://www.aapq.org (in French)

Association of Professional Landscape Designers, www.apld.org
Brick Industry Association, www.gobrick.com
British Columbia Society of Landscape Architects, www.bcsla.org
Builder Online, www.builderonline.com
Building Online, www.buildingonline.com
Canadian Green Building Council, www. cagbc.org
Canadian Society of Landscape Architects, www.csla-aapc.ca/csla-aapc
Canadian Wood Council, www.cwc.cacom
Chain Link Manufacturers Institute, www.chainlinkinfo.org (United States, Canada, Mexico)
Concrete Reinforcing Steel Institute, www.crsi.org
Construction Specifications Institute, www.csiresources.org
Council of Educators in Landscape Architecture, www.thecela.org
European Council of Landscape Architecture Schools, www.eclas.org/universities.php
European Forum for Agricultural and Rural Advisory Services, www.eufras.eu
European Landscape Architecture Student Association, www.elasa.org
General Services Administration, www.gsa.gov (for federal specifications; a vast site, but search by topic and practice area, e.g., https://www.gsa.gov/acquisition/purchasing-programs/requisition-programs/gsa-global-supply/supply-standards/index-of-federal-specifications-standards-and-commercial-item-descriptions)
Forest Stewardship Council, www.fsc.org
Greenroofs.com, www.greenroofs.com (based in Alpharetta, GA, a suburb of Atlanta)
Greenroofs for Healthy Cities, www.greenroofs.org (based in Toronto)
Illuminating Engineering Society, www.ies.org
International Conference of Building Officials, https://global.ihs.com (source for international building code publications)
International Federation of Landscape Architects, www.iflaonline.org
International Society of Arboriculture, www.isa-arbor.com
Julius Blum & Co., www.juliusblum.com (stock components for architectural metal work in aluminum, bronze, stainless steel, and nickel silver, including tubings, bars, shapes, and traditional railings and brackets)
Livingroofs.org, www.livingroofs.org (one of the leading independent European green roof organizations)
Manitoba Association of Landscape Architects, www.mala.net
National Association of Home Builders, www.nahb.org
National Association of the Remodeling Industry, www.nari.org
National Audubon Society, www.audubon.org
National Frame Building Association, www.postframe.org

Nature Conservancy, www.nature.org
North American Wholesale Lumber Association, www.nawla.org
National Concrete Masonry Association, www.ncma.org
National Landscape Nurserymen's Association, www.nlae.org
Ontario Association of Landscape Architects, https://www.oala.ca
Painting Contractors Association, https://pdca.org
Passive House Institute, https://passivehouse.com
Perlite Institute, www.perlite.org
Programme for the Endorsement of Forest Certification, https://www.pefc.org and https://www.pefc.co.uk
Saskatchewan Association of Landscape Architects, www.sala.sk.ca
Southern Forest Products Association, www.sfpa.org
Southern Nursery Association, www.sna.org
Sweets Architectural Catalog, www.sweets.construction.com (organized by products, manufacturers, catalogs, CAD, BIM (Building Information Modeling), specs)
Transportation Research Board, www.trb.org
United States Green Building Council, www.usgbc.org
Vermiculite Association, www.vermiculite.org
Walpole Woodworkers, https://www.walpolewoodworkers.com
Western Wood Products Association, www.wwpa.org (represents lumber manufacturers in twelve western states and Alaska)
WoodWorks, www.woodworks.org (for nonresidential construction)
World Green Roofs Congress, http://greenroofworld.com
World Green Roof Infrastructure Network Congresses, www.worldgreenroof.org/own-congresses.html

APPENDIX

Projects and Contact Information

The following is a list of addresses, e-mail addresses, web sites, and phone numbers for the contributors to the book. Contact information is provided first for Steven L. Cantor, the author, and Vicky Chan, and Richard Alomar, the principal illustrators, for the book. Thereafter, all information is organized by chapter, in the order it appears in the book. It is not possible to include every person or firm that played a major role in each project discussed or presented. However, you may wish to contact the organizations or individuals listed here in order to obtain additional information or to reach a particular consultant or specialist.

Steven L. Cantor, RLA, ASLA, StevenLCantor@gmail.com

Vicky Chan, RA, AIA, LEED AP, Founder, Avoid Obvious Architects, www.aoarchitect.us

Richard Alomar, RLA, Associate Professor, Landscape Architecture, and Associate Director, Office of Urban Extension and Engagement, Rutgers University

CHAPTER 1 INTRODUCTION

Shop Drawings
LDGN Landscape Architect DPC
Jeff Dragan, Principal
www.ldgn-ny.com

CHAPTER 2 PROJECT MANAGEMENT AND OFFICE ADMINISTRATION

Olympic Sculpture Park, created and operated by the Seattle Art Museum
9-acre park at 2901 Western Avenue, Seattle, WA 98121
Designed by Marion Weiss and Michael Manfredi, www.weissmanfredi.com
Richard Serra's sculpture *Wake* is one of many featured sculptures
https://www.seattleartmuseum.org/Documents/OSP%20Map%20and%20Guide.pdf

Floriculture Greenhouse Outdoor Living Lab Rutgers University
Rain Garden (see Chapter 5)

Solar panels and green roofs in Germany; also Neubauer and Leipzig greenroofs, in Chapter 7
ZinCo GmbH
Heidrun Eckert, Business Unit Manager
Lise-Meitner-Strasse 2
72622 Nuertingen
Germany
www.zinco-greenroof.com
heidrun.eckert@zinco-greenroof.com
ZinCo USA, Inc.
401 VFW Drive
Rockland, MA 02370
(866) 766 3155
info@zinco-usa.com
ZinCo Canada Inc.
P.O. Box 29
Carlisle, ON, L0R 1H0
(905) 690 1661
greenroof@zinco.ca

CHAPTER 3 CONTRACTS

Net Zero, Montpelier, Vermont
Richard Alomar, Registered Landscape Architect, Urban Field Studio
Vicky Chan, Registered Architect, Avoid Obvious Architects
Erik Madsen, Professional Structural Engineer, Madsen Consulting Engineering
Ching-Yu Lin, Lighting Consultant, CosmoC Design

Peter Vanhage, Computational Design Expert/Architect
Leo Lei, Branding and Marketing Expert, Matisse Design
Size: 20 city blocks
Year: Competition in 2016

CHAPTER 4 MARKETING AND HUMAN RESOURCES

Net Zero, Montpelier, Vermont (see Chapter 3)

CHAPTER 5 DRAWINGS

Jongmyo Park Mountain concept (Seun City Walk), South Korea
Architecture and Planning: Vicky Chan, Krystal Lung, DaeWook Lee, Alex Mailloux, Melissa Chan, and Ava Chow
Marketing and Branding: Matisse Design Ltd
Lighting: CosmosC Design Ltd
Landscape Consultant: Sookyung Shin
Size:1 km long
Year: Competition in 2015

Wetown Conceptual Model, near Vancouver, Canada
Design Team: Vicky Chan, Emily Manasc, Richard Alomar, Erik Madsen, Leo Lei
Client: Auguston Town Development Inc.—Angela Au, Allen Au, Ian Renton
Size: 1,287,235 m^2 GFA
Year: To be completely built by 2030

2nd Street Park, Perth Amboy, NJ
Center for Urban Environmental Sustainabilty
Rutgers Bloustein School of Planning and Public Policy
City of Perth Amboy
Middlesex County Improvement Community Conceptual Design, 2014.

Floriculture Greenhouse Outdoor Living Lab at Rutgers University, New Brunswick, NJ
Neil J. Werket, student in landscape architecture curriculum at Rutgers University
Richard Alomar, Rutgers Floriculture Greenhouse Living Lab Design
Nicolette Graf, Research Farm Supervisor

Carl Schurz Park, New York City
Two projects within New York City Department of Parks and Recreation.
Design by Banford Landscapes LLC, New York City
www.banfordlandscapes.com
info@banfordlandscapes.com
Both projects funded by the Carl Schurz Park Conservancy

CHAPTER 6 BUILDING CODE, SPECIFICATIONS, AND COST ESTIMATES

Water Feature, Queens Botanical Garden
http://www.greenroofs.com
linda@greenroofs.com
See the garden at 43-50 Main Street, Flushing, New York 11355
Specific information at https://queensbotanical.org/?galleries=sustainable-landscape-biotope-bioswales
See also http://www.greenroofs.com/projects/queens-botanical-garden-visitor-and-administrative-center

CHAPTER 7 AREAS OF PRACTICE

Sun Trust Plaza, downtown Atlanta, GA, at Peachtree, West Peachtree, and Baker Streets; photographed from adjacent hotel at June 4, 2009, 6:00 p.m.

Beijing Vanke, China
Firm: Zhang (Beijing) Landscape Design Studio
Room 0918, Richen International Center, 13 Nongzhanguan South Road, Chaoyang District, Beijing 100125, China
www.zlhdesign.com
(+86) 10-64096003
info@zlhdesign.com
Design Team: Lihong Zhang (Design Director), Xiaodong Zhao (Design Manager), Yuzhi Qin (Project Manager), and Pei Yang (Project Landscape Architect).

Barlow-Hill Residence, Easthampton
Steven L. Cantor, Registered Landscape Architect, New York City, SCantorRLA@aol.com
Christina Lynch, Registered Landscape Architect, Pasadena, California,

fluteclr@gmail.com
Stephen Mahoney, Mahoney Farm, 162 Long Lane, East Hampton, NY 11937
(631) 324-6574
stephen@mahoneyfarm.com
On-site drip irrigation: Marders, 120 Snake Hollow Road, P.O. Box 1261, Bridgehampton, NY 11932
info@marders.com
General Contractor: Wright & Co. Construction Inc., Bridgehampton, NY
Saskas Surveying Company, East Hampton, NY
https://www.linkedin.com/company/saskas-surveying-co
House Mover: Guy Davis, Davis Construction Building Movers, Blue Point, NY

Browns Race, Rochester, NY
LDGN Landscape Architect DPC
Jeff Dragan, Principal www.ldgn-ny.com

KLM Equities Inc. Sun Roof, New York, NY
Sanford M. Berger, AIA, S. M. Berger Architecture PC
P.O. Box 222113
Great Neck, NY
Mark Davies, Higher Ground Horticulture
470 West 24th Street, #14E
New York, NY 10011
(212) 691-3633
higherground@verizon.net
highergroundhorticulture.com

The Grange, at the Brooklyn Navy Yard
Completed 2012, New York City Department of Environmental Protection's green infrastructure grant, the first and (at that time) largest grant of almost $593,000 for 1.56-acre (0.6 ha) green roof at Building 3, Brooklyn Navy Yard
Brooklyn Navy Yard Development Corporation
James Corley, Vice President Construction, DEP's Office of Green Infrastructure

USPS Morgan Processing & Delivery Center, Manhattan
Completed in 2009 on building constructed in 1933 of almost 2.5 acres (1 ha); part of total roof rehab of $7 million
Architect of Record: The Office of (the late) Dilip C. Khale, AIA, Architect
Elizabeth Kennedy, Landscape Architect, PLLC, MBE, WBE (both projects)

Brooklyn Navy Yard Building 275 Room 202
63 Flushing Avenue
Brooklyn, NY 11205
(718) 596-8837 (ph) (718) 709-5885 (fax)
www.eklastudio.com

Neubauer and Leipzig greenroofs, ZinCo (see Chapter 2)
Rushmore Environmental Impact Assessment, New York
Raymond J. Heimbuch, PE (died 2017), President, Clarke + Rapuano, Inc.
Steven L. Cantor, RLA, Clarke + Rapuano, Inc.
Robert T. Zappalorti, herpetologist, www.herpetologicalassociates.com
Town of Woodbury, Orange County, New York
Many other consultants

Freshkills Park, Staten Island, NY
Eloise Hirsh, Freshkills Park Administrator
freshkillspark@parks.nyc.gov
(212) 602-5372

Walk DVRC Hong Kong
Design Team: Vicky Chan, Gianfranco Galagar, Angie Kan, Krystal Lung, Emil Manasc, and Wijdene Kaabi
Client: Walk DVRC
Area: 60,000 m^2
Completion Date: November 2019 to open

CHAPTER 8 CONSTRUCTION MATERIALS AND DETAILS

Emory Knolls Farm, MD
Ed and Lucy Snodgrass
3410 Ady Road
Street, MD 21154
www.greenroofplants.com

Bamboo Light Walk, Taiwan
(for Yuejin Lantern Festival 2014)
Designer: CosmoC Design Ltd
eth@cosmoc-design.com
Location: Yuejin Harbor Water Park, Yanshui Dist, Tainan City, Taiwan (ROC)

Fiberglass liner and wood planter
Mark Davies (see Chapter 7)
Michael Rubin Architect PLLC
www.michaelrubinarchitects.com,
Fiberglass Engineering, Midland, VA 22728
(540) 788-4800
fibrglas@earthlink.net

Ornamental gate for church and fiberglass liner for Cor-Ten planter
Gate pictures are from the children's playground at St. Philip's Church in the Highlands, Garrison, NY
Designed and forged by Dean Anderson and Amy Lahey
Planters and jackets, pictures from Super Square studio in Newburgh, NY
Fabricated in Cor-Ten by Dean Anderson
Super Square Corporation
547 Broadway, Newburgh, NY 12550
Mailing address: P.O. Box 636, Beacon, NY 12508
(845) 565-3539
supersquare2@aol.com

Wood deck at residence
LDGN Landscape Architect DPC
Jeff Dragan, Principal
www.ldgn-ny.com

CHAPTER 9 ETHICS AND SUSTAINABILITY

UpStream Art, Bentonville, Arkansas
Jane Maginot, UpStream Art
2536 N. McConnell Ave.
Fayetteville, AR 72704
www.uaex.edu/nwastomrwater
jmaginot@uaex.edu
(479) 444-1755

Horsetopia, South Korea
Architecture and Planning: Vicky Chan, Krystal Lung, DaeWook Lee, Alex Mailloux, Melissa Chan, and Ava Chow
Marketing and Branding: Matisse Design Ltd
Lighting: CosmosC Design Ltd

Landscape Consultant: Sookyung Shin
Size: 1,474,883 m²
Year: Competition in 2016

Boa'an G107 Highway, China
Architecture, Urban Planning, Environment: Avoid Obvious Architects (USA), TETRA Architects & Planners (Hong Kong)
AOA: Vicky Chan, Melissa Chan, Krystal Lung, Eve Hocheng, Daewook Lee, and Aime Vailes-Marcie
TETRA: Allen Poon, Gilbert Yeung, Andrew Kinoshita, Gary Chan, Brian Mok, and Clive Chan
Associate Architects and Traffic Consultant: BCCI (China)
Landscape and Water Management: Richard Alomar, Urban Field Studio (USA)
Structural Engineering: Erik Madsen, Madsen Engineering (USA)
Drone Consultant: AEE (China)
Lighting Design: Ching-Yu Lin, CosmoC Design (Taiwan)
Chinese Calligraphy: Isabella Sim (Hong Kong)
English Narrative: Andrew Montante (USA)
Size: 30 km long city plan along G107 Highway
Year: Competition in 2016

K-farm, Urban farming, Hong Kong
Design Architect: Avoid Obvious Architects, Vicky Chan, Melissa Chan, Gianfranco Galagar, and Krystal Lung
Executive Architect and Engineer: David S. K. Au & Associates Ltd, David S. K. Au
Joseph C. H. Lee, Philip K. S. Chan, Eugene Shum, Thomas T. Y. Tse, Jessie K. Y. Lau, and Duncan S. C. Yiu
Farming Curator: Farmacy HK Limited, Raymond Mak, Sanford Liu, Calvin Nip
NGO: Rough C Limited
Size: 2000 m²
Year: To open in November 2019

CHAPTER 10 RESOURCES

Biotention and bioswale
Downtown Little Rock, Arkansas
The Creative Corridor, LID Development in Downtown Little Rock, AR
Arkansas Natural Resources Commission

101 East Capitol, Suite 350
Little Rock, AR 72201
(501) 682-1611
Design Engineering Firm: Crafton Tull
10825 Financial Centre Parkway, Suite 300
Little Rock, AR 72211
(501) 664-3245

INDEX

Figures are indicated by an italic *f* following the page number.

Abbey, Edward, 5, 74–75
addendum
   definition of, 7
Alomar, Richard
   construction details, 353*f*, 357*f*, 363*f*, 365*f*, 373*f*, 375*f*, 395*f*, 396*f*, 397*f*, 406*f*, 454*f*,
   Floriculture Green House Outdoor Living Lab at Rutgers University (with Neil J. Werket), 38*f*, 39*f*, 131*f*, 134*f*, 138*f*, 139*f*, 140*f*, 141*f*, 154*f*
   Net Zero competition, Vermont, 63*f*, 77*f*
   Perth Amboy 2nd Street Park, 122*f*, 123*f*
   project management illustrations, 20*f*, 27*f*, 28*f*, 36*f*
   zoning and building code illustrations, 177*f*, 184*f*
American Institute of Architects, (AIA), 64, 188, 193, 336
American Society of Landscape Architect (ASLA), 62, 336, 434
American Society for Testing and Materials (ASTM), specification standards, 191, 192, 199
American Standards for Nursery Stock, 166
Americans with Disabilities Act of 1990, 180–83
   access, 181–82
   detectable warnings, 182
   history and impacts of, 180–81
   other factors, 183
   parking, 182
   public works, 253
   requirements for landscape architecture, 180–83
Anderson, Dean, Super Square Incorporated (and Lahey, Amy)
   church gate, 398*f*, 399*f*
   Cor-Ten planters, 391*f*
Army Corps of Engineers, 223
Au, David, S.K. & Associates Ltd., Hong Kong, 22*f*
AutoCAD, 1, 7, 49, 51–53

Banford Landscapes LLC
   Carl Schurz Park, New York City, 161, 162*f*, 164*f*
Barkskins (Proulx), 402
bid, 188–89
bid form, definition of, 7–8
bidding, contract, 81, 188–89
bidding requirements, 188–89
billing
   ancillary costs, 47
   delays in payment, 46
   for each phase of work, 45
   for reviews and approvals, 47–48
   importance of, 43–44
   timely, 45
BOCA National Building Code, 180
bond
   definition of, 8, 189–90
   labor and materials payment, 189
   payment, 189–90
   performance, 189
   surety, 189

brainstorming, 20f
Brenneisen, Stefan, 298
brick, *see* pavement; walls
Building Codes Illustrated, A Guide to Understanding the International Building Code (Ching and Winkel), 179
Building codes, impacts on
 driveways, garages and carports, 176
 easements, 175
 fences and gates, 177–78
 floor area, height, number of stories, 175
 history, 179–80
 invasive species, 178–79
 irrigation and lawns, 179
 recreational activities, 178
 right-of-way, 175–76
 runoff, storm water, 174–75
 setbacks, 172, 173f
 specialized requirements, 176–77, 184f
 swimming pools, 177
 tree protection, 177
 utilities, 172–74
 variance, 170–72
 wells, 174–75
 zoning, 172

Cantor, Steven L.
 Barlow-Hill Residence, Easthampton, NY, 236, 237f–47f
 construction details and photos where noted throughout
carbon footprint, 434, 439
Chan, Melissa (not related), *see* Chan, Vicky
Chan, Vicky (Avoid Obvious Architecture)
 Boa'an G107 Highway, Shenzhen, China, 447, 454f–55f
 construction details, 353f, 354f, 357f, 361f, 362f, 363f, 365f, 373f, 374f, 375f, 377f, 395f, 396f, 397f, 398f, 406f
 Horsetopia, Yeongcheon, South Korea, 438–39, 440f–46f, 448f–53f
 Jongmyo Park Mountain and Seun City Walk, Seoul, South Korea, 111–12, 113f–20f
 Net Zero competition, Vermont, 63f, 77f
 Urban farming, Hong Kong, 447, 456f, 457f
 Walk DVRC, Hong Kong, 338, 339f–41f
 Wetown, Abbotsford (near Vancouver), British Columbia, Canada, 112, 121f
 zoning and building code illustrations, 173f, 184f
change order, definition of, 8
Ching, Francis, D.K and Winkel Steven R, 179–80
City of New York Parks and Recreation Freshkills Park, 322, 323–30f
Clarke + Rapuano, Inc.
 Rushmore Environmental Impact Statement, 307–12
 preferred alternatives, 313f–14f
 phasing diagrams, 315f–19f
client, definition of, 8
construction details, *see* type of construction
construction services, contract, 73, 80
Construction Specifications Institute (CSI) Format standard, 188, 194, 196, 198
consultant, definition of, 8
contract
 administration, 70–71
 bids, 70, 72
 calendar of work, 67
 construction services, 72–73, 412–17
 contingencies, 71
 coordination with other disciplines, 71
 definition of, 8, 62
 digital records, 69
 disputes, 70
 Dos and Don'ts, 74
 exclusions, 65
 existing conditions and documents, 69–70
 extra services, 66
 hourly rates, 66–67
 inspections for completion, 67
 key personnel, 69
 LEED, 70
 limits on meetings, 66
 liquidated damages, 11, 73
 multiple sites, 68–69

payments, 66
permits and approvals, 68
project documentation, 72
scope of work, 64, 65
scope development, 62, 65
special conditions, 70
support services, 65
contractor, definition of, 9
Corbusier le, 216
cost estimate, 201–5
   break-down into convenient categories, 202
   Cost-Estimate requirements, 203
   lump sum items, 203
   mental arithmetic (common sense), 204
   not included items, 203
   permits and approvals, 203–4
   preliminary cost estimate example, residential, 204–10
   specialty items, 202–3
critical path, 21, 22f–23f

Daugherty, Edward L., 28
Davies, Mark, see Higher Ground Horticulture
dead load, 273–74
definitions
   addendum, 7
   bid form, 7–8
   bond, 8
   change order, 8
   client, 8
   concrete, 350–51
   consultant, 8
   contract, 8 (see also Chapter 3)
   contract documents, 9
   contractor, 9
   disability, 9
   ethics, 9 (see also Chapter 9)
   green roof, 287
   human resources, 88
   imagination, 9
   insurance, 9–10
   joints (pavement), 355–56, 357f
   joint venture, 10
   landscape architect, 10
   landscape designer, 10
   Leadership in Energy and Environmental Design (LEED), 10–11, 436
   lien, 11
   liquidated damages, 11, 190
   Minority-owned business enterprise (MBE), 11
   mortar mixes, 351
   owner, 11
   project management, art of, 16
   project manager, role of, 18–19
   public works, 251–52
   shop drawing, 11
   specifications, 12
   survey, 100
   sustainability (see Chapter 9)
   sustainable design, 12, 433, 434 (see also Chapter 9)
   unit prices, 12
   variance, 12
   Women-owned business enterprise (WBE), 13
Desert Solitaire (Abbey), 5, 74
design library of details, 348
design principles, 217
details, see type of construction or material
disability, definition of, 9
Dos and Don'ts
   brick pavements, 352
   construction drawings/contract drawings
      all drawings, 125–28
      demolition, 128–30
      detail drawings, 139
      grading and storm drainage, 135–37
      layout, 132–33
   contracts, 74
   cost estimates, 211
   definition of, 4
   design drawings
      conceptual, schematic, design development, and master plan, 109
      conceptual and schematic drawings, 110
      design development, 110–11
      master plan, 111
   emails and texting, 97–99
   environmental assessment, 320–21
   green roofs, 303–4
   inspections, site (see site inspections)
   international work, 342–43

Dos and Don'ts (*cont.*)
  irrigation, 382
  lighting, 378–79
  metals and plastics, 393–94
  minutes of meetings, 54–56
  pavements (including concrete), 358–60
  planting design, 155–61 (*see also* planting)
  plastics (*see* metal; plastics)
  public works projects, 261–62
  residential design (design considerations) 232–36
  residential design (project management) 229–32
  resources, 460–62
  roof gardens, 278–79
  site furniture, 409–10
  site planning, 218–19
  specifications, 199–200
  stepping-stones, 364
  steps, 366–67
  surveys, 100–6
  sustainability, 436–37
  walls, dry-laid, 372
  walls (general), 371–72
  water features, 386
  wood, 404–5
Downing, Andrew Jackson, 224
Dragan, Jeff
  Brown's Race, Rochester, NY, 262–63, 264*f*, 266*f*, 268*f*, 269*f*
  shop drawing example, 14*f*
  shop drawing stamp, 12*f*
  survey, 107*f*
  wood deck, 407*f*, 408*f*

EKLA Studio, *see* Kennedy, Elizabeth J.
email, *see* office standards
environmental assessment, 305–16
  Dos and Don'ts, 320–21
  fee proposals, 312
  Freshkills Park (*see* City of New York Parks and Recreation)
  history, 305
  requirements and process, 306–7
  terms, 306
  typical table of contents, 308–12
ethics
  definition of, 9, 422–24
  environmental, 422–24
  legal aspects of, 422–24
  questions for discussion, 424–33
  sustainable design, evolving from, 422–24 (*see also* sustainability)
Evangelista, Philip
  KLM Equities Sun Roof Garden, 280, 281–86*f*
Evans, Gary, 2*f*, 384*f*

Facebook, 90–91, 336. *See also* office standards
Federal Emergency Management Administration (FEMA), 223
French drain, 152
Freshkills Park, *see* City of New York Parks and Recreation

Gabello, William, *see* Evans, Gary
Gedge, Dusty, 298
green roof. *See* also roof gardens
  benefits, 288–98
  costs, 287–88, 298
  definition, 287
  Dos and Dont's, 303–4
  extensive vs intensive, 287
  Forschungsgesellschaft Landschaftsentwicklung Landschaftsbau (FLL), Guidelines for the Planning, Execution and Upkeep of Green-Roof Sites, 298–99
  Grange, the (*see* Kennedy, Elizabeth J.)
  layers and components, 287
  LEED ratings required for some buildings, 289
  modular systems, 298–99
  semi-intensive, 287
  stormwater overflow storage, 288
  United States Morgan Processing & Delivery Center (*see* Kennedy, Elizabeth J.)

Halprin, Lawrence, 24
Heimbuch, Raymond, PE, 307
Higher Ground Horticulture (Mark Davies)
  KLM Equities Sun Roof, 280, 281–86*f*
  planter (with Mark Rubin), 392*f*
Holism and Evolution (Smuts), 434
human resources
  benefits comparisons, 89–90
  consistent policies on documents, 90

definition of, 88
emails and texting, 91
FaceBook, Skype, Twitter, etc., 91
range of diverse activities, 89
self employment vs working for others, 88–89
smart phones and social media, 90–92
(see also subsequent questions)

imagination, definition of, 9
insurance
  certificates of, 9–10
  definition of, 9–10
  liability, 9
  property, 9
  workmen's compensation, 9–10
International Building Code 2000, 5, 180, 337
international landscape architecture practice
  artificial intelligence software, 336
  basic requirements, 332–33
  Beijing, Vanke, China (Zhang), 220, 221f, 222f, 338
  broadening perspective, 333–34
  "cookie cutter" designs, 337
  feng shui, 333
  health insurance, 334–35
  language fluency and liaison, 336–37
  legal protections, 337
  networking opportunities, 336–37
  Peace Corps model, 335–36
  potential cultural conflicts, 334–35
  speed, 335–36
  total immersion, 335–36
  See also Chan, Vicky; David S. K. Au, David S. K. & Associates Ltd.; Lin Ching-yu, Zhang, Lihong
irrigation
  basic requirements of zoning or building code, 380
  controllers and rain sensors, 380
  design vs specify, 380
  Dos and Don'ts, 382
  major manufacturers, 381
  manual watering offseason, 381
  response of the industry to drought and climate change, 381

joint venture, definition of, 10

Kellert, Stephen, 433–34
Kellert, Stephen R., see sustainable design
Kennedy, Elizabeth J., Landscape Architect, EKLA Studio
  the Grange, Brooklyn Navy Yard, 288–89, 290f–95f
  United States Morgan Processing & Delivery Center, New York City, 299, 300–2f
KLM Equities Sun Roof Garden, NYC, see Evangelista, Philip

Lahey, Amy, see Anderson, Dean, Super Square Incorporated
landscape architect/ landscape designer, definition of, 10
landscape architecture firms
  multidisciplinary, 76–77
landscape architecture practice
  types of, 213
Leadership in Energy and Environmental Design (LEED)
  definition of and certification process, 10–11
  GreenFormat Specifications, 195
Lee, Daewook, see Chan, Vicky
LEED, see Leadership in Energy and Environmental Design
LEED, contract, 70, 195, 436
Lei, Leo, see Chan, Vicky
lien, definition of, 11
lighting
  Bamboo Light Walk, Taiwan (Lin) 376, 377f
  Dos and Don'ts, 378–79
  evolution in landscape architecture, 376
  include with electrical requirements, 376
  types of fixtures, 376
Lin, Ching-yu
  Bamboo Light Walk, Taiwan, 376, 377f
liquidated damages
  contract, when to use, 73
  definition of, 11
live load, 273–74
load, dead, live, seismic, wind, 273
Louisiana Museum, Denmark, 35
Lung, Krystal, see Chan, Vicky
Lynch, Christina, 236
Lynch, Kevin, 215

marketing principles
  cover letter or narrative, 79
  design sketches, 81
  diversity, 83–84
  fee breakdown, 80–81
  open-endedness, 82
  pro bono efforts, 84
  project management timeline, 80
  project summaries, 79–80
  red-flags, 82–83
  required elements, 82
  resumes, 79
marketing trends
  specialization, 76–79
  story line, 79
  use of social media, 78
  videoconferencing, 78–79
metals
  aluminum, 388
  brass, 388
  cast iron vs wrought iron, 387, 389
  copper, 388
  details, 184f, 365f, 391f–92f, 395f–99f
  Dos and Don'ts, 393–94
  fastening, 389
  finishes, 388
  painting, 388
  processes, 389
  shapes, 389
  stainless steel, 388
  weights, 388
metric system
  TABLE: Metric and Imperial Measurements, 58
  compared to imperial system, 57–59, 60f–61f
minority owned business enterprise (MBE), definition of, 11

office standards
  AutoCAD, 49, 51–53
  Dos and Don'ts for smartphones, apps, emails and texting, 97–99
  email, 90–92
  Excel, 49
  folders, set-up, 50
  metric system, 57–59
  minutes of meetings, 53–56
  social media, 90–92 (*see also* questions at end of Chapter 4)
  Word, 49
Olmsted, Frederick Law, 224
Olympic Sculpture Park, Seattle, Washington, USA, 17f
owner, definition of, 11

passive design, 235
pavements
  asphalt, 350
  brick, 351, 353f, 354f
  concrete, 350–51
  Dos and Don'ts for brick pavements, 352
  Dos and Don'ts for Pavements (Including Concrete), 358–60
  Dos and Don'ts for Stepping-Stones, 364
  finishes, 356
  freezing weather, avoid, 356
  Hastings pavers, 352
  joints, 357f, 355–56
  mortar mixes, 351
  pedestrian vs. vehicular, 350
  reinforcement, 355
  sealants, 355–56
  stamped, 356–58
  stepping-stones, 360, 361f
  stone, 355, 362f
  traditional vs contemporary materials, 352
  unit pavers, 351
payment
  for extra services, 30
  for landscape architect, 62
Peace Corps, policies in international programs, 335–36
permits and approvals, 68, 199, 203, 205, 256–57, 260, 262, 279, 413
planting design and construction
  budget time for meetings, 159–60
  cost-cutting. 158
  details, 142
  Dos and Dont's, 155–61
  drainage, 149, 152
  fertilizers, 151
  guarantee, 152
  herbicides and pesticides, 151

inspections and records, 157, 160
invoicing, 160
irrigation, 156
landscape contractors and nurseries, local expertise, 158
maintenance, 153
method of labeling on plans, 144
mulching, 150
new plant materials, testing, 159
odd/ even number of plants, 150
phasing, 156
plan graphics, 144
plant lists, 146–47
plant materials, geographical range and local sources, 158–59, 166, 200
potential conflicts of interpretation, 159
pre-qualifications of contractors, 160
scale, 143
sequence of spaces, 143
shop drawings, 157
soil testing, 151–52
staking, 149–50
street trees, 149
structural planting elements, 143
structural soil, Cornell University and NYC, 149
substitutions, 158
tagging or pre-growing trees, 150–51
plastics
   Dos and Don'ts (see Dos and Don'ts for Metal and Plastics)
   uses, 390
Powerpoint, 47
practical considerations, 3
problem statements and exercises, x, 3–4
professional considerations, 3
professional practice, definition of, 1
project management, principles of
   adaptability, 28–29
   common sense, 27
   communication, 18–19
   control expectations, 32
   critical path, 22f
   delegation, 26
   don't reinvent the wheel, 35–37, 36f, 38f, 39f
   flexibility, 28–29
   honesty, 29

instant design, 40
listening, 19
look before you leap, 40
organization, 21–23
payroll review, 31–32
practicality, 27
publicize, 34
respect, 29
responsibility, 26–27
seek out others more knowledgeable, 33–34
separation of design/ contractual matters, 33
share credit, 34
tenacity, 29–30
timing, 30–31
trust, 24–25
Proulx, Annie, 402–3
public works
   bid process, 257, 258
   Brown's Race, Rochester, NY (see Dragan, Jeff)
   cost estimate, 257–58
   Dos and Don'ts, 261–62
   extra services/ additional work, 256–57
   fees, 259–61
   Freshkills Park (see City of New York Parks and Recreation)
   permits and approvals, 256–57
   role of the landscape architect, 252–53
   special challenges, 255
   specifications and bid process, 258–59
   time frame, extended, 256
   unique specifications, 253–54

Queens Botanical Garden, New York City, 171f

residential landscape design
   American history, 224
   Barlow-Hill residence, Easthampton, New York (Cantor), 236, 237–47f
   contemporary practice, 224–25
   Dos and Don'ts (Design Considerations), 232–36
   Dos and Don'ts (Project Management), 229–32

residential landscape design (*cont.*)
  sequence of services, 225–26
  survey, 226–28
  sustainable design, 229
resources
  bibliography, 462–67
  developing a library, 348
  DOS and DONT's, 460–62
  miscellaneous reference, 467–68
  trade associations and organizations, 468–72
role-play, 3–4, 250
roof gardens
  access, 275
  construction materials for, 276–77
  design and building code challenges, 275–76
  design process for, 276
  Dos and Don'ts, 278–79
  drains and irrigation, 277
  KLM Equities Sun Roof Garden (*see* Philip Evangelista)
  lightweight soil mixes, 277
  parapet height, 278
  paving on pedestal systems, 276
  pergolas, 275, 278
  planters, 277
  railings and guardrails, 275–76
  steps, 277
  structural reinforcement, 274–75
  water features, 278
RSVP Cycles (Halprin), 24
rumor mongering, 40–41
Rutgers University Center for Environmental Sustainability, *see* Alomar, Richard

Schubert, Franz, 25*f*
Serra, Richard (*Wake*), 17*f*
shop drawing, definition of, 11, 12*f*, 14*f*
site furniture
  custom-designed and mass produced, 409
  Dos and Don'ts, 409–10
site inspections
  construction details, 415
  demolition, 414
  duration and number, 412–13
  fee calculation or estimate, 72–73, 412–13
  final inspection, 417

  general, 412
  grading, 415
  layout, 415
  phases or tasks, 413–19
  planting design, 416
  pre-final inspection, 417
  special elements, 416
  start-up meeting, 413–14
site planning
  definition of, 213–15
  design principles, 217–18
  design process, 215
  Dos and Don'ts, 218–19
  programming, 216–17
  site analysis, 216
  study costs, 218
Site Planning (Lynch), 215
Smuts, Jan Christiaan, 434
social media, *see* office standards
solar power, California Energy Commission, 289
Southern Building Code, 180
specifications
  content, 196–97
  CSI format, 188
  definition of, 181, 190–91, 197
  descriptive, 190, 191
  Dos and Don'ts, 199–200
  examples, 191–92
  general conditions, 188
  government agencies, 196
  interrelationship between drawings and specifications, 192
  organization of, 192–94
  performance, 190–91
  prewritten software packages, 198–99
  proprietary, 191
  public works, 258–59
  reference, 191–92
  style, 196, 197
  writing methods, 197–98
standardized details, 35
steps
  descriptions, 184*f*, 353*f*, 364, 365*f*
  Dos and Don'ts, 366–67
stone pavements, *see* pavement
stone walls, *see* walls
structural soil, 149
surveys
  definition, 100

Dos and Don'ts, 100–6
  residential design, 107f, 226–28
sustainability, see sustainable design
sustainable design
  American Society of Landscape
    Architects, definition, 434
  bioretention and bioswales, Little
    Rock, Arkansas, USA, 461f
  Chan, Vicky (Avoid Obvious
    Architecture)
    Boa'an Highway, Shenzhen,
      China, 447, 454f–55f
    Horsetopia, Yeongcheon, South
      Korea, 438–39, 440f–46f
    Jongmyo Park Mountain and Seun
      City Walk, Seoul, Korea, 111–12,
      113f–20f
    Urban farming, Hong Kong, 447,
      456f, 457f
  correlation to whole building
    design, 434
  definition of, 12, 433, 434
  Dos and Don'ts, 436–37
  drainage structure, painted,
    Bentonville, Arkansas, USA, 423f
  Gandhi, Mahatma, 433
  Kellert, Stephen R. 433
  Leadership in Energy and
    Environmental Design (LEED), 436
  residential design, in, 229
  restorative environmental design,
    433–34
  tapestry of processes and design
    approaches, 433
  tracking design processes and
    safeguards, 435

Uniform Building Code, 180

unit prices, definition of, 12
variance, definition of, 12
Vaux, Calvert, 224

walls
  capstone critical in design, 368
  concrete block, 369–70
  corners, 371
  details, 368, 373–75f
  Dos and Don'ts, General, 371–72
  Dos and Don'ts for Dry-Laid Walls, 372
  dry-laid, 349f, 369
  footing drains, 370
  footings, 369
  most importance in design, 368
  prequalification of mason or
    contractor, 368
  review by engineer, 368, 370
  stone veneer, 369
  weepholes, 370
water features
  consultants or in-house, 383
  details, 384f, 385f
  Dos and Don'ts, 386
  maintain attractive appearance
    when not in use, 383
  recirculation of water, 383
Werket, Neil, see Alomar, Richard
whole building design, 434
Wilson, Edward O., 305, 422
women owned business
  enterprise (WBE)
  definition of, 13
wood
  actual vs nominal size, 400–1
  ASTM standards, 400
  avoid use of endangered or threatened
    species, 400, 401
  Barkskins (Proulx), 402
  contrast to vinyl for outdoor
    structures, 400, 404
  details, 400, 406f, 407f, 408f
  Dos and Don'ts, 404–5
  forestry practices, certification, 403
  forestry practices in the United
    States, 401
  painting and staining, 403–4
  planters, 404
  plywood, 404
  pressure treating, 403
  specified sizes smaller over time, 400–1
  staple of landscape architecture
    vocabulary, 400

Zhang, Lihong
  Beijing Vanke, China, 220, 221f,
    222f, 338
Zinco
  metric/imperial equivalents, 60f–61f
  Neubauer and Leipzig, Germany,
    296f–97f

www.ingramcontent.com/pod-product-compliance
Ingram Content Group UK Ltd.
Pitfield, Milton Keynes, MK11 3LW, UK
UKHW021255180426
11947UKWH00010B/793